Lecture Notes in Computer

Edited by G. Goos, J. Hartmanis and J.

Advisory Board: W. Brauer   D. Gries

T0230172

Springer
Berlin
Heidelberg
New York
Barcelona
Budapest
Hong Kong
London
Milan
Paris
Santa Clara
Singapore
Tokyo

J.-M. Alliot   E. Lutton   E. Ronald
M. Schoenauer   D. Snyers (Eds.)

Artificial Evolution

European Conference, AE 95
Brest, France, September 4-6, 1995
Selected Papers

 Springer

Series Editors

Gerhard Goos, Karlsruhe University, Germany

Juris Hartmanis, Cornell University, NY, USA

Jan van Leeuwen, Utrecht University, The Netherlands

Volume Editors

Jean-Marc Alliot
Ecole Nationale de l'Aviation Civile
7 avenue Edouard Belin, F-31055 Toulouse Cedex, France

Evelyne Lutton
INRIA-Rocquencourt
B.P. 105, F-78153 Le Chesnay Cedex, France

Edmund Ronald
Marc Schoenauer
Ecole Polytechnique, CMAP, CNRS-URA 756
F-91128 Palaiseau, France

Dominique Snyers
TélécomBretagne, LIASC
B.P. 832, F-29285 Brest Cedex, France

Cataloging-in-Publication data applied for

Die Deutsche Bibliothek - CIP-Einheitsaufnahme

Artificial evolution : European conference ; selected papers /
AE 95, Brest, France, September 4 - 6, 1995. J.-M. Alliot ...
(ed.). - Berlin ; Heidelberg ; New York ; Barcelona ; Budapest ;
Hong Kong ; London ; Milan ; Paris ; Santa Clara ; Singapore ;
Tokyo : Springer, 1996
  (Lecture notes in computer science ; Vol. 1063)
  ISBN 3-540-61108-8
NE: Alliot, Jean-Marc [Hrsg.]; AE <1995, Brest>; GT

CR Subject Classification (1991): F.1, F.2.2, I.2.6, I.5.1, G.1.6, J.3

ISBN 3-540-61108-8 Springer-Verlag Berlin Heidelberg New York

This work is subject to copyright. All rights are reserved, whether the whole or part of the material is
concerned, specifically the rights of translation, reprinting, re-use of illustrations, recitation, broadcasting,
reproduction on microfilms or in any other way, and storage in data banks. Duplication of this publication
or parts thereof is permitted only under the provisions of the German Copyright Law of September 9, 1965,
in its current version, and permission for use must always be obtained from Springer-Verlag. Violations are
liable for prosecution under the German Copyright Law.

© Springer-Verlag Berlin Heidelberg 1996
Printed in Germany

Typesetting: Camera-ready by author
SPIN 10512732   06/3142 – 5 4 3 2 1 0   Printed on acid-free paper

A la mémoire de Joëlle Biondi ...

Preface

The Artificial Evolution conference was originally conceived as a forum for the French-speaking Evolutionary Computation community, but has acquired a more European audience. In future it will be held every two years, alternating with PPSN, which remains the main European conference. This volume collects extended versions of papers presented in the first and second AE conferences, held in 1994 and 1995.

In 1994 the organizing committee, preferring a secluded provincial setting to Parisian pomp, sited the conference at ENAC in Toulouse, and supplied the food and beverages which the Gallic temperament regards as necessary fuel for the shaping of scientific theories. This must have been a success, because in 1995, various "foreigners" from such non-francophone countries as Germany and Hungary submitted papers, came to the ENST campus near the coastal city of Brest and consumed cider.

In the morning, the participants suffered through their hangovers. In order to provide time for them to wake up, keynote speakers were invited. These speakers presented viewpoints divergent from the conventional bitstring Genetic Algorithms framework: Nick Radcliffe's oral presentation in Toulouse asked the question "Why binary isn't always best?", while Thomas Bäck and David Fogel introduced in Brest the two other pillars of the EC building, namely Evolution Strategies and Evolutionary Programming.

The papers reproduced in this volume have been grouped into seven sections.

1. **Invited Papers** by Thomas Bäck, and David B. Fogel and Lawrence J. Fogel, present the state of the art in ES, and in EP; see above.

2. **Theoretical contributions**: Raphaël Cerf in a complex technical paper creates a probabilistic framework for the analysis of GA dynamics, and Gilles Venturini supplies a highly readable overview of the relationship between GA–deceptive and hill-climbing–difficult problems.

3. **GA Techniques**: In this section, authors investigate various novel "mutations" and extensions of the basic GA. Márk Jelasity and József Dombi present a new niching technique, Cyril Godart and Martin Krüger introduce steady-state reproduction inside a heterogeneous parallel implementation, Caroline Ravisé, Michèle Sebag, and Marc Schoenauer use inductive learning to supervise the GA, and Rémi Viennet, Christian Fonteix, and Ivan Marc exploit diploidy for multi-objective optimization.

4. **Coevolution**: Alternative evolutionary paradigms are introduced in this section. Paul Bourgine and Dominique Snyers study the coevolution dynamics of interacting species, Renaud Dumeur's symbiotic algorithm is inspired

by the recent ideas of Lynn Margulis, Nicolas Meuleau and Claude Lattaud investigate the influence of selection on the iterated prisoner dilemma and Spyros Xanthakis, Constantinos Karapoulios, Régis Pajot and Ahmed Rozz explore a biological analogy between fault tolerant computing and the immune system.

5. **Neural Nets**, when genetically trained, allow a representation of active behaviors. Frédéric Gruau synthesizes nets that simulate various walking gaits in hexapodal automata. Olivier Michel drives a robot through a simulated or real-world maze. Marc Schoenauer and Edmund Ronald investigate the automatic tuning of net parameters, and present a probabilistic analysis of the time required to train a net by GA.

6. **Image Processing:** Guillaume Cretin, Evelyne Lutton, Jacques Levy-Vehel, Philippe Glevarec, and Cedric Roll solve the inverse IFS fractal representation problem by GP, Jeanine Graf and Wolfgang Banzhaf follow Richard Dawkins in incorporating interactive selection in a GA, Evelyne Lutton and Patrice Martinez apply GAs to graphic primitive detection, and Jean Louchet effects object identification in motion analysis.

7. **Applications:** This section demonstrates the successful applicability of GAs in a broad range of problem domains: Jean–Marc Alliot and Nicolas Durand – evolution of Othello playing programs, Thomas Bäck, Martin Schütz and Sami Khuri – constraint handling techniques on the set covering problem, Jin-Kao Hao and Raphaël Dorne – radio frequency assignment in cellular telephone networks, Vittorio Gorrini and Marco Dorigo – packing and routing problems for a real–world robot, Couro Kane and Marc Schoenauer – specific crossover operators for structural optimization, Frédéric Medioni, Nicolas Durand, and Jean–Marc Alliot – air traffic collision avoidance, and Mohammed Slimane, Gilles Venturini, Jean–Pierre Asselin de Beauville, T. Brouard, and A. Brandeau – the optimization of hidden Markov chains in pattern recognition.

At this point, we would like to extend our thanks to Jean-Louis Latieule, Ghislaine LeGall, and Isabelle Parcoit for their help in making the conferences run smoothly.

Finally, we would like to thank both EA94 and EA95 program committees for their work in ensuring the high scientific content of the papers presented. The names of the individuals, who found time to do the refereeing, are listed on the following page. The following additional referees generously donated their time: J.-M. Ahuactzin, S. Augier, L. Gambardella, V. Gorrini, O. Lebeltel, C. Ravisé, and E. G. Talbi.

The organizing committee:

Jean-Marc Alliot (ENAC Toulouse) – Evelyne Lutton (INRIA Rocquencourt)
Edmund Ronald (CMAP Palaiseau) – Marc Schoenauer (CMAP Palaiseau)
Dominique Snyers (TELECOM Bretagne)

Evolution Artificielle 94
19 - 23 September 1994

ENAC, TOULOUSE

Organising Committee

Jean-Marc Alliot (ENAC Toulouse) - Evelyne Lutton (ENRIA Rocquencourt)
Edmund Ronald (CMAP Palaiseau) - Marc Schoenauer (CMAP Palaiseau)

Program Committee

Robert Azencott (DIAM Cachan) - Hughes Bersini (VUB Brussels)
Pierre Bessiere (LIFIA Grenoble) - Joëlle Biondi (I3S Nice)
Paul Bourgine (CEMAGREF Anthony) - Bertrand Braunschweig (IFP Rueil)
Michel Cosnard (LIP Lyons) - Marco Dorigo (ULB Brussels)
Jean Erceau (ONERA Chatillon) - Jean-Arcady Meyer (ENS Paris)
Francesco Mondada (EPFL Lausanne) - Spyros Xanthakis (OPL Toulouse)

Evolution Artificielle 95
4 - 6 September 1995

Télécom Bretagne, BREST

Organising Committee

Jean-Marc Alliot (ENAC Toulouse) - Evelyne Lutton (ENRIA Rocquencourt)
Edmund Ronald (CMAP Palaiseau) - Marc Schoenauer (CMAP Palaiseau)
Dominique Snyers (TELECOM Bretagne)

Program Committee

Thomas Bäck (Dortmund University) - Pierre Bessiere (LIFIA Grenoble)
Joëlle Biondi (I3S Nice) - Paul Bourgine (CEMAGREF Anthony)
Bertrand Braunschweig (IFP Rueil) - Michel Cosnard (LIP Lyons)
Marco Dorigo (ULB Brussels) - Reinhardt Euler (UBO Brest)
Bernard Manderick (VUB Brussels) - Francesco Mondada (EPFL Lausanne)
Nick Radcliffe (EPCC Edinburgh) - Michèle Sebag (LMS Palaiseau)
Spyros Xanthakis (Riversoft Toulouse)

Contents

---

[1] This paper represents a synthesis of two papers presented at the
1994 and 1995 conferences
[2] This paper was presented at the 1994 conference

Neural Networks

Image-Processing

Applications

Invited Papers

Evolution Strategies:
An Alternative Evolutionary Algorithm

Thomas Bäck

Informatik Centrum Dortmund
Joseph-von-Fraunhofer-Str. 20
D-44227 Dortmund

Abstract. In this paper, *evolution strategies* (ESs) — a class of evolutionary algorithms using normally distributed mutations, recombination, deterministic selection of the $\mu > 1$ best offspring individuals, and the principle of self-adaptation for the collective on-line learning of strategy parameters — are described by demonstrating their differences to *genetic algorithms*. By comparison of the algorithms, it is argued that the application of canonical genetic algorithms for continuous parameter optimization problems implies some difficulties caused by the encoding of continuous object variables by binary strings and the constant mutation rate used in genetic algorithms. Because they utilize a problem-adequate representation and a suitable self-adaptive step size control guaranteeing linear convergence for strictly convex problems, evolution strategies are argued to be more adequate for continuous problems.
The main advantage of evolution strategies, the self-adaptation of strategy parameters, is explained in detail, and further components such as recombination and selection are described on a rather general level.
Concerning theory, recent results regarding convergence velocity and global convergence of evolution strategies are briefly summarized, especially including the results for (μ,λ)-ESs with recombination. It turns out that the theoretical ground of ESs provides many more results about their behavior as optimization algorithms than available for genetic algorithms, and that ESs have all properties required for global optimization methods. The paper concludes by emphasizing the necessity for an appropriate step size control and the recommendation to avoid encoding mappings by using a problem-adequate representation of solutions within evolutionary algorithms.

1 Optimization and Genetic Algorithms

In contrast to the title and the overall intention of this article to provide an overview of *evolution strategies* (ESs) [35, 36, 40, 44] (and, to make the reader curious of reading more about them), I take the freedom to start with a brief look at the global optimization problem and those evolutionary algorithms which are widely used to approximately solve this problem: *Genetic algorithms* (GAs) [21, 25]. Although these two different, independently developed branches of evolutionary computation are known since more than thirty years, genetic algorithms have gained much more interest during the past ten years than evolution strategies did. *Evolutionary programming* (EP) [18, 20], the third main stream evolutionary algorithm, has strong similarities to evolution strategies and could be discussed here as well — but this is more adequately done by David B. Fogel, whose contribution in this volume is a must for every reader interested in evolutionary computation.

In the following, I will assume some basic familiarity with the principles of organic evolution, namely, a *population* of *individuals* which are subject to processes of *mutation*, *recombination*, and *selection*. On a higher level of abstraction, the notion of different *species* and the interactions between species are also incorporated into the algorithms. Furthermore, I also take the freedom to assume that genetic algorithms in their canonical form — using a binary representation of individuals as fixed length strings over the alphabet $\{0, 1\}$, a mutation operator that occasionally inverts single bits with a small probability p_m, a crossover operator that exchanges substrings between individuals with crossover probability p_c, and a proportional selection operator that selects individuals probabilistically on the basis of their relative fitness $f_i / \sum_{j=1}^{\lambda} f_j$ — are known to the reader. If this is not the case, the following should be sufficiently general to provide at least an impression of the major principles and differences between the algorithms, or the reader might wish to consult a more technical overview article such as [10] or the in-depth introductions to evolutionary computation [4, 18].

Though it is often claimed that genetic algorithms are developed for solving *adaptation* rather than *optimization* problems (see e.g. [26, 27]), the distinction (if any) between both terms is neither obvious nor formally defined. Biologically, adaptation denotes a general advantage in efficiency of an individual over other members of the population, and the process of attaining this state (see [31], pp. 134–135). This implies an interpretation of adaptation as optimization in a changing environment — a problem which is handled well by GAs, ESs, and EP as well. In fact, most practical applications of GAs exploit their optimization capabilities (see the bibliography of Alander [1] for a collection of numerous applications of evolutionary algorithms, especially GAs). It is even more surprising to see that many practical applications of GAs do not aim at solving strictly pseudoboolean optimization problems

$$f : M \subseteq \{0, 1\}^\ell \to I\!\!R \quad , \tag{1}$$

(which arise for a variety of combinatorial problems such as knapsacks [30], scheduling [29], graph problems [7, 28], and others), but rather at solving problems of the form

$$f : M \subseteq I\!\!R^n \to I\!\!R \quad , \tag{2}$$

which are naturally defined over a continuous, real-valued search space. To apply GAs in their canonical form, a suitable encoding of real-valued vectors $\mathbf{x} \in I\!\!R^n$ as binary strings $\mathbf{y} \in \{0, 1\}^\ell$ is required, and typically this is achieved by subdividing $\mathbf{y}$ into n segments of equal length, decoding each segment to yield the corresponding integer value (either using the standard binary code or a Gray code), and mapping the integer value linearly to an interval $[u_i, v_i] \subset I\!\!R$ of real values. Figure 1 illustrates this decoding process for string segments of length nine bits (allowing for the representation of the integers $\{0, 1, \ldots, 511\}$), which are mapped to the interval $[-50, 50]$.

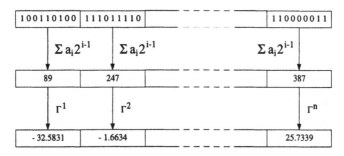

Fig. 1.: Decoding process used in canonical genetic algorithms for continuous search spaces. Γ^i denotes the linear mapping of an integer value $k \in \{0, \ldots, 2^\ell - 1\}$ to the interval $[u_i, v_i]$.

Notice that this representation requires the specification of so-called *box constraints* $[u_i, v_i]$ for the search space. Furthermore, it introduces a discretization of the original search space, and the resolution of the resulting grid of points depends critically on the number of bits used to encode single object variables (I am always critical about applications where less than ten bits are used per object variable — but sometimes, when the precision of the object variables is a priori known to be limited, as in the case of some high energy physics experiments, this might also be an invaluable advantage [23]). The discretization of the search space introduced by the binary code implies that the genetic algorithm might fail to locate an optimal solution exactly just because this solution is not represented by any of the binary strings.

An other disadvantage (or advantage ?) of the binary code is that by mutation of a single bit, depending on its significance in the code, an arbitrarily large modification of the corresponding integer value might result. No preference of small phenotypic changes over large ones — as common in nature — is realized in canonical GAs for parameter optimization tasks. Only few properties of the binary versus Gray code and its impact on the search in combination with the mutation and recombination operators are known and provide some hints towards an explanation of the fact that canonical GAs are sometimes quite useful and the Gray code is typically observed to yield better results than the standard binary code [14, 24]:

- The Gray code and bitwise mutation introduces a probability density function in the decoded object variable space that prefers smaller modificiations over large ones and therefore has some basic similarity to the normal distribution in ESs [4, 24].

– The Gray code does not introduce additional multimodality on the level of the decoding function, as the standard code typically does [3].

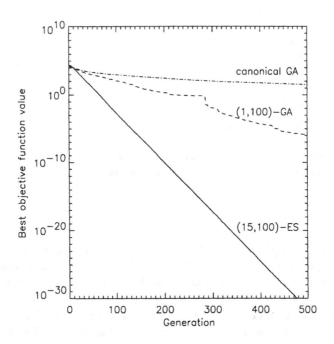

Fig. 2.: Comparison of the convergence velocity of a (15,100)-ES with $n_\sigma = 1$, a (1,100)-GA with a mutation rate $p_m = 0.001$, and a canonical GA with $p_m = 0.001$, $p_c = 0.6$, two-point crossover, and proportional selection.

To look for the pure effect of mutation on binary strings in a simple genetic algorithm, I present some experimental results on an objective function which is so simple that nobody would ever apply an evolutionary algorithm to optimize it — but it serves extremely well for experimental investigations regarding the *convergence velocity* of such algorithms: The sphere model $f(\mathbf{x}) = \sum_{i=1}^{n} x_i^2$. The experiment compares the convergence velocity, measured in terms of the best objective function value f_t' occurring in the population as a function of the generation counter t, for a canonical GA with population size $\lambda = 100$, a simple (15,100)-ES, and a (1,100)-GA *without recombination*. The notation (15,100) describes the selection scheme, indicating that 15 parent individuals generate 100 offspring individuals, and the 15 best of the offspring individuals are deterministically selected as parents of the next generation. The evolution strategy works with a self-adaptive control of the mutation step size (see the next section for an explanation of this), while the genetic algorithm uses a constant mutation rate $p_m = 1/\ell$ and a string length $\ell = 30 \cdot n$ in order to obtain a search grid resolution comparable to that of the evolution strategy. A search space dimension $n = 30$ is used in the experiment. The results, averaged over 20 runs per experiment, are shown in figure 2.

Notice that the canonical GA tends to stagnate after a number of generations, while the modified GA, using mutation only, clearly shows a *linear order of convergence*, i.e., $\ln f_t'/f_0' \propto t$. In fact, although this linear convergence is not comparable to the convergence velocity of an evolution strategy, it may serve as a counterargument against the claim that recombination is the most important search operator in genetic algorithms and mutation plays only a minor role [21, 25] (minimum requirements for evolution are mutation and selection, not recombination !). Using proportional selection rather than (1,100)-selection, however, the canonical genetic algorithm is not able to exploit the innovative, but undirected power of mutation, and its convergence velocity decreases as the optimum is approached, because a favorable mutation becomes more and more unlikely and, at the same time, the preservation of favorable genotypes is quite weak. In fact, in order to approach the optimum further and further, *the GA has to adjust its "step size" of mutation implicitly* by concentrating bit inversion on bits of smaller and smaller significance, while the step size adaptation in an ES works by evolutionary processes of trial and error.

The order of convergence is an important measure for any kind of optimization method, because a sufficiently large order of convergence assures that the algorithm will at least yield a local optimum within a reasonable time — and time is a critical factor for any practical application. While it was recently claimed by Voigt, Mühlenbein and Cvetković that linear convergence is the best one can achieve for evolutionary algorithms [48], Bäck, Rudolph and Schwefel have shown that evolution strategies and evolutionary programming are in fact able to achieve this convergence order [8].

In addition to a linear order of convergence, which assures a sufficiently large velocity, the property of *global convergence* is the second, at least theoretically important requirement for any global optimization algorithm: Given unlimited time, the algorithm has to find a globally optimal solution with probability one, i.e.,

$$\mathcal{P}(\lim_{t \to \infty} \mathbf{x}_t' = \mathbf{x}^*) = 1 \qquad (3)$$

($\mathbf{x}_t'$ denotes the best solution occurring at generation t, and $\mathbf{x}^*$ denotes a globally optimal solution, i.e., $f(\mathbf{x}^*) \leq f(\mathbf{x}) \ \forall \mathbf{x} \in M$). For *elitist* evolutionary algorithms, where the best parental solution is guaranteed to survive when no offspring individual represents an improvement of the currently best parent, global convergence with probability one holds under rather general conditions, provided the mutation step size is larger than zero (see [37] for GAs, [8] for ESs and EP), but the canonical GA lacks this property [37]. For evolution strategies, Rudolph recently demonstrated that the non-elitist $(1,\lambda)$-strategy converges globally with probability one if the objective function is stricly convex at least in a neighborhood around the globally optimal point [38].

For practical applications, however, it is not the global convergence with probability one that one is most interested in, but rather the capability of the algorithm to find a solution which is better than the solution known so far. In some sense, it is the *practical convergence reliability* of an evolutionary algorithm that we are most interested in.

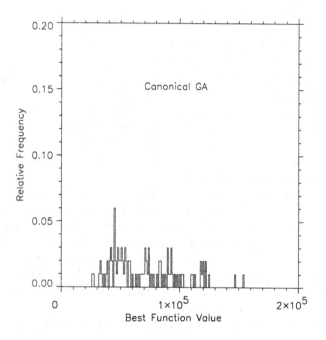

Fig. 3.: Histogram of the final best objective function value obtained from a canonical GA on the function after Fletcher and Powell.

It is almost impossible to assess the general convergence reliability of evolutionary algorithms, and any attempt to do so is necessarily restricted to a limited number of experiments on a limited number of practical problems or test problems. An interesting test problem which was introduced by Fletcher and Powell [16] as a typical representative of nonlinear parameter estimation problems will serve here as an example that reflects my personal experience of the comparative convergence reliability of evolution strategies and genetic algorithms — this assessment is based on personal experience, not on any measurable, general quantity. The Fletcher and Powell function is highly multimodal, and its optima are randomly distributed according to two random $n \times n$ matrices $\mathbf{A} = (a_{ij})$ and

$\mathbf{B} = (b_{ij})$ $(a_{ij}, b_{ij} \in [-100, 100])$, where

$$f_{FP}(\mathbf{x}) = \sum_{i=1}^{n} (\mathbf{A}_i - \mathbf{B}_i)^2 \tag{4}$$

$$A_i = \sum_{j=1}^{n} (a_{ij} \sin \alpha_j + b_{ij} \cos \alpha_j) \tag{5}$$

$$B_i = \sum_{j=1}^{n} (a_{ij} \sin x_j + b_{ij} \cos x_j)$$

and the global optimum equals the vector $\boldsymbol{\alpha}$, with $f_{FP}(\boldsymbol{\alpha}) = 0$. Up to 2^n extrema are located in the search region $|x_i| \leq \pi$. For the details of the matrices $\mathbf{A}$ and $\mathbf{B}$ as well as the vector $\boldsymbol{\alpha}$, the reader is referred to [4].

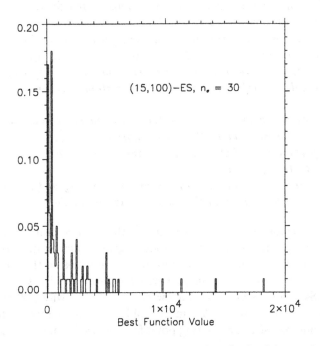

Fig. 4.: Histogram of the final best objective function value obtained from a (15,100)-ES with $n_\sigma = 1$ on the function after Fletcher and Powell.

Since different runs of an evolutionary algorithm on multimodal functions typically stagnate in different local optima, some insight into the reliability to obtain good solutions can be gained from performing a large number of independent runs and looking at the histogram of final solution quality, i.e., the frequency distribution of the best objective function value found after a certain number of generations. I performed 100 runs over 2000 generations, each, with a canonical GA and the (15,100)-ES (with n self-adaptive step sizes), and the resulting histograms are shown in figure 3 for the genetic algorithm, and in figure 4 for the evolution strategy. Notice that most of the runs of the ES yield an objective function value close to zero, the globally optimal one, and only few outliers with final objective function values above 10^4 can be identified. On the contrary, the canonical GA yields a distribution which is much more spreaded between best values of about $2.5 \cdot 10^4$ and worst values of about $1.5 \cdot 10^5$ (notice that the abscissa axis of figure 3 covers a range of values ten times as large as in figure 4).

As I indicated before, these results are representative of my personal experience, and there is no evidence that evolution strategies would generally outperform genetic algorithms on multimodal problems. To be fair, I have to emphasize that the ES certainly benefits from a larger degree of freedom by working with n different self-adaptive standard deviations per individual in contrast to a single mutation rate in genetic algorithms, exogenously predefined and constant over the complete evolution process.

My conclusions, up to this point, however, are certainly a bit provocative, because they attempt a clear distinction of the preferred application domains of GAs and ESs: *For continuous parameter optimization problems, evolution strategies should be preferred over genetic algorithms, because*

- *the binary search space representation might introduce additional multimodalities and therefore make the search more difficult* [1],
- *the self-adaptation of strategy parameters provides a larger flexibility of the evolution strategy, and*
- *the evolution strategy combines convergence velocity and convergence reliability in a more robust way than genetic algorithms do.*

Presently, some researchers circumvent the coding problem in genetic algorithms by recently developed real-coded "genetic algorithms" (e.g., the real-coded GAs of Davis, which adapt operator probabilities on the basis of the past usefulness of these operators in previous generations [15]), such that the gap between ESs and GAs becomes narrower. Like Michalewicz [32] and Davis [15], I prefer the point of view that the most problem-adequate representation of individuals should be used as a starting point for the development of an evolutionary heuristic.

Besides the representation, step size adaptation over the search process is a topic of major importance. Linear convergence order can be achieved only by an appropriate adjustment of step sizes, and ESs use an elegant way to achieve this: Evolutionary learning, called self-adaptation. This principle, which is not that simple to understand at first glance, is the topic of the next section.

---

[1] I did not discuss this statement in sufficient detail here, but refer the reader to [3], where the multimodality problem is analyzed.

2 The Tricky Step: Self-Adaptation

In addition to the object variables $x_i \in I\!R$ themselves, also the strategic meta-parameters — e.g., variances and covariances of a normal distribution for mutation — can be learned by evolutionary processes during the search. This is the simple, but strikingly powerful idea of Schwefel[2] [40], to search the space of solutions and strategy parameters in parallel and to enable a suitable adjustment *and* diversity of mutation parameters under arbitrary circumstances. The natural model consists in the existence of repair enzymes and mutator genes coded on the DNA, thus providing partial control of the DNA over its own mutation probability ([22], pp. 269–271).

Technically, this so-called *self-adaptation* principle combines the representation of a solution and its associated strategy parameters within each individual, and the strategy parameters are subject to mutation and recombination just as the object variables. Selection exploits the implicit link between fitness and strategy parameters, thus favoring useful strategy parameters due to their advantageous impact on object variables. In case of evolution strategies, mutation works by adding a normally distributed random vector[3] $z \sim N(0, C)$ with expectation 0 and covariance matrix C^{-1}, where variances and covariances are the strategy parameters. In order to avoid the overly technical details, I will not discuss the most general case of *correlated mutations*, where the full covariance matrix is subject to self-adaptation (see e.g. [10]), but restrict attention to the self-adaptation of $n_\sigma \in \{1, n\}$ variances σ_i^2. For $n_\sigma = n$, each object variable x_i is mutated according to a normally distributed random number $z_i \sim N(0, \sigma_i'^2)$, i.e.,

$$x_i' = x_i + z_i \quad , \tag{6}$$

and the variance $\sigma_i'^2$ is obtained from mutating σ_i according to a logarithmic normal distribution, i.e.,

$$\sigma_i' = \sigma_i \cdot \exp(s') \cdot \exp(s_i) \quad , \tag{7}$$

where $s' \sim N(0, \frac{1}{2n})$ and $s_i \sim N(0, \frac{1}{2\sqrt{n}})$ (s' is identical for the complete individual, while s_i is sampled anew for each component). For $n_\sigma = 1$, the modification of σ reduces to

$$\sigma' = \sigma \cdot \exp(s_0) \tag{8}$$

where $s_0 \sim N(\frac{1}{n})$. The values $\tau'^2 = \frac{1}{2n}$, $\tau_i^2 = \frac{1}{2\sqrt{n}}$, and $\tau_0^2 = \frac{1}{n}$ of the variances were proposed by Schwefel (see [40], p. 168) and result from partially theoretical investigations.

The choice of a logarithmic normal distribution for the mutation of mutation step sizes σ_i is motivated by just empirical but nevertheless "naturally reasonable" arguments:

---

[2] And, independently, of David B. Fogel, who reinvented the method in EP [17].

[3] The notation $x \sim N(\zeta, \sigma^2)$ indicates x to be a realization of a random variable X which is normally distributed with expectation ζ and variance σ^2.

- A multiplicative modification preserves positive values.
- The median of a multiplicative modification should equal one to guarantee that, on average, a multiplication by a value c occurs with the same probability as a multiplication by $1/c$.
- Small modifications should occur more often than large ones.

Extensive experimental investigations by Schwefel [41, 42, 44] clarified that this mechanism for self-adaptation is extremely robust with respect to the setting of the meta-parameters (or *learning rates*) τ, τ', and τ_0, respectively. Presently, a variety of further work clearly demonstrates that the general principle also works for the adaptation of other parameters such as crossover exchange probabilities [33], mutation rates in canonical genetic algorithms [2], mutation rates in evolutionary programming for the evolution of finite state machines [19], mutation rates for discrete object variables in a hybrid algorithm of ES and GA [9, 39], and momentum adaptation in ESs [34].

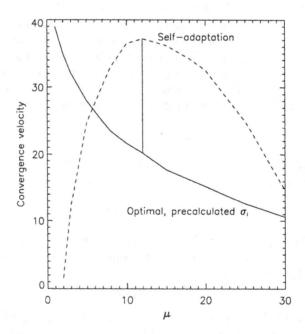

Fig. 5.: Progress rates for the $(\mu,100)$-ES using optimally, precalculated σ_i and the self-adaptive strategy.

To illustrate some of the insights regarding the self-adaptation of $n_\sigma = n$ variances in ESs, I will briefly discuss one of Schwefel's well-designed experiments, using the objective function $F(\mathbf{x}) = \sum_{i=1}^{n} i \cdot x_i^2$ [43]. For this function, each variable is differently scaled, such that self-adaptation requires to learn the scaling[4] of n different σ_i. Furthermore, the optimal settings of standard deviations $\sigma_i^* \propto 1/\sqrt{i}$ are known in advance, such that self-adaptation can be compared to an ES using optimally adjusted σ_i for mutation. The experiment compares the convergence velocity, measured in terms of a logarithmic scale $\ln f_t'/f_0'$, of $(\mu,100)$-ESs with self-adaptation and the optimal strategy, varying μ within the range $\{1, \ldots, 30\}$. The convergence velocity for both strategies is shown in figure 5 as a function of μ.

For the strategy using optimal standard deviations σ_i, the convergence rate is maximal for $\mu = 1$, because this setting exploits the perfect knowledge in an optimal sense. The self-adaptive strategy reaches optimal performance for a value of $\mu = 12$, and both larger and smaller values of μ cause a loss of convergence speed. About 12 imperfect parents collectively yield a performance by means of self-adaptation which almost equals the performance of the perfect (1,100)-strategy and clearly outperforms the collection of 12 perfect individuals.

Clearly, self-adaptation is a mechanism that requires the existence of a knowledge diversity (a diversity of *internal models*), i.e., a number of parents larger than one, and benefits from a collective intelligence phenomenon. Further experiments have shown that the limited life span of individuals (to allow for forgetting misadapted strategy parameters), the creation of a surplus $\lambda > \mu$ of offspring individuals, and the application of recombination also on strategy parameters seem to be important prerequisites for a successful self-adaptation.

Of course, more experimental and especially theoretical work needs to be done in order to understand the self-adaptation process in detail, but the fact that it works in a variety of implementations and for large ranges of parameter settings for the "learning rates" (determining the speed of self-adaptation) provides clear evidence that the basic principle is extremely useful and widely applicable.

3 Completing the Evolution Strategy

After describing the structure of individuals — a vector $\mathbf{x} \in I\!\!R^n$ of object variables and a vector σ of variances (and, for correlated mutations, a vector α of rotation angles representing the covariances; see [4]) — and the principle of self-adaptation, especially regarding the mechanism of the mutation operator, only few remains to complete a standard (μ,λ)-ES algorithm.

Because the population sizes of parent and offspring population are different, the reduction from λ to μ individuals by selection (performed by just determ-

---

[4] Similarly, Schwefel utilized the sphere model for investigations regarding the self-adaptation of $n_\sigma = 1$ standard deviation, and the function

$$F'(\mathbf{x}) = \sum_{i=1}^{n} \left(\sum_{j=1}^{i} x_j^2 \right) \qquad (9)$$

for correlated mutations [43].

inistically choosing the μ best offspring as new parents) has to be compensated by a repeated application of either mutation or recombination to create λ individuals. In ESs, this is achieved by recombination, which generates one offspring individual from ϱ ($1 \leq \varrho \leq \mu$) randomly selected parent individuals per application. Typically, $\varrho = 2$ or $\varrho = \mu$ (so-called *global* recombination) are chosen. The λ-fold application of recombination to the parent population yields λ individuals, each of which is mutated and enters the offspring population. The objective function values of the offspring individuals are calculated, selection chooses the μ best to survive, and the recombination-mutation-evaluation-selection cycle is repeated until a termination criterion is fulfilled. The initial population is either generated at random or by means of mutation from a single, predefined starting point.

In contrast to GAs, recombination in ESs is a more flexible operator, incorporating between 2 and μ parents for the creation of a single descendant. Recombination types on object variables and strategy parameters are usually different, and typical examples are *discrete recombination* (random exchanges between parents) for x_i and *intermediary recombination* (arithmetic averaging) for σ_i (see e.g. [10] for details).

That's all to complete the ES — a lot of details have been added, but they are not mandatory for a working strategy (the details are described e.g. in [44], including reasons why they are incorporated into the algorithm). The most recent proposal for a "contemporary ES" incorporates some additional ideas from GAs (such as the recombination probability and multi-point crossover), multirecombination with an arbitrary number ϱ of parents, a life span κ between the two extremes one ((μ,λ)-selection) and infinity (($\mu+\lambda$)-selection, where the parents are choosen out of offspring *and* old parents), and thus introduces a variety of new possibilities for experimental investigations of evolutionary principles and their effectiveness for the purpose of optimization (see [46] for details regarding the contemporary ES).

4   Theory

Traditional GA-theory, concentrating on the notion of schemata, building blocks, and the schema-theorem giving an upper bound on the expected number of instances of a schema at generation $t + 1$ [21, 25] does not yield any kind of constructive result regarding optimization properties such as global convergence or convergence velocity — and is therefore simply useless for such applications. On the contrary, ES-theory has always focused on these properties, and already Rechenberg calculated optimal step size and maximal convergence velocity of a (1+1)-ES for the corridor and sphere model [35].

Here, I will neither present theoretical results for the simple (1+1)-ES (which does not use a population but just one individual) or any other strategy based on elitist selection, nor discuss any of the mathematical derivations, but I will just provide an overview of the results for (μ,λ)-ESs on the sphere model

$$f(\mathbf{x}) = \|\mathbf{x} - \mathbf{x}^*\|^2 = r^2 \quad , \tag{10}$$

where r equals the Euclidean distance between a search point x and the optimum x^*. For this model, the *relative progress* at generation t is defined by

$$P = \frac{r_t - r_{t+1}}{r_t} \quad , \tag{11}$$

and for $n_\sigma = 1$ and $n \gg 1$, its expectation for a $(1,\lambda)$-ES is given by the quadratic equation

$$E[P_{(1,\lambda)}] = \frac{2\hat{\sigma}c_{1,\lambda} - \hat{\sigma}^2}{2n} \quad , \tag{12}$$

using the normalized, dimensionless variable $\hat{\sigma} = \sigma \cdot n/r$. Notice that the equation reflects a *progress gain* $\hat{\sigma}c_{1,\lambda}/n$ as well as a *progress loss* $\hat{\sigma}^2/2n$, and these two terms have to be kept in balance (either by increasing the gain or by decreasing the loss) in order to achieve positive convergence velocity. The constant $c_{1,\lambda}$ denotes the expectation of the maximum of λ stochastically independent standardized normally distributed random variables and reflects the strong relationship between (μ,λ)-selection and the theory of *order statistics* (see [5] for details).

From equation (12), the optimal standard deviation

$$\sigma^* = c_{1,\lambda} \cdot \frac{r}{n} \tag{13}$$

and the resulting maximal relative progress

$$E[P_{1,\lambda}^*] = \frac{c_{1,\lambda}^2}{2n} \tag{14}$$

are easily obtained. The *convergence velocity* φ (or its normalized counterpart $\hat{\varphi} = \varphi \cdot n/r$) is related to the relative progress according to $E[P_{1,\lambda}] = \varphi/r = \hat{\varphi}/n$, such that, using the asymptotic result $c_{1,\lambda} \approx \sqrt{2\ln\lambda}$ [8], one finally obtains

$$\hat{\varphi}_{(1,\lambda)}^* \approx \ln\lambda \quad . \tag{15}$$

For the details of this derivation, the reader is referred to [45].

In case of the (μ,λ)-ES without recombination, the progress coefficients $c_{\mu,\lambda}$ are generalized to reflect the expectation of the average of the μ best offspring, the convergence velocity is given by the equation

$$\hat{\varphi}_{(\mu,\lambda)} = \hat{\sigma}c_{\mu,\lambda} - \frac{1}{2}\hat{\sigma}^2 \quad , \tag{16}$$

and one obtains the results

$$\hat{\sigma}^* = c_{\mu,\lambda} \tag{17}$$

and

$$\hat{\varphi}_{(\mu,\lambda)}^* = \frac{1}{2}c_{\mu,\lambda}^2 \quad , \tag{18}$$

which asymptotically yields

$$\hat{\varphi}_{(\mu,\lambda)}^* \sim \ln\frac{\lambda}{\mu} \tag{19}$$

$(c_{\mu,\lambda} \approx \mathcal{O}(\sqrt{\ln \lambda/\mu})$; see [11] for details). This result reflects the dominating impact of the *selective pressure* λ/μ for the convergence velocity.

Taking also global recombination (i.e., with $\varrho = \mu$ parents involved in the creation of a single offspring) into account, Beyer and Rechenberg recently derived the equation

$$\hat{\varphi}_{(\mu,\lambda),\varrho=\mu_I} = \hat{\sigma} c_{\mu/\mu,\lambda} - \frac{1}{\mu} \frac{\hat{\sigma}^2}{2} \quad , \tag{20}$$

resulting in the optimal step size

$$\hat{\sigma}_I^* = \mu c_{\mu/\mu,\lambda} \tag{21}$$

and the corresponding convergence velocity

$$\hat{\varphi}_{(\mu,\lambda),\varrho=\mu_I}^* = \frac{\mu}{2} \cdot c_{\mu/\mu,\lambda}^2 \tag{22}$$

for intermediary recombination on the sphere model [11, 36]. With the asymptotical result $c_{\mu/\mu,\lambda} \approx \sqrt{\ln \lambda/\mu}$ (the relation $0 \leq c_{\mu/\mu,\lambda} \leq c_{\mu,\lambda} \leq c_{1,\lambda} < \sqrt{2 \ln \lambda}$ holds for the various progress coefficients), a μ-fold speedup of intermediary recombination as opposed to no recombination is obtained:

$$\hat{\varphi}_{(\mu,\lambda),\varrho=\mu_I}^* \approx \mu \ln \frac{\lambda}{\mu} \quad . \tag{23}$$

Surprisingly, for discrete recombination Beyer obtained the same result for the optimal convergence velocity $\hat{\varphi}^*$ from the progress equation

$$\hat{\varphi}_{(\mu,\lambda),\varrho=\mu_D}^* = \sqrt{\mu} \hat{\sigma} c_{\mu/\mu,\lambda} - \frac{1}{2} \hat{\sigma}^2 \quad , \tag{24}$$

but the optimal step size

$$\hat{\sigma}_D^* = \sqrt{\mu} c_{\mu/\mu,\lambda} \tag{25}$$

turns out to be a factor of $\sqrt{\mu}$ smaller.

Resulting from his mathematical analysis, Beyer is able to explain the beneficial effect of recombination by the fact that recombination reduces the progress loss term in the quadratic equation for $\hat{\varphi}$. In other words, recombination serves as a statistical error correction method, reducing the impact of the harmful part of mutations. This effect, called *genetic repair* by Beyer [13], clarifies that the conjecture of the building block hypothesis that the combination of good partial solutions explains the advantage of recombination does not hold for ESs. Also for discrete recombination, Beyer concludes that it works by implicitly performing a genetic repair, such that (by analogy with uniform crossover in GAs [47]), it is very likely that his results will also fundamentally change the direction of further theoretical research regarding GAs (see [12] for a summary).

5   So What ?

The purpose of this article is not to compare ESs and GAs and to claim that the former or the latter are in some sense better or worse — both classes of evolutionary algorithms have their advantages, and both certainly have their preferable domain of application problems. As a rule of thumb, I personally claim that ESs should be applied in case of continuous problems whereas GAs serve most useful in case of pseudoboolean problems. Obviously, hybridizations of both algorithms are promising for the application to mixed-integer problems involving both discrete and continuous object variables (see [9] for an application of such a hybrid to optical filter design problems).

Furthermore, GAs and ESs offer many possibilities for cross-fertilization by testing certain principles of one algorithm within the context of the other. The most promising (and, at the same time, most complicated to understand) potential for cross-fertilization can be identified in the technique of self-adaptation of strategy parameters, which certainly offers a powerful alternative to the common utilization of a constant mutation rate in GAs — in fact, I have shown that for a simple pseudoboolean objective function, the optimal mutation rate critically depends on both search space dimension ℓ and distance to the optimum [3].

Concerning the theoretical understanding of ESs, it is undoubtedly true that, from the very beginning, much theoretical effort was invested to derive results concerning global convergence and convergence velocity, such that presently a relatively deep understanding of their working principles is available. Most recent investigations regarding recombination demonstrate the validity of its interpretation as a genetic repair operator rather than the superposition effect of good partial solutions as claimed by the building block hypothesis of GAs.

GAs are known more widely than ESs, and the number of researchers using GAs for practical applications is much larger than for ESs, which are still basically used nowhere except in Germany. Their usefulness, however, has been clearly demonstrated not only by an extensive comparison with a variety of traditional search methods [44], but also by a large number of practical applications in domains such as biology, chemistry, computer aided design, physics, medicine, production planning, etc. [6].

Acknowledgements

The author gratefully acknowledges financial support by the project EVOALG, grant 01 IB 403 A from the German BMBF. EVOALG is a joint research project of the Informatik Centrum Dortmund (ICD), the Humboldt-Universität zu Berlin, and Siemens AG Munich.

The author would also like to thank Marc Schoenauer for the invitation to present this talk at Evolution Artificielle 1995.

References

1. J. T. Alander. An indexed bibliography of genetic algorithms: Years 1957–1993. Art of CAD Ltd, Espoo, Finland, 1994.
2. Th. Bäck. Self-Adaptation in Genetic Algorithms. In F. J. Varela and P. Bourgine, editors, *Proceedings of the First European Conference on Artificial Life*, pages 263–271. The MIT Press, Cambridge, MA, 1992.
3. Th. Bäck. Optimal mutation rates in genetic search. In S. Forrest, editor, *Proceedings of the Fifth International Conference on Genetic Algorithms*, pages 2–8. Morgan Kaufmann, San Mateo, CA, 1993.
4. Th. Bäck. *Evolutionary Algorithms in Theory and Practice*. Oxford University Press, New York, 1995.
5. Th. Bäck. Generalized convergence models for tournament- and (μ,λ)-selection. In L. Eshelman, editor, *Proceedings of the 6th International Conference on Genetic Algorithms*, pages 2–8. Morgan Kaufmann Publishers, San Francisco, CA, 1995.
6. Th. Bäck, F. Hoffmeister, and H.-P. Schwefel. Applications of evolutionary algorithms. Report of the Systems Analysis Research Group SYS–2/92, University of Dortmund, Department of Computer Science, February 1992.
7. Th. Bäck and S. Khuri. An evolutionary heuristic for the maximum independent set problem. In *Proceedings of the First IEEE Conference on Evolutionary Computation*, pages 531–535. IEEE Press, 1994.
8. Th. Bäck, G. Rudolph, and H.-P. Schwefel. Evolutionary programming and evolution strategies: Similarities and differences. In D. B. Fogel and W. Atmar, editors, *Proceedings of the Second Annual Conference on Evolutionary Programming*, pages 11–22. Evolutionary Programming Society, San Diego, CA, 1993.
9. Th. Bäck and M. Schütz. Evolution strategies for mixed-integer optimization of optical multilayer systems. In *Proceedings of the 4th Annual Conference on Evolutionary Programming*, 1995.
10. Th. Bäck and H.-P. Schwefel. An overview of evolutionary algorithms for parameter optimization. *Evolutionary Computation*, 1(1):1–23, 1993.
11. H.-G. Beyer. Towards a theory of 'evolution strategies' — Results from the N-dependent (μ,λ) and the multi-recombinant $(\mu/\mu,\lambda)$-theory. Report of the Systems Analysis Research Group SYS–5/94, University of Dortmund, Department of Computer Science, 1994.
12. H.-G. Beyer. How GAs do NOT work. Understanding GAs without Schemata and Building Blocks. Report of the Systems Analysis Research Group SYS–2/95, University of Dortmund, Department of Computer Science, 1995.
13. H.-G. Beyer. Toward a theory of evolution strategies: On the benefits of sex — the $(\mu/\mu, \lambda)$-theory. *Evolutionary Computation*, 3(1):81–111, 1995.
14. R. A. Caruna and J. D. Schaffer. Representation and hidden bias: Gray vs. binary coding for genetic algorithms. In J. Laird, editor, *Proceedings of the 5th International Conference on Machine Learning*, pages 153–161. Morgan Kaufmann Publishers, San Mateo, CA, 1988.
15. L. Davis, editor. *Handbook of Genetic Algorithms*. Van Nostrand Reinhold, New York, 1991.
16. R. Fletcher and M. J. D. Powell. A rapidly convergent descent method for minimization. *Computer Journal*, 6:163–168, 1963.
17. D. B. Fogel. *Evolving Artificial Intelligence*. PhD thesis, University of California, San Diego, CA, 1992.

18. D. B. Fogel. *Evolutionary Computation: Toward a New Philosophy of Machine Intelligence.* IEEE Press, Piscataway, NJ, 1995.

19. L. Fogel, D. B. Fogel, and P. J. Angeline. A preliminary investigation on extending evolutionary programming to include self-adaptation on finite state machines. *Informatica,* 18:387–398, 1994.

20. L. J. Fogel, A. J. Owens, and M. J. Walsh. *Artificial Intelligence through Simulated Evolution.* Wiley, New York, 1966.

21. D. E. Goldberg. *Genetic algorithms in search, optimization and machine learning.* Addison Wesley, Reading, MA, 1989.

22. W. Gottschalk. *Allgemeine Genetik.* Georg Thieme Verlag, Stuttgart, 3 edition, 1989.

23. S. Hahn, K. H. Becks, and A. Hemker. Optimizing monte carlo generator parameters using genetic algorithms. In D. Perret-Gallix, editor, *New Computing Techniques in Physics Research II — Proceedings 2nd International Workshop on Software Engineering, Artificial Intelligence and Expert Systems for High Energy and Nuclear Physics,* pages 255–265, La Londe-Les-Maures, France, January 13–18 1992. World Scientific, Singapore, 1992.

24. R. Hinterding, H. Gielewski, and T. C. Peachey. On the nature of mutation in genetic algorithms. In L. Eshelman, editor, *Genetic Algorithms: Proceedings of the 6th International Conference,* pages 65–72. Morgan Kaufmann Publishers, San Francisco, CA, 1995.

25. J. H. Holland. *Adaptation in natural and artificial systems.* The University of Michigan Press, Ann Arbor, MI, 1975.

26. K. A. De Jong. Are genetic algorithms function optimizers ? In R. Männer and B. Manderick, editors, *Parallel Problem Solving from Nature 2,* pages 3–13. Elsevier, Amsterdam, 1992.

27. K. A. De Jong. Genetic algorithms are NOT function optimizers. In D. Whitley, editor, *Foundations of Genetic Algorithms 2,* pages 5–17. Morgan Kaufmann Publishers, San Mateo, CA, 1993.

28. S. Khuri and Th. Bäck. An evolutionary heuristic for the minimum vertex cover problem. In J. Kunze and H. Stoyan, editors, *KI-94 Workshops (Extended Abstracts),* pages 83–84. Gesellschaft für Informatik e. V., Bonn, 1994.

29. S. Khuri, Th. Bäck, and J. Heitkötter. An evolutionary approach to combinatorial optimization problems. In D. Cizmar, editor, *Proceedings of the 22nd Annual ACM Computer Science Conference,* pages 66–73. ACM Press, New York, 1994.

30. S. Khuri, Th. Bäck, and J. Heitkötter. The zero/one multiple knapsack problem and genetic algorithms. In E. Deaton, D. Oppenheim, J. Urban, and H. Berghel, editors, *Proceedings of the 1994 ACM Symposium on Applied Computing,* pages 188–193. ACM Press, New York, 1994.

31. E. Mayr. *Toward a new Philosophy of Biology: Observations of an Evolutionist.* The Belknap Press of Harvard University Press, Cambridge, MA, and London, GB, 1988.

32. Z. Michalewicz. *Genetic Algorithms + Data Structures = Evolution Programs.* Springer, Berlin, 1994.

33. J. Obalek. Rekombinationsoperatoren für Evolutionsstrategien. Diplomarbeit, Universität Dortmund, Fachbereich Informatik, 1994.

34. A. Ostermeier. An evolution strategy with momentum adaptation of the random number distribution. In R. Männer and B. Manderick, editors, *Parallel Problem Solving from Nature 2,* pages 197–206. Elsevier, Amsterdam, 1992.

35. I. Rechenberg. *Evolutionsstrategie: Optimierung technischer Systeme nach Prinzipien der biologischen Evolution.* Frommann–Holzboog, Stuttgart, 1973.
36. I. Rechenberg. *Evolutionsstrategie '94*, volume 1 of *Werkstatt Bionik und Evolutionstechnik.* frommann–holzboog, Stuttgart, 1994.
37. G. Rudolph. Convergence analysis of canonical genetic algorithms. *IEEE Transactions on Neural Networks, Special Issue on Evolutionary Computation*, 5(1):96–101, 1994.
38. G. Rudolph. Convergence of non-elitist strategies. In Z. Michalewicz, J. D. Schaffer, H.-P. Schwefel, D. B. Fogel, and H. Kitano, editors, *Proceedings of the First IEEE Conference on Evolutionary Computation*, pages 63–66. IEEE Press, 1994.
39. M. Schütz. Eine Evolutionsstrategie für gemischt-ganzzahlige Optimierungsprobleme mit variabler Dimension. Diplomarbeit, Universität Dortmund, Fachbereich Informatik, 1994.
40. H.-P. Schwefel. *Numerische Optimierung von Computer-Modellen mittels der Evolutionsstrategie*, volume 26 of *Interdisciplinary Systems Research.* Birkhäuser, Basel, 1977.
41. H.-P. Schwefel. Evolutionary learning optimum–seeking on parallel computer architectures. In A. Sydow, S. G. Tzafestas, and R. Vichnevetsky, editors, *Proceedings of the International Symposium on Systems Analysis and Simulation 1988, I: Theory and Foundations*, pages 217–225. Akademie-Verlag, Berlin, September 1988.
42. H.-P. Schwefel. Imitating evolution: Collective, two-level learning processes. In U. Witt, editor, *Explaining Process and Change — Approaches to Evolutionary Economics*, pages 49–63. The University of Michigan Press, Ann Arbor, MI, 1992.
43. H.-P. Schwefel. Natural evolution and collective optimum–seeking. In A. Sydow, editor, *Computational Systems Analysis: Topics and Trends*, pages 5–14. Elsevier, Amsterdam, 1992.
44. H.-P. Schwefel. *Evolution and Optimum Seeking.* Sixth-Generation Computer Technology Series. Wiley, New York, 1995.
45. H.-P. Schwefel and Th. Bäck. Evolution strategies ii: Theoretical aspects. In J. Périaux and G. Winter, editors, *Genetic Algorithms in Engineering and Computer Science*, chapter 7. Wiley, Chichester, 1995.
46. H.-P. Schwefel and G. Rudolph. Contemporary evolution strategies. In F. Morán, A. Moreno, J. J. Merelo, and P. Chacón, editors, *Advances in Artificial Life. Third International Conference on Artificial Life*, volume 929 of *Lecture Notes in Artificial Intelligence*, pages 893–907. Springer, Berlin, 1995.
47. G. Syswerda. Uniform crossover in genetic algorithms. In J. D. Schaffer, editor, *Proceedings of the 3rd International Conference on Genetic Algorithms*, pages 2–9. Morgan Kaufmann Publishers, San Mateo, CA, 1989.
48. H.-M. Voigt, H. Mühlenbein, and D. Cvetković. Fuzzy recombination for the breeder genetic algorithm. In L. Eshelman, editor, *Genetic Algorithms: Proceedings of the 6th International Conference*, pages 104–111. Morgan Kaufmann Publishers, San Francisco, CA, 1995.

An Introduction to Evolutionary Programming

David B. Fogel
Lawrence J. Fogel
Natural Selection, Inc.
3333 North Torrey Pines Ct., Suite 200
La Jolla, CA 92037
fogel@sunshine.ucsd.edu

Abstract

Evolutionary programming is a method for simulating evolution that has been investigated for over 30 years. This paper offers an introduction to evolutionary programming, and indicates its relationship to other methods of evolutionary computation, specifically genetic algorithms and evolution strategies. The original efforts that evolved finite state machines for predicting arbitrary time series, as well as specific recent efforts in combinatorial and continuous optimization are reviewed. Some areas of current investigation are mentioned, including empirical assessment of the optimization performance of the technique and extensions of the method to include mechanisms to self-adapt to the error surface being searched.

1 Introduction

Recent years have seen a rapid increase in the use of simulated evolution to address difficult problems in machine learning. These efforts within the field of *evolutionary computation* have generally followed three main lines of investigation: (1) *genetic algorithms*, (2) *evolution strategies*, or (3) *evolutionary programming*. These techniques are broadly similar. Each relies on a population of contending trial solutions which are subjected to random alterations and compete to be retained as parents of successive progeny. The differences between the methods concern the level in the hierarchy of evolution being modeled: the chromosome, the individual, or the species.

Genetic algorithms model evolution as a succession of changing gene frequencies, with contending solutions being analogous to chromosomes. The space of possible solutions is searched by applying transformations to the trial solutions as observed in the chromosomes of living organisms: crossover, inversion, point mutation, and so forth (Holland 1975). Solutions are typically made to propagate into future generations with a likelihood that is proportional to their fitness relative to all other existing solutions. In this manner, it is hoped that the population will recombine building blocks of independently discovered solutions with above-average fitness and thereby generate new, improved solutions over successive generations (Goldberg 1989).

In contrast, evolution strategies (Rechenberg 1965; Schwefel 1995) and evolutionary programming (Fogel 1962; Fogel et al. 1966) (also sometimes collectively referred to as *evolutionary algorithms*, as opposed to genetic algorithms; Mühlenbein 1992; Fogel 1993a; Goldberg 1994) model evolution as a process of adaptive behavior of individuals or species, respectively, rather than of adaptive genetics. Populations of trial solutions are evolved, but the solutions are modified such that there is a continuous range of possible new behaviors while at the same time maintaining a strong behavioral link between parents and offspring. For example, if the solutions are represented as real-valued vectors, a typical mutation operation is to add a multi-variate zero mean Gaussian random variable to each parent. Selection operates, either deterministically or probabilistically, respectively, to eliminate the worst solutions in the population rather than promote copies of those with above-average fitness.

Any review of evolutionary computation that does not cover all three lines of investigation, as well as additional efforts in simulating the evolutionary dynamics of populations (Conrad and Rizki 1989; Galar 1991; Ray 1991) is incomplete. Yet, due to the nature of the authors' invitation to participate, this paper will focus solely on evolutionary programming, with brief comparisons to evolution strategies and genetic algorithms made as appropriate. Readers interested in a review of all of these techniques (as well as their derivatives, e.g., *genetic programming*) are directed to Fogel (1994a, 1995a), Bäck and Schwefel (1993), Bäck (1995), and Koza (1992).

2 Background on Evolutionary Programming

The original motivation for evolutionary programming centered on generating an alternative approach to artificial intelligence. Rather than emulate or simulate humans, either in their neurophysiological structure or their particular behaviors, natural evolution was modeled as a process that generates organisms of increasing intellect over time. Intelligence was defined as the ability of an organism to achieve goals in a range of environments (Fogel et al. 1966) and intelligent behavior was regarded as requiring the ability to predict future environmental occurrences coupled with a translation of those predictions into suitable responses.

Simple evolutionary prediction experiments were conducted (Fogel et al. 1966) in which finite state machines were used to represent the regularity that underlies the observed environment (Fig. 1). As offered by Fogel et al. (1966), evolutionary programming operated on a population of finite state machines as follows: The logical form of an environment was chosen to represent a sequence of observations, some portion of this sequence having been observed. Initial finite state machines were created at random over a prespecified number of states. Each of these machines was tested on the observed sequence of symbols with respect to how well each could predict as yet unobserved symbols, this with respect to an arbitrary payoff function relating the worth of each possible correct prediction and each possible error. Offspring machines were created from these parents (usually one machine per parent) by random mutation across five possible modes: (1) add a state, (2) delete a state, (3) change the start state, (4) change an output symbol, and (5) change a state transition. Typically, the probabilities for each of these mutations were set equal, and the number of mutations to apply was determined *a priori*. Each offspring machine was tested in a manner similar to their parents and the best half of the population was retained to become parents of the next generation. If a prediction of the next symbol was required, the best machine in the population was used to generate the prediction and the new symbol was then included in the observed environment.

For example, Fogel et al. (1966) considered the nonstationary sequence of symbols generated by classifying each of the increasing possible integers as being prime (represented by the symbol 1) or nonprime (represented by the symbol 0). The environment

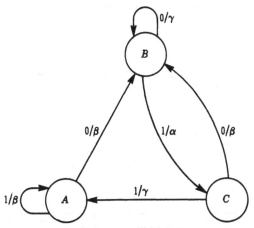

Fig. 1. A finite state machine is defined over an input and output alphabet. For each state, every possible input symbol has a corresponding output symbol and a next-state transition. The machine is presumed to start in state A unless otherwise noted. In the figure, input symbols {0,1} are shown to the left of the virgule ("/") and output symbols {α,β,γ} are shown to the right of the virgule. The machine responds to a sequence of input symbols with a sequence of output symbols determined by the initial state, the input-output relationships of each state and the path through the states as defined by the next-state transitions.

consisted of the sequence 01101010001 ..., wherein each symbol depicts the primeness of the numbers 1, 2, 3, 4, 5, 6, 7, 8, 9, 10, 11, ..., respectively. When using a population of five parent finite state machines undergoing a single mutation per offspring and 10 generations before making each next prediction (introducing the next symbol), Fig. 2 shows the cumulative percent correct, where each machine was scored based on its percentage of correct predictions and a penalty for complexity of 0.01/state, this favoring smaller machines (Occam's razor).

The fundamental principles of the original evolutionary programming of Fogel (1962; Fogel et al. 1966) have remained mostly unchanged. There are two noteworthy additions.

1. Stochastic competition is now typically employed to determine which parents to keep and which to cull from the population. A round-robin tournament is constructed such that each solution must compete for "wins" with a preselected number of randomly chosen members of the population. In each encounter, if the solution being conditioned upon is at least as good as its opponent, it receives a win. Those solutions with the most wins are retained to become parents of the succeeding generation. By varying the number of competitions, selection can range from being stringent (i.e., deterministic, as before) to quite relaxed.

2. Self-adaptive parameters are often included within the coding structure. These parameters indicate the distribution or step size of mutation to be applied to each parameter of the solution vector. For example, if the task is to find the real-valued vector x associated with the minimum of a functional $f(x)$, then the coding structure could take the form (x,σ), where x is the solution to be implemented, σ is the vector of standard deviations for Gaussian mutation, and new solutions (x',σ') are generated by:

$$x' = g(x,\sigma)$$
$$\sigma' = h(\sigma)$$

More detail on this procedure is offered below in the section on future directions.

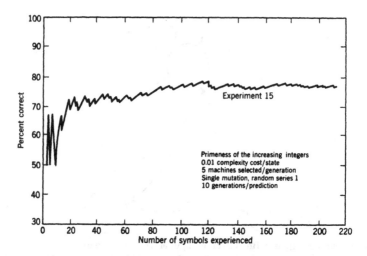

Fig. 2. The cumulative percent correct when using evolutionary programming to evolve finite state machines to predict the primeness of the increasing integers as the function of the previous sequence of primes and non-primes. The figure is taken from Fogel et al. (1966, p. 37).

3 Other Applications of Evolutionary Programming

In evolutionary programming, the representation used follows directly from the problem at hand. There is no suggestion to code all problems as binary strings, logical trees, or any other particular representation. As such, when the technique has been applied to other problems, a wide variety of representations and operators have been implemented. A few different implementations will be highlighted by examples.

3.1 The Traveling Salesman Problem

The traveling salesman problem (TSP) has received considerable attention from the evolutionary computation community. The problem is interesting because (1) it has become a benchmark test case, (2) it is easily stated yet difficult to solve, and (3) it is broadly applicable to a number of routing and networking problems. The task is to take n cities and construct a minimum length tour such that each city is visited once and only once, with a return to the original city completing the circuit. The problem is NP-hard and the number of possible solutions increases as $(n-1)!/2$. As such, it is often less of interest to find the optimum routing than to quickly find reasonably good solutions.

One natural representation of the problem is simply a listing of the cities to be visited in order. Possible solutions are then permutations of this listing. Under evolutionary programming, the construction of a suitable mutation operation should take into consideration two facets: (1) maintaining a strong behavioral link between each parent and its offspring, and (2) providing a (nearly) continuous range of potential new behaviors. One possible mutation operation that meets these requirements for the TSP is an inversion operation: Select two cities in the tour and reverse the order of the segment defined by those two cities. The mutation operation that would offer the generally smallest functional change would be to simply swap two adjacent cities. This is an inversion of length two. As the inversion length is increased, the expected functional change between a parent tour and its offspring is also increased, up to a maximum inversion length of $n/2$, due to the symmetry of the tour. Note that this inversion operation is not chosen because it has any analogous operator in biota; it simply provides a method of variation that will allow for suitable evolution in this problem.

Specifically, the method can be implemented as follows. A population of random tours are scored with respect to their Euclidean length. Each tour is mutated using the inversion operation defined above to generate an offspring. All of the offspring are then scored and a stochastic competition including all parents and offspring is conducted to determine which tours to purge from the population, and as a consequence which tours remain to generate the next succession of progeny.

This method has been implemented on several uniform TSPs (i.e., problems in which the cities are distributed in accordance with a uniform distribution) (Fogel 1993b,c). Fig. 3 shows an evolved solution to a 1,000-city uniform TSP after 4×10^7 function evaluations in Fogel (1993b). The optimization of the best tour in the population is quite rapid. The routing depicted was estimated to be within five to seven percent of the expected optimal tour length. Fogel (1993c) estimated that the number of tours that must be examined in order to achieve solutions which are on average 10 percent worse than the expected best in uniform TSPs increases as the square of the number of cities. Previous efforts by Ambati et al. (1991) with similar procedures estimated that solutions that are on average 25 percent worse than the expected best could be discovered in $O(n \log n)$.

In earlier research, Fogel (1988) implemented a different mutation operation which selected a city at random and replaced it in a random location. Ambati et al. (1991) demonstrated that more effective mutation operations could be constructed, yet the method of Fogel (1988) was observed to outperform some genetic algorithm techniques that relied on recombination (e.g., PMX) (Goldberg and Lingle 1985).

3.2 Evolving Neural Networks

Neural networks are general nonlinear mapping functions which are defined by a number of processing nodes and the manner in which these nodes are connected. Speculation that such networks could be optimized using simulated evolution goes back at least to Bremermann (1966). More recently there have been many efforts to apply evolutionary programming to the design and optimization of neural networks in pattern recognition and control systems.

The most common network involved in these studies has been the mutilayer perceptron (MLP) (Fig. 4). If the network architecture is predefined, the problem becomes one of trying to find optimal settings for the weight and bias terms. A natural representation for this is a real-valued vector where each component corresponds to a weight or bias. For such continuous-valued representations, the common method of mutation is to add a multi-variate zero mean Gaussian random variable to all components, thereby in this case changing all of the weights and biases simultaneously. The behavior of each offspring network is strongly related to its parent and there is a continuous range of possible new behaviors.

The variability between each parent and its offspring can be controlled directly by specifying the individual variances for each Gaussian perturbation (or covariances). One common method for accomplishing this is to set the step size (standard deviation) to be proportional to the mean squared error of the parent network. In this manner, as better solutions are discovered, the step size is reduced and the search effort is concentrated around the parent, and conversely, a larger variance is used for parents with rela-

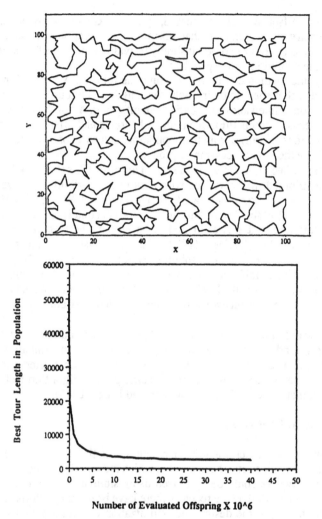

Fig. 3. (a) The final best evolved tour for a 1,000-city TSP in which the cities were distributed at random in accordance with a uniform distribution in both the x and y dimensions. (b) The rate of optimization of the best solution in the population as a function of the number of tours generated. The figure is from Fogel (1993c).

tively poor performance. As noted in the introduction, more sophisticated methods of self-adapting the variability of the permutations have been considered and are discussed below in the section on future efforts.

Some of the first efforts at applying evolutionary programming to optimizing neural networks can be found in Fogel et al. (1990) and Fogel (1991a). More recent research has involved simultaneously evolving both the structure and weights of feed forward and recurrent networks (McDonnell and Waagen 1993, 1994; Angeline et al. 1994). Some attention has also been given to evolving fuzzy neural networks in which classification of input patterns are made with respect to their fuzzy membership in evolved clusters (Brotherton and Simpson 1995).

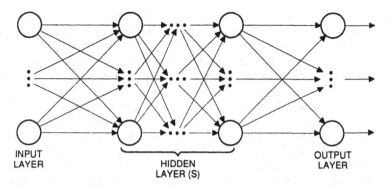

Fig. 4. A multilayer feedforward perceptron consists of a number of input, output and hidden nodes. Each node typically performs a weighted sum of the inputs to that node, subtracts off a variable bias, and then passes the result through a nonlinear filter.

3.3 Evolution and Games

One of the more interesting avenues of research in evolutionary computation involves the co-evolution of solutions in gaming, wherein the appropriateness of a solution depends on its interactions with other solutions in the population. Among these investigations are the *Tierra* simulations of Ray (1991), the competitive sorting research of Kinnear (1993), and many other simulations involving the iterated prisoner's dilemma (Axelrod 1987; Fogel 1991b, 1993d; Harrald and Fogel 1995).

The prisoner's dilemma is a two-person game in which each of the players can adopt one of two behavioral policies: cooperate or defect. *Cooperation* implies increasing the reward of both participants, while *defecting* implies increasing one's own reward at the expense of the other player. A typical payoff matrix for the game is shown in Table 1 (see Fogel (1993d) for a general description of the game). If the game is played for a single iteration, the dominant move is to defect. Defection is also rational if the game is

Table 1. The payoff matrix for typical experiments with the iterated prisoner's dilemma. The matrix was used in Axelrod (1987), Fogel (1991, 1993d), and others. The payoffs indicate the reward to each player for adopting a particular strategy in light of the other player's strategy. Mutual cooperation generates three points for each player. Mutual defection generates only one point for each player. Defecting against a cooperator yields five points for the defector and no points for the cooperator.

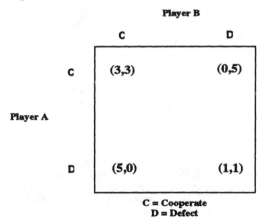

iterated over a series of plays under conditions in which both player's decisions are not affected by previous plays; the game degenerates into a series of independent plays. But if the players' strategies can depend on the results of previous iterations, mutual cooperation can become a rational policy. Evolutionary simulations have generally focused on the dynamics of a population of players interacting according to the payoff matrix in Table 1, or variants of this matrix.

Within the framework of evolutionary programming, the iterated prisoner's dilemma has been simulated by using both finite state machines (Fogel 1991b, 1993d, 1995b) and neural networks (Harrald and Fogel 1995). Fig. 5 shows a typical finite state machine used in such simulations, while Fig. 6 shows the design of a neural network. Finite state machines have been used for conditions in which there are discrete choices between cooperating and defecting, while neural networks have been used to offer a continuum of possible behaviors between these two extremes. In general, the results of experiments with these two constructions have indicated that mutual cooperation is more likely to occur when behaviors are limited to the extremes; when a continuum of behaviors exists, it becomes easier to slip toward mutual defection. Certainly, these results and conclusions are preliminary and readers interested in pursuing research on the prisoner's dilemma should also review Axelrod (1984), Lindgren (1991), Stanley et al. (1994), and others.

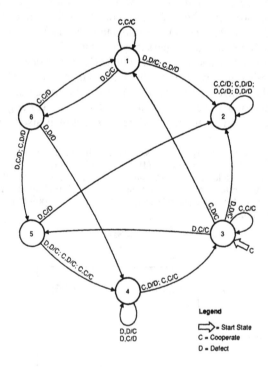

Fig. 5. A typical finite state machine evolved in evolutionary programming experiments with the iterated prisoner's dilemma from Fogel (1993d). The input symbols are {(C,C), (C,D), (D,C), (D,D)}, corresponding to the previous move (cooperate or defect) of the finite state machine and its opponent. The output symbols are {C, D}, corresponding to the possible behavior resulting from the previous pair of moves and the current state.

4 Future Efforts

Much of the current effort in evolutionary programming and other methods of evolutionary computation involves gaining a better understanding of the mathematical properties of these techniques as optimization algorithms. Most of the useful results in convergence have been offered by the German researchers in evolution strategies (Bäck
and Schwefel 1993). For example, it can be shown that when optimizing on a strongly
convex function, evolutionary programming and evolution strategies will (1) converge
geometrically to globally optimal solutions, and (2) demonstrate a speed up of order
log(λ) for using λ offspring from a single parent rather than one offspring. Similar
results with genetic algorithms have been difficult to obtain and this had led to many
empirical comparisons. Some of these have illustrated cases favoring certain recombination methods (Schaffer and Eshelman 1991). But when the objective criterion is a
highly interactive nonlinear function of the parameters to vary, the results have often

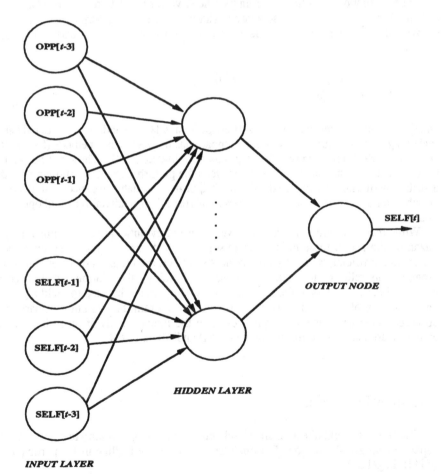

Fig. 6. A multilayer perceptron used for evolving continuous behaviors in the iterated prisoner's
dilemma. There are six input nodes corresponding to the three previous moves of the neural network and its opponent. The number of hidden nodes is variable. The output node is scaled to range
from [-1,1] where -1 indicates complete defection and +1 indicates complete cooperation.

been generally favorable to methods of evolutionary programming and evolution strategies (Bäck and Schwefel 1993; Bäck 1993; Fogel and Atmar 1990; Fogel 1994b; Fogel and Stayton 1994; and others). The careful construction of comparisons between different methods of evolutionary computation and other optimization algorithms, as well as the mathematical analysis of these procedures, remains an open area of research.

Another interesting line of investigation regards the use of self-adaptation in determining how each parent should generate offspring. Rather than predetermine a step size for each dimension of concern, each solution can evolve this information along with the parameters involved in the objective function. There have been two main proposals for such a procedure within the evolutionary programming and evolution strategies communities. Fogel et al. (1991) suggested a method:

$$x_i' = x_i + N(0,\sigma_i)$$
$$\sigma_i' = \sigma_i + \alpha N(0,\sigma_i)$$

where each component of the offspring parameter vector x is determined by a Gaussian perturbation on a self-adaptive standard deviation which is itself perturbed by a scaled normally distributed random variable. In contrast, Schwefel (1981) had earlier proposed a method:

$$\sigma_i' = \sigma_i \cdot \exp(\tau N_i(0,1) + \tau' N(0,1))$$
$$x_i' = x_i + N(0,\sigma_i)$$

where the main difference is that the self-adaptive standard deviations are updated using scaled lognormal perturbations. Initial comparisons between these methods (Saravanan and Fogel 1994) have favored the proposals of Schwefel (1981). These methods have been extended to include correlations between the perturbations, thus allowing for the distribution of trials in arbitrary directions. It appears reasonable to expect that it may be possible to further refine these techniques to increase their potential for rapid optimization.

While these two areas directly involve the use of evolutionary algorithms for optimization, the use of these procedures to study the patterns of complex systems, both natural and artificial, is equally important. As always, the progression of evolutionary programming relies on the careful observation of biological evolution and striving toward facilitating models of the natural process which might be used to yield a greater understanding of the conditions that promote or constrain the evolution of particular observed patterns. Readers more interested in the fundamentals of modeling natural evolution are encouraged to see Fogel (1995a) and Atmar (1994).

5 Acknowledgments

The authors would like to thank M.Schoenauer for the opportunity to contribute this paper. Thanks are also owed to E. Ronald and E. Lutton for facilitating the participation of D.B. Fogel at EA95.

6 References

B.K. Ambati, J. Ambati, and M.M. Mokhtar (1991) "Heuristic combinatorial optimization by simulated Darwinian evolution: a polynomial time algorithm for the traveling salesman problem," *Biological Cybernetics*, Vol. 65, pp. 31-35.

P.J. Angeline, G.M. Saunders and J.B. Pollack (1994) "An evolutionary algorithm that constructs recurrent neural networks," *IEEE Transactions on Neural Networks*, Vol. 5, pp. 54-65.

W. Atmar (1994) "Notes on the simulation of evolution," *IEEE Transactions on Neural Networks*, Vol. 5, pp. 130-148.

R. Axelrod (1987) "The evolution of strategies in the iterated prisoner's dilemma," *Genetic Algorithms and Simulated Annealing*, L. Davis (ed.), Pitman, London, pp. 32-42.

R. Axelrod (1984) *The Evolution of Cooperation*, Basic Books, NY.

T. Bäck and H.-P. Schwefel (1993) "An overview of evolutionary algorithms for parameter optimization," *Evolutionary Computation*, Vol. 1:1, pp. 1-24.

T. Bäck (1995) *Evolutionary Algorithms in Theory and Practice*, IOP Press, Philadelphia, PA, in press.

H.J. Bremermann (1966) "Numerical optimization procedures derived from biological evolution processes," *Cybernetic Problems in Bionics*, H.L. Oestreicher and D.R. Moore (eds.), Gordon and Breach, London, pp. 543-562.

T.W. Brotherton and P.K. Simpson (1995) "Dynamic Feature Set Training of Neural Nets for Classification," *Evolutionary Programming IV: Proceedings of the Fourth Annual Conference on Evolutionary Programming*, J.R. McDonnell, R.G. Reynolds, and D.B. Fogel (eds.), MIT Press, Cambridge, MA, 1995, pp. 83-94.

M. Conrad and M.M. Rizki (1989) "The artificial worlds approach to emergent evolution," *BioSystems*, Vol. 23, pp. 247-260.

D.B. Fogel (1988) "An evolutionary approach to the traveling salesman problem," *Biological Cybernetics*, Vol. 60, pp. 139-144.

D.B. Fogel (1991a) "An information criterion for optimal neural network selection," *IEEE Transactions on Neural Networks*, Vol. 2, 1991, pp. 490-497.

D.B. Fogel (1991b) "The evolution of intelligent decision making in gaming," *Cybernetics and Systems*, Vol. 22, pp. 223-236.

D.B. Fogel (1993a) "On the philosophical differences between evolutionary algorithms and genetic algorithms," *Proceedings of the Second Annual Conference on Evolutionary Programming*, D.B. Fogel and W. Atmar (eds.), Evolutionary Programming Society, La Jolla, CA, pp. 23-29.

D.B. Fogel (1993b) "Applying evolutionary programming to selected traveling salesman problems," *Cybernetics and Systems*, Vol. 24, pp. 27-36.

D.B. Fogel (1993c) "Empirical estimation of the computation required to discover approximate solutions to the traveling salesman problem using evolutionary programming," *Proceedings of the Second Annual Conference on Evolutionary Programming*, D.B. Fogel and W. Atmar (eds.), Evolutionary Programming Society, La Jolla, CA, pp. 56-61.

D.B. Fogel (1993d) "Evolving behaviors in the iterated prisoner's dilemma," *Evolutionary Computation*, Vol. 1, pp. 77-97.

D.B. Fogel (1994a) "An introduction to simulated evolutionary optimization," *IEEE Transactions on Neural Networks*, Vol. 5:1, pp. 3-14.

D.B. Fogel (1994b) "Applying evolutionary programming to selected control problems," *Comp. Math. Applic.*, Vol 27:11, pp. 89-104.

D.B. Fogel (1995a) *Evolutionary Computation: Toward a New Philosophy of Machine Intelligence*, IEEE Press, Piscataway, NJ.

D.B. Fogel (1995b) "On the relationship between the duration of an encounter and the evolution of cooperation in the iterated prisoner's dilemma," *Evolutionary Computation*, in press.

D.B. Fogel, L.J. Fogel and V.W. Porto (1990) "Evolving neural networks," *Biological Cybernetics*, Vol. 63, pp. 487-493.

D.B. Fogel and L.C. Stayton (1994c) "On the effectiveness of crossover in simulated evolutionary optimization," *BioSystems*, Vol 32:3, pp. 171-182.

D.B. Fogel and J.W. Atmar (1990) "Comparing genetic operators with Gaussian mutations in simulated evolutionary processing using linear systems," *Biological Cybernetics*, Vol. 63, pp. 111-114.

D.B. Fogel, L.J. Fogel and J.W. Atmar (1991) "Meta-evolutionary programming," *Proc. of the Asilomar Conf. on Signals, Systems and Computers*, R.R. Chen (ed.), Maple Press, San Jose, CA, pp. 540-545.

L.J. Fogel (1962) "Autonomous automata," *Industrial Research*, Vol. 4, pp. 14-19.

L.J. Fogel, A.J. Owens and M.J. Walsh (1966) *Artificial Intelligence through Simulated Evolution*, John Wiley, NY.

R. Galar (1991) "Simulation of local evolutionary dynamics of small populations," *Biological Cybernetics*, Vol. 65, pp. 37-45.

D.E. Goldberg (1989) *Genetic Algorithms in Search, Optimization and Machine Learning*, Addison-Wesley, Reading, MA, 1989.

D.E. Goldberg (1994) "Genetic and evolutionary algorithms come of age," *Communications of the ACM*, Vol. 37, pp. 113-119.

D.E. Goldberg and R. Lingle (1985) "Alleles, Loci, and the Traveling Salesman Problem," *Proceedings of an International Conference on Genetic Algorithms and Their Applications*, J.J. Grefenstette (ed.), pp. 154-159.

P.G. Harrald and D.B. Fogel (1995) "Evolving continuous behaviors in the iterated prisoner's dilemma," *BioSystems*, in press.

J.H. Holland (1975) *Adaptation in Natural and Artificial Systems*, Univ. of Michigan Press, Ann Arbor, MI.

K. Kinnear (1993) "Evolving a sort: lessons in genetic programming" *IEEE International Conference on Neural Networks 1993*, IEEE Press, Piscataway, NJ.

J.R. Koza (1992) *Genetic Programming*, MIT Press, Cambridge, MA.

K. Lindgren (1991) "Evolutionary phenomena in simple dynamics," *Artificial Life II*, C.G. Langton, C. Taylor, J.D. Farmer and S. Rasmussen (eds.), Addison-Wesley, Reading, MA, pp. 295-312.

J.R. McDonnell and D. Waagen (1993) "Neural network structure design by evolutionary programming," *Proceedings of the Second Annual Conference on Evolutionary Programming*, D.B. Fogel and W. Atmar (eds.), Evolutionary Programming Society, La Jolla, CA, pp. 79-89.

J.R. McDonnell and D. Waagen (1994) "Evolving recurrent perceptrons for time-series modeling," *IEEE Transactions on Neural Networks*, Vol. 5, pp. 24-38.

H. Mühlenbein (1992) "Evolution in time and space — the parallel genetic algorithm," Foundations of Genetic Algorithms, G.J.E. Rawlins (ed.), Morgan Kaufmann, San Mateo, CA, pp. 316-337.

T. Ray (1991) "An approach to the synthesis of life," *Artificial Life II*, C.G. Langton, C. Taylor, J.D. Farmer and S. Rasmussen (eds.), Addison-Wesley, Reading, MA, pp. 371-408.

I. Rechenberg (1965) "Cybernetic solution path of an experimental problem," Royal Aircraft Establishment, Library Translation No. 1122, August.

N. Saravanan and D.B. Fogel (1994) "Learning strategy parameters in evolutionary programming: an empirical study," *Proc. of the Third Annual Conference on Evolutionary Programming*, A.V. Sebald and L.J. Fogel (eds.), World Scientific, River Edge, NJ, pp. 269-280.

J.D. Schaffer and L. Eshelman (1991) "On crossover as an evolutionarily viable strategy," *Proc. of the Fourth Intern. Conf. on Genetic Algorithms*, R.K. Belew and L.B. Booker (eds.), Morgan Kaufmann, San Mateo, CA, pp. 61-68.

H.-P. Schwefel (1981) *Numerical Optimization of Computer Models*, Chichester, UK.

H.-P. Schwefel (1995) *Evolution and Optimum Seeking*, John Wiley, NY.

E.A. Stanley, D. Ashlock and L. Tesfatsion (1994) "Iterated prisoner's dilemma with choice and refusal of partners," *Artificial Life III*, C.G. Langton (ed.), Addison-Wesley, Reading, MA, pp. 131-175

Evolutionary Computation Theory

An Asymptotic Theory for Genetic Algorithms

Raphaël CERF

Université de Montpellier*

Abstract. The Freidlin–Wentzell theory deals with the study of random perturbations of dynamical systems. We build several models of genetic algorithms by randomly perturbing simple processes. The asymptotic dynamics of the resulting processes is analyzed with the powerful tools developed by Freidlin and Wentzell and later by Azencott, Catoni and Trouvé in the framework of the generalized simulated annealing. First, a markovian model inspired by Holland's simple genetic algorithm is built by randomly perturbing a very simple selection scheme. The convergence toward the global maxima of the fitness function becomes possible when the population size is greater than a critical value which depends strongly on the optimization problem. In the bitstring case, the critical value of the population size is smaller than a linear function of the chromosome length, provided that the difficulty of the fitness landscape remains bounded. We then use the concepts introduced by Catoni and further generalized by Trouvé to fathom more deeply the dynamics of the two operators mutation–selection algorithm when the population size becomes very large. A new genetic algorithm is finally presented. A new selection mechanism is used, which has the decisive advantage of preserving the diversity of the individuals in the population. When the population size is greater than a critical value, the delicate asymptotic interaction of the perturbations ensures the convergence toward a set of populations which contain all the global maxima of the fitness function.

1 Overview

In the realm of stochastic optimization, attention has focused essentially on two techniques during the last decade: simulated annealing [1] and evolutionary algorithms [11, 13].

The theory of the simulated annealing is now extensively developed and a great number of results describing the dynamics of this kind of algorithms in various settings are available. As far as we know, the most accurate work in this area has been achieved by Catoni, in the spirit of the Freidlin–Wentzell theory [2, 3, 4]. Nevertheless, the simulated annealing presents a fundamental drawback: it is sequential in nature.

In an attempt to investigate the theoretical aspects of the parallelization of this algorithm, Trouvé carried out a systematic study (initiated by Hwang and

---

* Current address: Université Paris–Sud, Mathématique, bâtiment 425, 91405 Orsay.
  E–mail: cerf at stats.matups.fr

Sheu [12]) of a broader class of algorithms he baptized generalized simulated annealing [16, 17, 18]. As it turns out, the general framework of the Markov chains with exponentially vanishing transition probabilities is also well adapted for evolutionary algorithms and allows to prove several convergence results. We put forward the existence of critical population sizes.

Up to now, there have been several analysis of genetic algorithms as Markov chains which have not lead to satisfactory results [9, 14, 15]: either they are very restrictive or they describe situations where convergence toward the global maxima fails.

We first propose a markovian model inspired by Holland's simple genetic algorithm which is built by randomly perturbing a very simple selection scheme [5]: mutations and crossovers are considered as vanishing random perturbations. We prove that the convergence toward the global maxima of the fitness function becomes possible when the population size is greater than a critical value (which depends strongly on the optimization problem). Surprisingly, the crossover is not fundamental to ensure this convergence: the crucial point is the delicate asymptotic interaction between the local perturbations of the individuals (i.e. the mutations) and the selection pressure. We show how our general model specializes to the bitstring case, where the state space is $\{0, 1\}^N$. A rough upper bound of the critical value of the population size implies that it is smaller than a linear function of the chromosome length, provided that the difficulty of the fitness landscape remains bounded.

We then use the concepts introduced by Catoni [2, 3, 4] and further generalized by Trouvé [16, 17, 18] to fathom more deeply the dynamics of the two operators mutation–selection algorithm when the population size becomes very large [6]. The key result lies in the structure of the trajectories of populations joining two uniform populations: a small group of individuals sacrifice themselves in order to create an ideal path which is then followed by all other individuals. As a consequence, the various quantities associated with the algorithm (such as the communication cost, the virtual energy, the communication altitude) are affine functions of the population size and the hierarchy of cycles on the set of the uniform populations stabilizes. Furthermore, if the mutation kernel is symmetric, the limiting distribution is the uniform distribution over the set of the global maxima of the fitness function.

We introduce eventually two major modifications to the previous schemes [7]. The crossover is now integrated in the unperturbed process and is not considered any more as a random vanishing perturbation: although this operator is not essential to ensure the desired convergence, it certainly increases the efficiency of the algorithm. Three conditions of biological relevance are imposed on the crossover:
• When two identical individuals mate, they produce offsprings identical to themselves.
• There is always a non–zero probability that nothing happens during a crossover.
• The two individuals of the mating pair play symmetric roles (the populations are asexual).

Amazingly enough, the first two conditions (especially the first one) are crucial for our algorithm to work. Furthermore, we propose a new selection mechanism which has the decisive advantage of preserving the diversity of the individuals in the population. The analysis of this algorithm follows the road opened by Freidlin and Wentzell [10]. Unlike the situations studied in [5, 6], the structure of the set of the attractors of the unperturbed process is very rich: these are particular subsets of populations and they stand in one to one correspondence with the equi–fitness subsets of the state space whose cardinality is less than the population size. When the population size is large enough, there is a unique ideal attractor whose populations contain all the maxima of the fitness function. We study the communication cost between attractors: the costs of bad transitions (either those which decrease the maximal fitness of the population or those which lose some peak fitness individuals) increase linearly with the population size, whereas the costs of good transitions (those which create some new peak fitness individuals) remain bounded. As a consequence, when the population size is greater than a critical value, the minimum of the virtual energy corresponds to the ideal attractor previously described. Therefore the sequence of the stationary measures (associated with a fixed level of intensity of the perturbations) concentrates on this attractor as the perturbations vanish. The remaining problem is to adapt carefully the rate of decreasing of the perturbations in order to obtain an inhomogeneous Markov chain with the same limiting law. Besides, it is possible to ensure a stronger convergence: we may force the process to be forever trapped in the attraction basin of the ideal attractor after a finite number of transitions. Furthermore, when the population size is large, the cycles which do not contain the ideal attractor are reduced to one single attractor and the optimal convergence exponent increases faster than an affine function of the population size.

Our algorithm finds simultaneously all the global maxima of the objective function in finite time and thus solves completely the optimization problem: it seems to be the first of this kind. The cornerstone of this cooperative search procedure is the delicate asymptotic interaction between the mutations and our enhanced selection mechanism; the process explores simultaneously and without respite the neighbourhoods of the best points found so far (instead of focusing on a particular point). Moreover, we hope to have found the right way of using the crossover operator.

We only outline here the main results of [5, 6, 7] without any proof. Each section starts with a rough non technical summary. The precise formulations are given thereafter.

2   Asymptotic convergence of genetic algorithms

We propose a general model of a simple genetic algorithm [5]. We first consider a process (X_n^∞) which is a pure selection process: at each time step it selects the best fit individuals of the population and destroy the others. This process is quickly trapped in a uniform population (that is a population containing only copies of the same individual) where it remains forever. We then build a perturbed process (X_n^l) (the parameter l controls the perturbation level) using three operators: mutation, crossover, selection. As l goes to infinity the transition mechanism of (X_n^l) converges to the transition mechanism of (X_n^∞): the mutations and the crossovers vanish whereas the selection pressure increases. In the homogeneous case (i.e. where the intensity of the perturbations is kept constant in time), we study the asymptotic behavior of the equilibrium state of the process (X_n^l) as l goes to infinity (the equilibrium state is, roughly speaking, the probabilistic law of the process after it has run a very long time). Using the Freidlin–Wentzell theory [10], we prove that, under certain conditions which are not automatically fulfilled, the probability to end up with a population made of global maxima converges to one, whatever the initial population. The most interesting of these conditions is the existence of a critical population size which depends heavily upon the optimization problem. In the inhomogeneous case, we try to find an increasing sequence $l(n)$ such that the same result holds directly for the process $(X_n^{l(n)})$. This can be achieved under certain conditions on the rate of decreasing of the perturbations. We also show how our model specializes to the bitstring case and we give an upper bound for the critical population size.

We start now with the precise formulation of this program. Let E be a finite set and f a positive function defined on E (the fitness function). The aim is to determine the set f^* of the global maxima of f. Let m be a population size. For $x = (x_1, \cdots, x_m)$ an element of E^m and i a point of E, we define the maximal fitness reached in the population x,

$$\widehat{f}(x) = \widehat{f}(x_1, \cdots, x_m) = \max\{\, f(x_k) : 1 \le k \le m \,\},$$

the set of the individuals of the population x taking this maximal value,

$$\widehat{x} = \{\, x_k : 1 \le k \le m,\ f(x_k) = \widehat{f}(x) \,\},$$

the number of the individuals of the population identical to i,

$$x(i) = \mathrm{card}\{\, k : 1 \le k \le m,\ x_k = i \,\},$$

and the set of all points of the search space appearing at least once in the population i,

$$[x] = \{\, x_k : 1 \le k \le m \,\}.$$

The set S is the set of equi–fitness populations, that is populations whose individuals have the same fitness:

$$S = \{\, x \in E^m : f(x_1) = \cdots = f(x_m) \,\}.$$

Let (X_n^∞) be the Markov chain on E^m having for transition matrix (this is the matrix indexed by E^m whose coefficient (y, z) is the probability of jumping from the population y to the population z)

$$P\left(X_{n+1}^\infty = z / X_n^\infty = y\right) = \prod_{k=1}^m \frac{1_{\widehat{y}}(z_k)\, y(z_k)}{\mathrm{card}\,\widehat{y}}$$

that is, the individuals of the population X_{n+1}^∞ are chosen randomly (under the uniform distribution) and independently among the elements of $\widehat{X}_n^\infty$ which are the best individuals of X_n^∞ according to the fitness function f. The Markov chain (X_n^∞) is instantaneously trapped in S and after a finite number of steps it is absorbed in a uniform population (i.e. a population containing m copies of the same individual). We build a sequence of Markov chains (X_n^l) by randomly perturbing (X_n^∞). The parameter l controls the intensity of the perturbations. We decompose the transition mechanism of (X_n^l) in three stages:

$$X_n^l \xrightarrow{\text{mutation}} U_n^l \xrightarrow{\text{crossover}} V_n^l \xrightarrow{\text{selection}} X_{n+1}^l.$$

$X_n^l \longrightarrow U_n^l$: **mutation.** The mutation operator is modeled by random independent perturbations of the individuals of the population X_n^l. The transition probabilities from X_n^l to U_n^l are then given by

$$P\left(U_n^l = u/X_n^l = x\right) = p_l(x_1, u_1) \cdots p_l(x_m, u_m)$$

where p_l is a markovian kernel on E, that is a function defined on $E \times E$ with values in $[0, 1]$ such that

$$\forall i \in E \qquad \sum_{j \in E} p_l(i, j) = 1.$$

The quantity $p_l(i, j)$ is the probability to transform i into j by mutation at perturbation level l. We suppose that p_l satisfies

Hypothesis H_p: There exist a function α defined on $E \times E$ with values in $\mathbb{R}^+$ which is an irreducible kernel i.e.

$$\forall i, j \in E \quad \exists i_0, i_1, \cdots, i_r \quad i_0 = i, \quad i_r = j, \quad \prod_{0 \le k \le r-1} \alpha(i_k, i_{k+1}) > 0$$

and a positive real number a such that, for all $s \ge a$, p_l admits the expansion

$$p_l(i, j) = \begin{cases} \alpha(i, j) l^{-a} + o\left(l^{-s}\right) & \text{if } i \ne j \\ 1 - \alpha(i, j) l^{-a} + o\left(l^{-s}\right) & \text{if } i = j \end{cases}$$

$U_n^l \longrightarrow V_n^l$: **crossover.** The crossover operator is modeled by random independent perturbations of the couples formed by consecutive individuals of the population (X_n^l). The transition probabilities from U_n^l to V_n^l are

$$P\left(V_n^l = v/U_n^l = u\right) = \delta_m(u_m, v_m) \prod_{1 \le k \le m/2} q_l\left((u_{2k-1}, u_{2k}), (v_{2k-1}, v_{2k})\right)$$

where $\delta_m(i, j) = \delta(i, j)$ if m is odd (the last individual remains unchanged), $\delta_m(i, j) = 1$ if m is even, and q_l is a markovian kernel on E satisfying

Hypothesis H_q: There exists a kernel β over $E \times E$ and a positive real number b such that, for all $s \ge b$, q_l has the expansion

$$q_l((i_1, j_1), (i_2, j_2)) = \begin{cases} \beta\left((i_1, j_1), (i_2, j_2)\right) l^{-b} + o\left(l^{-s}\right) & \text{if } (i_1, j_1) \ne (i_2, j_2) \\ 1 - \beta\left((i_1, j_1), (i_2, j_2)\right) l^{-b} + o\left(l^{-s}\right) & \text{otherwise} \end{cases}$$

$V_n^l \longrightarrow X_{n+1}^l$: **selection.** The transition probabilities from V_n^l to X_{n+1}^l are

$$P\left(X_{n+1}^l = x/V_n^l = v\right) = \prod_{r=1}^{m} \frac{v(x_r)\exp(cf(x_r)\ln l)}{\sum_{k=1}^{m}\exp(cf(v_k)\ln l)}$$

(where $c \in \mathbb{R}_*^+$ is a scaling parameter).

As l grows toward infinity the perturbations become smaller and smaller and with overwhelming probability, the chain (X_n^l) behaves as (X_n^∞) would do: it is attracted by the set S. Using the Freidlin–Wentzell theory, we carry out a precise study of the asymptotic dynamics of the chain when the perturbations disappear. Let T_n be the instant of the n-th visit to the set S and let $Z_n^l = X_{T_n}^l$. The study of the Markov chain (Z_n^l) of the successive visits to the set of the attractors S is equivalent to the study of the whole process (X_n^l). We have then

$$P\left(Z_{n+1}^l = z/Z_n^l = y\right) \sim \widetilde{C}(y,z)\, l^{-\widetilde{V}(y,z)}$$

as l goes to infinity, where $\widetilde{C}$ and $\widetilde{V}$ are two positive kernels on S. We then compute estimates of the invariant measure of the Markov chain Z_n^l with the help of the representation formula using Freidlin–Wentzell graphs. The crucial quantity is the virtual energy W, defined for x in S by

$$W(x) = \min\left\{ \sum_{(y\to z)\in g} \widetilde{V}(y,z) : g \in G(x) \right\}$$

where $G(x)$ is the set of the x–graphs over S. We consider two situations. In the homogeneous case, we study the asymptotic behavior of the stationary measure of (X_n^l). We derive sufficient conditions to ensure that

$$\forall x \in E^m \qquad \lim_{l\to\infty}\ \lim_{n\to\infty}\ P\left([X_n^l] \subset f^*/X_0^l = x\right) = 1.$$

We show that our conditions are always fulfilled for a sufficiently large population size. In the inhomogeneous case, where l is an increasing function of n, we derive several conditions on the population size and on the rate of increasing of the sequence $l(n)$, some necessary and some sufficient, to ensure the settling in f^* in finite time i.e.

$$\forall x \in E^m \qquad P\left(\exists N\ \forall n \geq N\ [X_{T_n}] \subset f^*/X_0 = x\right) = 1.$$

Application to the bitstring case. We specialize our general model in order to apply our results to the case where $E = \{0,1\}^N$ for some integer N. A point i of E is a word of length N over the alphabet $\{0,1\}$ and is noted $i = i_1 \cdots i_N$ where $i_k \in \{0,1\}$. The Hamming distance $H(i,j)$ between two points i,j of E is the number of letters where i and j differ, that is $H(i,j) = \operatorname{card}\{k : 1 \leq k \leq m,\ i_k \neq j_k\}$. The kernel α defining the mutation kernel p_l is

$$\alpha(i,j) = \begin{cases} 0 & \text{if } H(i,j) > 1, \\ 1 & \text{if } H(i,j) = 1, \\ N & \text{if } H(i,j) = 0. \end{cases}$$

It is irreducible: the minimal number of transitions necessary to join two arbitrary points of E through the kernel α is N. In order to build the crossover operator, we define now a cutting operator T_k for k in $\{1 \cdots N-1\}$; T_k maps $E \times E$ onto $E \times E$ and for i, j in E, we put $T_k(i, j) = (i', j')$ where

$$i' = i_1 \cdots i_k j_{k+1} \cdots j_N, \qquad j' = j_1 \cdots j_k i_{k+1} \cdots i_N.$$

The kernel β associated to the crossover kernel q_l is then defined to be

$$\beta((i,j),(i',j')) = \text{card}\left\{ k : 1 \le k \le N-1, T_k(i,j) = (i',j') \right\}.$$

We choose two kernels p_l and q_l satisfying hypothesis H_p and H_q (for instance the kernels defined by the first terms of the expansions in H_p and H_q). We put

$$\Delta^\circ = \max_{i \in E \backslash f^*} \min \left\{ \max_{0 \le k < r} (f(i) - f(i^k)) : i^0 = i,\ r \le N,\ f(i^r) > f(i) \right\},$$

$$\Delta^* = \max_{i,j \in f^*} \min \left\{ \max_{0 \le k < r} (f(i) - f(i^k)) : i^0 = i,\ r \le N,\ i^r = j \right\},$$

where the sequences $i^0, \cdots, i^r$ are such that $H(i^0, i^1)H(i^1, i^2) \cdots H(i^{r-1}, i^r) > 0$. We let $\Delta^\circledast = \max(\Delta^\circ, \Delta^*)$ and $\delta = \min\{ |f(i) - f(j)| : i, j \in E,\ f(i) \ne f(j) \}$. Within this framework, all the convergence results do apply. We restate and comment only a rough upper bound of the critical population size, which has a more precise form in this context.

Theorem 1 *If the population size m is such that*

$$m > \frac{aN + c(N-1)\Delta^\circledast}{\min(a, b/2, c\delta)}$$

then the stationary measures of the processes (X_n^l) concentrate on f^* i.e.

$$\forall x \in E^m \qquad \lim_{l \to \infty} \lim_{n \to \infty} P\left([X_n^l] \subset f^* / X_0^l = x\right) = 1.$$

Our condition shows that the optimization problem may be solved with a sufficiently large population size. In addition, we are completely free for the choice of the parameters a, b and c: if we take a very large c, so that $\min(a, b/2) \le c\delta$, we see that δ does not intervene any more in the above condition. The parameter $\Delta^\circledast$ describes the difficulty for an individual to travel from one point to a better point. If $\Delta^\circledast$ is fixed and N becomes large, a population size of order CN for some constant C will always suffice to handle the problem. Finally, let us remark that our condition on the population size m is very rough: neither the fine structure of the optimization problem nor the possibility of using crossovers to travel between two populations have been taken into account to obtain it. The crucial point is that the perturbation mechanism allows the process to visit all the space, even when the random perturbations are very small. The role of the crossover is thus not fundamental (the algorithm without crossover corresponds to the case $b = \infty$): however this operator creates a lot of new possible transitions for the chain (X_n) and thus certainly decreases the value of the critical population size.

3 The dynamics of mutation–selection algorithms with large population sizes

An important result of the first paper is that the crossover operator is not essential in order to ensure the convergence of the algorithm toward the global maxima of the fitness function. In the second paper [6], we suppress the crossover in the previous model and focus on the simple mutation–selection algorithm. We first describe the perturbed process, which is essentially the one of the previous section but without crossover. We then show how to embed the study of this process into the abstract framework of the generalized simulated annealing. We state precisely the critical population size theorem, which asserts that convergence toward the global maxima of the fitness function occurs beyond a critical population size. We describe the dynamical behavior of the process. That is, we seek the typical structure of the trajectories of populations the process is most likely to follow when the perturbations vanish. This structure is completely elucidated when the population size is large. This is the essence of theorem 4. This result have far–reaching consequences concerning the dynamics of the process, which are only alluded to here. We end up with several applications of results coming from the theory of the generalized simulated annealing: a necessary and sufficient condition of convergence for the rate of decreasing of the perturbations and an estimation of the speed of convergence.

We thus study a Markov chain (X_n^l) built with the mutation and selection operators:

$$X_n^l \xrightarrow{\text{mutation}} Y_n^l \xrightarrow{\text{selection}} X_{n+1}^l.$$

$X_n^l \longrightarrow Y_n^l$: **mutation.** Let α be an irreducible markovian kernel on E. We set

$$\alpha_l(i,j) = \begin{cases} \alpha(i,j)\, l^{-a} & \text{if } i \neq j \\ 1 - \sum_{e \neq i} \alpha(i,e)\, l^{-a} & \text{if } i = j \end{cases}$$

The transition matrix from X_n^l to Y_n^l is defined by

$$P\left(Y_n^l = y / X_n^l = x\right) = \alpha_l(x_1, y_1) \cdots \alpha_l(x_m, y_m).$$

$Y_n^l \longrightarrow X_{n+1}^l$: **selection.** We use the same mechanism as in the previous model:

$$P\left(X_{n+1}^l = x / Y_n^l = y\right) = \prod_{r=1}^m \frac{y(x_r) \exp(cf(x_r) \ln l)}{\sum_{k=1}^m \exp(cf(y_k) \ln l)}$$

and the process (X_n^l) is still a perturbation of the process (X_n^∞), in the sense that for all y, z in E^m

$$\lim_{l \to \infty} P\left(X_{n+1}^l = z / X_n^l = y\right) = P\left(X_{n+1}^\infty = z / X_n^\infty = y\right).$$

We compute the asymptotic expansion of the transition probabilities as l goes to infinity:

$$P\left(X_{n+1}^l = z / X_n^l = y\right) \underset{l \to \infty}{\sim} q_1(y,z) \exp\left(-V_1(y,z) \ln l\right)$$

where $q_1(y, z)$ is a non–negative constant and $V_1(y, z)$ is the communication cost

$$V_1(y, z) = \min \left\{ a \operatorname{card} \{ k : 1 \leq k \leq m, y_k \neq x_k \} + c \sum_{k=1}^{m} (\widehat{f}(x) - f(z_k)) : \right.$$
$$\left. x \text{ such that } P\left(X_{n+1}^l = z, Y_n^l = x / X_n^l = y\right) > 0 \right\}.$$

We are now in the framework of the generalized simulated annealing studied by Trouvé. That is, the transition probabilities of the process (X_n^l) form a family of Markov kernels on the space E^m indexed by l which is admissible for the communication kernel q_1 and the cost function V_1 [16] [Definition 3.1]. We can therefore use all the tools developed by Catoni and Trouvé [2, 3, 4, 16, 17, 18]. The virtual energy W_1 associated to the communication cost V_1 is defined by

$$\forall x \in E^m \qquad W_1(x) = \min \left\{ \sum_{(x_1 \to x_2) \in g} V_1(x_1, x_2) : g \in G(x) \right\}$$

where $G(x)$ is the set of the x–graphs over E^m. Let W_1^* be the set of the global minima of W_1. The probability mass concentrates on W_1^* as l goes to infinity:

Proposition 1 (Freidlin and Wentzell) *For all x in E^m,*

$$\lim_{l \to \infty} \lim_{n \to \infty} P\left(X_n^l \in W_1^* / X_0^l = x\right) = 1.$$

We identify the uniform population $(i, \cdots, i)$ and the point i of E. The set W_1^* is included in the set of the stable attractors of the process (X_n^∞), i.e. the uniform populations, and can thus be seen as a subset of E. The main result of the first paper can be formulated as follows.

Theorem 2 (critical population size) *There exists a critical population size m^* depending on the space E, the function f, the mutation kernel α and the parameters a, c, such that*

$$m \geq m^* \qquad \Longrightarrow \qquad W_1^* \subset f^*.$$

The fundamental quantity for describing the dynamics of the algorithm is the communication altitude A_1 introduced by Trouvé and defined by

$$A_1(x, y) = \inf \left\{ \max_{1 \leq k < r} W_1(p_k) + V_1(p_k, p_{k+1}) : p_1 = x, p_r = y \right\}.$$

In the case of the sequential simulated annealing where $V(i, j) = (f(j) - f(i))^+$, the communication altitude is the minimal energy barrier one has to overcome to travel from x to y. All the information on the asymptotic dynamics of the algorithm is contained in the decomposition of the space into a hierarchy of particular subsets called the cycles. Roughly speaking, the cycles are the most attractive and stable sets we can build for the perturbed process. The hierarchy of the cycles can be built directly with the communication altitude and it is possible to define the heights of exit and of mixing (H_e, H_m) for arbitrary subsets of E. The height of exit of a set may be interpreted as the perturbation energy necessary to escape from the set.

We first replace the function V_1 by the function V defined by

$$V(x,y) = \min \left\{ V_1(p_1,p_2) + \cdots + V_1(p_{r-1},p_r) : p_1 = x, p_r = y \right\}$$

and we prove a result valid in the abstract framework of the generalized simulated annealing: the virtual energy W and the communication altitude A associated to V coincide with W_1 and A_1.

Our aim is to study the dynamics of the algorithm when m becomes large: a major difficulty is that the size of the state space E^m grows geometrically with m. The whole study relies on the following result, also valid in the abstract framework of the generalized simulated annealing.

Theorem 3 *Let H be a subset of E^m such that*

$$\forall x \in E^m \quad \exists y \in H \quad V(x,y) = 0.$$

We define a cost V_H over $H \times H$ by

$$V_H(x,y) = \inf \left\{ \sum_{k=1}^{r-1} V_1(p_k, p_{k+1}) : p_1 = x, p_r = y, p_k \notin H, 1 < k < r \right\}.$$

Let A_H and W_H be the communication altitude and the virtual energy over H associated to V_H. Then we have $W = W_H$ and $A = A_H$.

We apply this result to the set $H = U = \{ (i, \cdots, i) : i \in E \}$ of the uniform populations. This set U is in one to one correspondence with E and its cardinality does not depend on m. We then show that it is equivalent to build the whole hierarchy of the cycles over E^m, and then to restrict it to U (by keeping only the uniform populations in each cycle), or to build directly the hierarchy of the cycles over U with A_U. In a sense, the operations "building of the cycles" and "restriction to U" commute. We are thus lead to study the restriction of the dynamics to the set of the stable attractors U. We then examine the trajectories of minimal cost between two uniform populations. When the population size m is large, their structure may be described as follows: a small group of individuals sacrifice themselves in order to create an ideal path which is then followed by all other individuals.

Theorem 4 (linearity of the cost) *Let i,j be two points of E. When m is large enough, $V(i,j)$ is an affine function of m:*

$$\exists M \quad \forall m \geq M \quad V(i,j) = \theta(i,j) + m\Omega(i,j).$$

The coefficient $\theta(i,j)$ corresponds to the cost of the trajectories of the small group of individuals and $\Omega(i,j)$ is the cost of the ideal path. We give several properties of the coefficients $\theta(i,j)$ and $\Omega(i,j)$. We describe the behaviour of Ω in two extremal situations, when a/c is very small of very large compared to the variations of the function f.

All the quantities associated to the cycle decomposition are defined through V as the minimum or the maximum of a finite set of sums built with V and are thus affine functions of m when m is large: it is the case for the virtual energy and the communication altitude.

All these affine functions, which are in finite number, possess a finite number of points of intersection. Let M^* be an integer large enough beyond which there are no more changes between the relative positions of two of these affine functions. The hierarchy of the cycles over U is built with the help of equivalence classes of comparison relations linked with the communication altitude. These relations are frozen when m is greater than M^* so that the hierarchy of the cycles stabilizes and does not depend any more on m when $m \geq M^*$.

We study the structure of this limiting hierarchy. We have $W^* \subset f^*$. We prove that the function f is constant over the cycles which do not intersect W^*.

Theorem 5 (limiting distribution)
Suppose that the kernel α is symmetric (i.e. $\alpha(i,j) = \alpha(j,i)$) and $m \geq M^*$. Then $W^* = f^*$ and the limiting distribution is the uniform distribution on f^*:

$$\forall x \in E^m \quad \forall i \in f^* \quad \lim_{l \to \infty} \lim_{n \to \infty} P\left(X_n^l = (i, \cdots, i)/X_0^l = x\right) = \frac{1}{|f^*|}.$$

We eventually give several convergence results.

Theorem 6 (critical height H_1, [16]) *Suppose that α is symmetric and m is large enough to ensure that $W^* = f^*$. There exists a critical value H_1, called the critical height, which is a bounded function of the size m, such that, for every non–decreasing sequence $l(n)$ going to infinity, we have the equivalence*

$$\sup_{x \in E^m} P\left(X_n \notin f^*/X_0 = x\right) \xrightarrow[n \to \infty]{} 0 \quad \Longleftrightarrow \quad \sum_{n=0}^{\infty} l(n)^{-H_1} = \infty.$$

Unlike H_1, the height of exit H_e^* of the basin of attraction of f^* increases linearly with m. Thus, by tuning conveniently the sequence $l(n)$, we can obtain an algorithm which definitely contains points of f^* after a finite number of transitions.

Theorem 7 *Let $H_e^* = H_e(\{ x \in E^m : [x] \cap f^* \neq \emptyset \})$. Suppose that m is large enough to ensure that $W^* = f^*$ and $H_1 < H_e^*$. For every non–decreasing sequence $l(n)$ going to infinity, we have the equivalence*

$$\forall x \quad P\left(\exists N \ \forall n \geq N \ [X_n] \cap f^* \neq \emptyset/X_0 = x\right) = 1$$

$$\Longleftrightarrow \quad \sum_{n=0}^{\infty} l(n)^{-H_1} = \infty, \quad \sum_{n=0}^{\infty} l(n)^{-H_e^*} < \infty.$$

We study the optimal convergence exponent α_{opt} defined by Azencott, Catoni and Trouvé which characterizes the optimal speed of convergence of the algorithm in finite time.

Theorem 8 *There exist two positive constants R_1 and R_2 such that for all n*

$$\frac{R_1}{n^{\alpha_{opt}}} \leq \inf_{l(1) \leq \cdots \leq l(n)} \max_{x \in E^m} P\left(X_n \notin f^*/X_0 = x\right) \leq \frac{R_2}{n^{\alpha_{opt}}}.$$

Moreover the exponent α_{opt} is an affine increasing function of m.

The speed of convergence increases linearly with the population size. This last result shows that the mutation–selection algorithm is intrinsically parallel.

4 A new genetic algorithm

We present now succinctly the new genetic algorithm [7] which finds simultaneously all the global maxima of the fitness function. The idea is to start with a more sophisticated unperturbed process. We integrate the crossover in it. For the selection mechanism, we first perform a selection on the set of the fitness values taken in the population, regardless of how many individuals do take these values. Once a fitness value has been selected, an individual corresponding to this value is chosen in the population in a way that avoids the loss of the genetic diversity. The attractors of the unperturbed process have a richer structure than in the previous cases. The analysis is conducted along the same lines. We show that, beyond a critical size, the population settles in a particular subset of populations, the populations of which contain all the global maxima of the fitness function. Such an algorithm completely solves the optimization problem.

We consider a finite space of states E and a positive function f defined on E. We wish to find the whole set f^* of the global maxima of f. For a subset K of E^m (that is a set of populations) we define the subset of all the individuals present in K,

$$[K] = \{\, x_k : 1 \leq k \leq m, \, x = (x_1, \cdots, x_m) \in K \,\},$$

the maximal fitness reached in $[K]$,

$$\widehat{f}(K) = \max\{\, f(x_k) : 1 \leq k \leq m, \, x = (x_1, \cdots, x_m) \in K \,\},$$

the set of the individuals of $[K]$ taking this maximal fitness,

$$\widehat{K} = \{\, x_k : 1 \leq k \leq m, \, x = (x_1, \cdots, x_m) \in K, \, f(x_k) = \widehat{f}(K) \,\}.$$

If x is an element of E^m, we note $[x] = [\{x\}], \widehat{x} = \widehat{\{x\}}$. If F is a subset of E, $f(F)$ (or fF) denotes the set $\{\, f(i) : i \in F \,\}$. For $\lambda \in \mathbb{R}_+^*$, we put $f_\lambda = f^{-1}(\{\lambda\})$ which is the set of the points of the search space having fitness λ. For x in E^m, we let $x^\lambda = [x] \cap f_\lambda$ (the set of the individuals in the population x having fitness λ) and we number the elements of the set x^λ: $x^\lambda = \{\, x_1^\lambda, \cdots, x_{|x^\lambda|}^\lambda \,\}$ in the order they appear in the sequence $(x_k^\lambda, 1 \leq k \leq |x^\lambda|)$. For the special case $\lambda = \widehat{f}(x)$ (i.e. where λ is the maximal value of the fitness in the population x), we use the notation $\widehat{x} = \{\, \widehat{x}_1, \cdots, \widehat{x}_{|\widehat{x}|} \,\}$. Finally, for x in E^m, we denote by $\lambda_1^x, \cdots, \lambda_{|f[x]|}^x$ the $|f[x]|$ elements of the set $f[x]$, where again the indexation respects the order of appearance of the elements of $f[x]$ in the sequence $(f(x_k), 1 \leq k \leq m)$.

We first describe the ground process which drives the algorithm. When there is no random perturbation, the process under study is a Markov chain (X_n^∞) with state space E^m. The transition from X_n^∞ to X_{n+1}^∞ is decomposed in two stages:

$$X_n^\infty \xrightarrow{\text{crossover}} Z_n^\infty \xrightarrow{\text{selection}} X_{n+1}^\infty.$$

$X_n^\infty \longrightarrow Z_n^\infty$: **crossover.** The phenomenon of crossover is modeled as a random operation on the couples formed by consecutive individuals of the population X_n^∞. Let β be a markovian kernel on the space $E \times E$. The quantity $\beta((i_1, j_1), (i_2, j_2))$ is interpreted as the probability of producing the pair of individuals (i_2, j_2) by performing a crossover on the couple (i_1, j_1). We impose three conditions on β, which are of biological relevance:

• When two identical individuals mate, they produce offsprings identical to themselves

$$\forall i \in E \qquad \beta((i,i),(i,i)) = 1.$$

• There is always a non–zero probability that nothing happens during a crossover

$$\forall i, j \in E \qquad \beta((i,j),(i,j)) > 0.$$

• The two individuals of the mating pair play symmetric roles (the populations are asexual) i.e. for any i_1, j_2, i_2, j_2,

$$\beta((i_1, j_1), (i_2, j_2)) = \beta((j_1, i_1), (j_2, i_2)).$$

Amazingly enough, the first two conditions (especially the first one) are crucial for our algorithm to work. The transition probabilities from X_n^∞ to Z_n^∞ are

$$P\left(Z_n^\infty = z / X_n^\infty = x\right) = \delta_m(x_m, z_m) \prod_{1 \leq k \leq m/2} \beta\left((x_{2k-1}, x_{2k}), (z_{2k-1}, z_{2k})\right)$$

where $\delta_m(i,j) = \delta(i,j)$ if m is odd (the last individual has no mating partner and remains unchanged after crossover) and $\delta_m(i,j) = 1$ if m is even.

$Z_n^\infty \longrightarrow X_{n+1}^\infty$: **selection.** We propose here an enhanced version of the selection mechanism used in the previous models [5, 6] which has the decisive advantage of preserving the diversity of the individuals present in the population. We define a triangular array of integers $\tau(k,h)$, $0 \leq k \leq h+1$, $1 \leq h \leq m$, by

$$\forall h \in \{1 \cdots m\} \qquad \tau(0,h) = 1, \quad \tau(h+1,h) = m,$$

$$\forall k, h \quad 1 \leq k \leq h \leq m \quad \tau(k,h) = 2k \left\lfloor \frac{m}{2(h+1)} \right\rfloor + 1.$$

Suppose $Z_n^\infty = z$. To build the population $X_{n+1}^\infty = x$, we select randomly with the uniform distribution a permutation σ of $\mathfrak{S}(|\hat{z}|)$. Using the subdivision $\tau(\cdot, |\hat{z}|)$ of $\{1 \cdots m\}$, we divide the set of indices $\{1 \cdots m\}$ in $|\hat{z}| + 1$ parts. The components of the r–th part (for $1 \leq r \leq |\hat{z}|$) are set equal to $\hat{z}_{\sigma(r)}$, that is for the indices k such that $\tau(r-1, |\hat{z}|) \leq k < \tau(r, |\hat{z}|)$ we put $x_k = \hat{z}_{\sigma(r)}$. The components of the $(|\hat{z}|+1)$–th part (i.e. for the indices k such that $\tau(|\hat{z}|, |\hat{z}|) \leq k \leq m$) are chosen independently and uniformly on the set $\hat{z}$.

Notice that the first $|\hat{z}|$ parts have an even cardinality, so that the crossover cannot act on a pair of individuals belonging to two distinct parts. Since in addition each part contains only one type of individual, the first condition imposed on the crossover implies that it has no effect on the first $|\hat{z}|$ parts. The main interest of the $(|\hat{z}|+1)$–th part is to allow different individuals to mate together without restriction.

The transition mechanism implies that the process (X_n^∞) is instantaneously absorbed in particular subsets of populations that we call the attractors.

Proposition 2 (attractors of X_n^∞)

The attractors of the process (X_n^∞) are the sets of populations K such that

i) $[K] = \widehat{K}$

ii) a population x of E^m belongs to K if and only if

- $\forall r \in \{1 \cdots |\widehat{K}|\} \quad \forall k, h \quad \tau(r - 1, |\widehat{K}|) \leq k, h < \tau(r, |\widehat{K}|) \implies x_k = x_h,$
- $\{ x_k : 1 \leq k < \tau(|\widehat{K}|, |\widehat{K}|) \} = [K],$
- $\{ x_k : \tau(|\widehat{K}|, |\widehat{K}|) \leq k \leq m \} \subset [K].$

The set of all the attractors is denoted by $\mathcal{K}$ (hence $\mathcal{K} \subset \mathcal{P}(\mathcal{P}(E^m))$).

An attractor K is stable if and only if the crossover cannot create individuals having a fitness greater than $f(K)$ starting with two arbitrary individuals of $[K]$. The structure of the set $\mathcal{K}$ is much richer than in the preceding models. All the individuals present in the populations of a fixed attractor have the same fitness. Notice that the bracket operator [ ] provides a one to one correspondence between the set of all the attractors and the set $\{ F \subset E : |F| \leq m, \, \widehat{F} = F \}$. We denote by K^* the unique attractor (it exists for $m \geq |f^*|$) such that $[K^*] = f^*$. In particular, the populations of K^* contain the whole set f^*.

The previous Markov chain (X_n^∞) is then randomly perturbed by two distinct mechanisms. The first one acts directly upon the population and mimics the phenomenon of mutation. The second one consists in loosening the selection of the individuals. The intensity of the perturbations is governed by an integer parameter l: as l goes to infinity, the perturbations progressively disappear. The transition mechanism of the perturbed Markov chain (X_n^l) is decomposed in three stages:

$$X_n^l \xrightarrow{\text{mutation}} Y_n^l \xrightarrow{\text{crossover}} Z_n^l \xrightarrow{\text{selection}} X_{n+1}^l.$$

$X_n^l \longrightarrow Y_n^l$: **mutation.** The mutations are modeled by random independent perturbations of the individuals of the population X_n^l. Let α be an irreducible Markov kernel on the space E. Let α_l be the irreducible Markov kernel on E defined by

$$\alpha_l(i, j) = \begin{cases} \alpha(i, j) l^{-a} & \text{if } i \neq j \\ 1 - \sum_{e \neq i} \alpha(i, e) l^{-a} & \text{if } i = j \end{cases} \quad (\text{where } a \in \mathbb{R}_+^*)$$

The transition probabilities from X_n^l to Y_n^l are given by

$$P\left(Y_n^l = y / X_n^l = x\right) = \alpha_l(x_1, y_1) \cdots \alpha_l(x_m, y_m).$$

$Y_n^l \longrightarrow Z_n^l$: **crossover.** The crossover is not perturbed in any way: this stage is exactly the same as the passage from X_n^∞ to Z_n^∞.

$Z_n^l \longrightarrow X_{n+1}^l$: **selection.** The selection mechanism of the chain (X_n^∞) is perturbed by loosening randomly the selection pressure (individuals below peak fitness may survive).

Suppose $Z_n^l = z$ and we wish to build the vector $X_{n+1}^l = x$.

Let us summarize briefly and without formula this mechanism.

We first traverse $f[z]$; with each element λ of this set, we associate a sequence $\psi_1, \cdots, \psi_{n_\lambda}$ obtained by reordering randomly (all orders being equally probable) the set $f^{-1}(\{\lambda\}) \cap f[z]$. The population x is then built in the following way: for each component $x_k, 1 \le k \le m$, we draw a value λ under a distribution probability on the set $f[z]$ which is biased toward the high values. With this value λ, we had previously associated a sequence $\psi_1, \cdots, \psi_{n_\lambda}$. We divide the set $\{1 \cdots m\}$ in $n_\lambda + 1$ parts. If the index k under consideration belongs to the r–th part, where $1 \le r \le n_\lambda$, we set $x_k = \psi_r$. If k belongs to the $(n_\lambda + 1)$-th part, we choose x_k randomly and uniformly over the set $\{\psi_1, \cdots, \psi_{n_\lambda}\}$.

We now describe precisely the selection mechanism.

For each h in $\{1 \cdots |f[z]|\}$, we select independently and randomly with the uniform distribution a permutation σ^h belonging to $\mathfrak{S}(|z^{\lambda_h}|) = \mathfrak{S}(|[z] \cap f_{\lambda_h}|)$. The law of each component x_k $(1 \le k \le m)$ of x is defined as follows: we select randomly a value λ in the set $f[z]$ with the distribution

$$\forall h \in \{1 \cdots |f[z]|\} \qquad P\left(\lambda = \lambda_h\right) = \frac{\exp(c\lambda_h \ln l)}{\sum_{r=1}^{|f[z]|} \exp(c\lambda_r \ln l)} \qquad \text{(where } c \in \mathbb{R}_+^*\text{)}$$

The value of x_k is then chosen in the set z^λ according to the value of the index k:
- if $\tau(|z^\lambda|, |z^\lambda|) \le k \le \tau(|z^\lambda|+1, |z^\lambda|) = m$, then x_k is chosen with the uniform distribution over the set $z^\lambda = [z] \cap f_\lambda$ (k lies in the last part of the set of indices).
- if there exists r in $\{1 \cdots |z^\lambda|\}$ such that $\tau(r-1, |z^\lambda|) \le k < \tau(r, |z^\lambda|)$, then we put $x_k = z^\lambda_{\sigma^h(r)}$, where h is the unique integer in $\{1 \cdots |z^\lambda|\}$ satisfying $\lambda = \lambda_h$ (the index k lies in the r–th part of the set of indices, where $1 \le r \le |z^\lambda|$).

We have

$$\forall y, z \in E^m \qquad \lim_{l \to \infty} P\left(X_{n+1}^l = z/X_n^l = y\right) = P\left(X_{n+1}^\infty = z/X_n^\infty = y\right)$$

so that the process (X_n^l) is a perturbation of the process (X_n^∞). To study the asymptotic dynamics of the algorithm, we compute the asymptotic expansion of the transition probabilities:

$$P\left(X_{n+1}^l = v/X_n^l = u\right) \sim q_1(u,v) l^{-V_1(u,v)}$$

where q_1 is a non–negative kernel and V_1 is a communication cost corresponding to a discrete variational problem. We are thus in the framework of the generalized simulated annealing studied by Trouvé [16, 17, 18]. The crucial quantity is the virtual energy W, defined for x in E^m by $W(x) = \min\{\sum_{(y \to z) \in g} V_1(y,z) : g \in G(x)\}$ where $G(x)$ is the set of all x–graphs over E^m. As l goes to infinity, the sequence of the stationary measures of the chains $(X_n^l), l \in \mathbb{N}$, concentrates on the set W^* of the global minima of the virtual energy W.

Theorem 9 *There exists a critical population size m^* depending on the function f, the kernels α, β and the parameters a, c such that $W^* = K^*$ when m is greater than m^*, whence*

$$\forall m \ge m^* \quad \forall x \in E^m \qquad \lim_{l \to \infty} \lim_{n \to \infty} P\left([X_n^l] = f^*/X_0^l = x\right) = 1.$$

Furthermore, the limiting distribution is uniform over the ideal attractor K^*.

We give various conditions (which depend strongly on the optimization problem) to ensure $W^* = K^*$ and upper bounds on m^*. We finally apply Catoni and Trouvé's tools for the inhomogeneous case [2, 3, 4, 16, 17, 18] where $l = l(n)$ is an increasing function of n.

Theorem 10 *Suppose m is large enough to have $W^* = K^*$. There exists a positive constant H_1 such that, for all increasing sequences $l(n)$ going to infinity, we have the equivalence*

$$\sup_{x \in E^m} P(X_n \notin K^*/X_0 = x) \underset{n \to \infty}{\longrightarrow} 0 \quad \Longleftrightarrow \quad \sum_{n=0}^{\infty} l(n)^{-H_1} = \infty.$$

Proposition 3 *The critical height H_1 is bounded as a function of m.*

We may adapt $l(n)$ in order to be trapped in the basin of attraction of K^*.

Theorem 11 *Define $H_e^* = H_e(\{ x \in E^m : f^* \subset [x] \})$.*
Suppose m is large enough to have $W^* = \{K^*\}$ and $H_1 < H_e^*$.
For all increasing sequences $l(n)$, we have the equivalence

$$\forall x \in E^m \quad P(\exists N \quad \forall n \geq N \quad f^* \subset [X_n]/X_0 = x) = 1$$

$$\Longleftrightarrow \quad \sum_{n=0}^{\infty} l(n)^{-H_1} = \infty, \quad \sum_{n=0}^{\infty} l(n)^{-H_e^*} < \infty.$$

Proposition 4 *The optimal convergence exponent α_{opt} is bounded between two affine strictly increasing functions of m.*

Theorem 12 *When the population size m is sufficiently large, each cycle over the set of attractors not containing the attractor K^* is reduced to one single attractor.*

5 Conclusion

Some substantial convergence results have been obtained in the theoretical framework proposed in this work. Several research directions are now possible. First, one would like to carry out a more refined study (even an experimental one) of the crucial quantities appearing in these first models. An essential conclusion is that an efficient implementation of a genetic algorithm heavily depends on the optimization problem.

This work deals only with the simplest of the genetic algorithms. A bundle of variations and extensions built with more sophisticated genetic operators can be analyzed with the same techniques. More generally, one would like to evaluate the impact and the interest of each genetic phenomenon (redundancy, diploidy, sex, dominance, non–coding segments, ...) for the optimization process.

Acknowledgement. I thank Marc Schoenauer for useful comments that improved the readability of this paper.

References

1. E.H.L. Aarts–P.J.M. Van Laarhoven, *Simulated annealing: theory and applications*, D. Reidel, Dordrecht–Tokyo, 1987.
2. O. Catoni, *Large deviations for annealing*, PhD Thesis, University Paris XI (1990).
3. O. Catoni, *Rough large deviations estimates for simulated annealing, Application to exponential schedules*, The Annals of Probability 20 no. 3 (1992), 1109–1146.
4. O. Catoni, *Sharp large deviations estimates for simulated annealing algorithms*, Ann. Inst. Henri Poincaré Probab. Statist. 27 no. 3 (1991), 291–383.
5. R. Cerf, *Asymptotic convergence of genetic algorithms*, Preprint, submitted (1993).
6. R. Cerf, *The dynamics of mutation-selection algorithms with large population sizes*, Ann. Inst. Henri Poincaré Probab. Statist., 1996, to appear.
7. R. Cerf, *A new genetic algorithm*, Preprint, submitted (1993).
   The postscript files of the above three papers are available by anonymous ftp at ftp://blanche.polytechnique.fr/pub/eark/papers/Raphael.Cerf/xxx.ps.gz or with netscape http://blanche.polytechnique.fr/www.eeaax/eeaax.html.
8. R. Cerf, *Une théorie asymptotique des algorithmes génétiques*, Thèse, Université Montpellier II (1994). The postscript file of this thesis is available by anonymous ftp at ftp 129.199.96.12, pub/reports/thesis/atga.ps.Z.
9. T.E. Davis–J.C. Principe, *A simulated annealing like convergence theory for the simple genetic algorithm*, Proceedings of the 4th ICGA, San Diego, Morgan Kaufmann, 174–181, 1991.
10. M.I. Freidlin–A.D. Wentzell, *Random perturbations of dynamical systems*, Springer–Verlag, New York, 1984.
11. D. Goldberg, *Genetic algorithms in search, optimization and machine learning*, Addison–Wesley, 1989.
12. C.R. Hwang–S.J. Sheu, *Singular perturbed Markov chains and exact behaviours of simulated annealing process*, Journal of Theoretical Probability Vol. 5 No. 2 (1992), 223–249.
13. J.H. Holland, *Adaptation in natural and artificial systems*, The University of Michigan Press, Ann Arbor, 1975.
14. A.E. Nix–M.D. Vose, *Modeling genetic algorithms with Markov chains*, Annals of Mathematics and AI, 5(1):79–88, 1991.
15. G. Rudolph, *Convergence analysis of canonical genetic algorithms*, IEEE Trans on Neural Networks 5(1), 96–101, 1994.
16. A. Trouvé, *Cycle decompositions and simulated annealing*, Preprint (1993).
17. A. Trouvé, *Parallélisation massive du recuit simulé*, PhD Thesis, University Paris XI (1993).
18. A. Trouvé, *Rough large deviation estimates for the optimal convergence speed exponent of generalized simulated annealing algorithms*, Ann. Inst. Henri Poincaré Probab. Statist., 1996, to appear.

Towards a Genetic Theory of Easy and Hard Functions

Gilles Venturini

Laboratoire d'Informatique,

Ecole d'Ingénieurs en Informatique pour l'Industrie,

Université de Tours,

64, Avenue Jean Portalis, Technopôle Boîte No 4,

37913 Tours Cedex 9 FRANCE

Tel: (+33)-47-36-14-33, Fax: (+33)-47-36-14-22

Email : venturi@lri.fr or venturini@univ-tours.fr

Abstract. According to the literature that deals with the difficulty of functions with respect to genetic algorithms (GA), the so-called GA-hard functions are usually hard for other methods. In this paper, we firstly show that a gradient easy function can be fully deceptive, and thus hard for a GA to optimize while being unimodal. More generally, we show that the global search method introduced by (Das and Whitley 1991) to optimize GA-easy functions can be simply adapted to solve GA-hard functions. The resulting algorithm, called GSC1, generates a set of binary strings and outputs the string that wins the first order schemas competitions as well as its binary complement. According to the theory of deceptiveness in GAs, this method solves GA-easy and GA-hard problems efficiently, as shown effectively in the reported experiments. This method is however only well suited for these functions, and does not deal with partially deceptive functions. It is then shown how it could be combined with a GA.

1 Introduction

Genetic algorithms were initially designed as procedures for adapting rule-based systems to various learning tasks (Holland 1975). These algorithms are now mostly used as function optimizers, and the term "genetic algorithms for function optimization" (GAFOs) used in (De Jong 1992) (De Jong et al. 1994) fits them truly well. As with any other search procedures, it is essential to know which functions are hard or easy to optimize by GAFOs, because the problem of solving an arbitrary function is NP-complete (Hart and Belew 1991). One cannot expect GAFOs to perform well on all possible functions but only on a subset of these functions. Also GAFOs and other evolution based algorithms can be simply described (Spears et al. 1993), their behavior is much more difficult to analyze. Hence, characterizing GA-easy or GA-hard functions is also a difficult task.

In the last few years, an extension of schema analysis has lead to the study of the so-called deceptive problems (Goldberg 1987 and 1989). These problems, or functions, mislead in theory the GAFOs in the fitness landscape. The actual theory of GA-easy and GA-hard functions is based on the notion of schema (Liepins and Vose 1990) (Whitley 1990). If the competitions between schemas always lead to the optimum, then the function is GA-easy. If these competitions always lead away from the optimum, then the function is GA-hard. It is interesting to know whether these functions are easy or hard to optimize by other search methods like gradient methods (GM), hill climbing (HC) or global search (GS). This helps in comparing GAFOs with other techniques, and also in understanding how GAFOs work. Several researchers have given some insight about such comparison: GA-easy functions are not necessarily HC-easy (Wilson 1991), GA-hard functions are HC-hard (Whitley 1990), functions with zero epistasy are GA-easy (Davidor 1990) (Manela and Campbell 1992), GA-easy functions are GS-easy (Das and Whitley 1991).

GA-hard functions given in the literature are usually hard for other techniques. For instance, the GS method introduced in (Das and Whitley 1991) can optimize GA-easy functions by solving the competitions of order 1, but this method cannot solve GA fully deceptive functions. This is the reason why many researchers try to improve GAs or other methods from that point of view, and test their algorithm on GA-hard functions like for instance in (Whitley 1990) (Goldberg et al. 1991) (Muhlenbein 1992). This seems reasonable if GA-hard functions are also difficult for other methods and represent challenging problems. However, we have found recently an example of GA-hard function which is unimodal and thus easy for a gradient search (Venturini 1994). This function is presented in section 3. GA-hard functions are thus not necessarily hard for other search methods. In section 4, we go one step further in that direction by showing that any GA-hard function, or more generally any consistently deceptive function, is easy for a global search method. The next section presents an overview of GA-easy and GA-hard functions, and also the relations that exist to other search methods.

2   GA-easy, GA-hard functions, and other search methods

2.1   Basic notations

The standard problem of optimizing a binary string can be stated as follows: given a search space $S = \{0, 1\}^l$, and a fitness function f from S to $\mathcal{R}^+$, find a string s^* that maximizes f. The GAFO considered in this paper is the canonical GA, which uses a roulette wheel selection and a one point crossover (Goldberg 1989). Schema analysis states that the GA processes not only binary strings but also hyperplanes of S, called schemas, represented by strings from $\{0, 1, *\}^l$. The order $o(h)$ of a schema h is the number of fixed bits in h. The fitness $f(h)$ of a schema h is defined by the mean of fitnesses of binary strings in h. A schema can be viewed as a search direction or a hyperplane of S.

2.2 GA-easy functions

One way to model the GA behavior is to define competitions between schemas. The GA solves such competitions, and concentrates in one area of the search space according to the winners of these competitions. For instance, the GA may have to solve the competition between the four following schemas ($l = 3$):

$$h_1 = 00*, \; h_2 = 01*, \; h_3 = 10*, \; h_4 = 11*$$

A competition at order N is thus defined by the comparison of the fitnesses of all schemas that have $*$ (or fixed bits) at the same positions. For instance, the previous schemas define a competition of order 2. The winner of a competition is the schema with the highest fitness. This model suggests that the GA combines blocks of bits from low order schemas in order to find the optimum s^*. This combination will work if the optimum s^* always belongs to the winning schemas. For instance, in the Onemax problem (Syswerda 1989) where $f(s)$ is equal to the number of 1 in s, any competition of any order N leads to the optimum s^*. This is the case for instance of all the first order competitions ($l = 3$):

$$C_1 : \; f(0**) = 1, \; f(1**) = 2$$
$$C_2 : \; f(*0*) = 1, \; f(*1*) = 2$$
$$C_3 : \; f(**0) = 1, \; f(**1) = 2$$

where any schema containing $s^* = 111$ wins the competition it is involved in. Such a function is said to be GA-easy. A function f is GA-easy whenever the winners of the competitions of any order N always lead to a global maximum of f. It is important to notice that optimizing such functions can be performed by solving the first order competitions only (Das and Whitley 1991), because the winners of these competitions indicate directly s^*. For instance, the Onemax problem can be solved in this manner: s^* is assigned a 1 at the first location because $f(1**) > f(0**)$, a 1 at the second location because $f(*1*) > f(*0*)$, and a 1 at the third location because $f(**1) > f(**0)$. This property will be important for GS methods.

2.3 GA-hard functions

The notion of deception in a competition occurs when the winner does not contain s^*. For instance, consider the following function (Whitley 1990):

$$f(000) = 28, \; f(001) = 26, \; f(010) = 22$$
$$f(011) = 0, \; f(100) = 14, \; f(101) = 0$$
$$f(110) = 0, \; f(111) = 30$$

The global maximum is 111, but a competition of order 1 is such that for instance $f(0**) > f(1**)$. Solving this competition would lead the search process away from s^* because a 0 would be assign to the first location. Such a function is called a deceptive function: there exists at least one competition of any order N less than l where the winner does not contain s^*. However, the GA may still be able to find the optimum, because there may be some other competitions of order N that are not deceptive. The GA may not have to solve the deceptive competition by following a path of nested "easy" competitions that leads to s^* (Liepins and Vose 1990).

More difficult functions can be defined: a function f defined on l bits is fully deceptive when the winners of all the competitions between schemas of order less than l lead to a global attractor s^- which is different than the optimum s^*. s^- is in fact not only different than s^* but must be the binary complement of s^* denoted by $\bar{s}^*$ (Whitley 1990). Such functions are GA-hard because all the competitions that the GA will have to solve lead away from s^*. For instance, the deceptive function given previously is fully deceptive of order 3.

A function is consistently deceptive of order l when all competitions of order less than l lead away from s^* (Whitley 1990). In this case, at least the competitions of order 1 lead to the complement of s^*. A fully deceptive function is consistently deceptive. Such functions are also GA-hard because the GA usually exploits information given by low order schemas, which will be misleading here.

A fundamental property of consistently deceptive functions for the GS method developped in section 4 is that solving the first order competitions leads directly to $\bar{s}^*$. For instance, the fully deceptive function given in this section is such that:

$$f(0**) > f(1**),\ f(*0*) > f(*1*),$$

$$f(**0) > f(**1)$$

Solving the first order competitions leads to the string 000, which is the complement of $s^* = 111$.

2.4 Relation to hill climbers

Wilson has shown that GA-easy functions are not necessarily easy for hill climbers (Wilson 1991) by giving a GA-easy function which contains local optima in the Hamming space. This function can be generalized to any dimension $l = 3k$.

Whitley has shown that GA fully deceptive functions necessarily contain a local optima in the Hamming space (Whitley 1990). These functions are thus hard for hill climbers.

One should not conclude from this that hill climbers should not be used in cooperation with evolutionary techniques (Mulhenbein 1992). By "hill climbing hard", people usually mean that there may exist at least one local optimum in the Hamming space. However there may exists another path that leads to the optimum without getting trapped in a local optimum.

2.5 Relations to global search methods

GA-easy functions can be optimized by a simple and straightforward method introduced in (Das and Whitley 1991). This method is a global search algorithm (denoted by GS1 in the following) which solves directly the first order competitions. It uses the following principle:

1. Generate a set S_1 of strings in $\{0,1\}^l$,

2. Evaluate each string of S_1,

3. Compute the fitnesses of all the first order schemas,

4. Generate the output s_1^* by solving the competitions of order 1 (as mentioned in section 2.2 or 2.3).

More precisely, GS1 uses a t-test to determine whether it should generate more points or not. If the competitions are solved independently, then this algorithm can only deal with GA-easy functions. GS1 may solve partially deceptive functions in the following way: solve first the first order competition with the greatest fitness ratio to determine one bit of s_1^*, and apply GS1 recursively to the remaining bits. In the following, we consider only the first version, which solves the competitions independently.

If f is GA-easy, the authors show that GS1 performs significantly better than a GA. In fact, GS1 represents the best behavior a GA could have when optimizing a GA-easy function. Any consistently deceptive function of order 2 or more misleads GS1. When f is partially deceptive, GS1 may ot find the optimum.

2.6 Relation to epistasis measure and fitness/distance correlation

The epistasis measure is useful to determine the correlations that may exists between binary genes (Davidor 1990). If no such correlations exist, then the influence of each bit value on the fitness function is independent from the values of the other bits. Thus, functions with 0 epistasis measure are GA-easy and can be optimized also by solving the first order competitions.

The fitness/distance correlation measures the correlation that may exists between the fitness value of a string s and its Hamming distance to the closest optimum s^* (Jones and Forrest 1995). This function is computed with a sample of points in S but requires that the global optima of f are known. When this

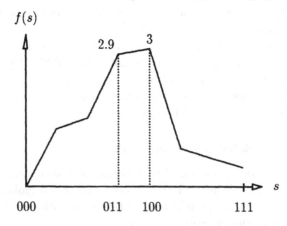

Figure 1: A fully deceptive function of order 3 which is gradient-easy.

correlation is close to -1, then the closer you get to one global optimum s^* the higher the fitness is. This is the case for instance of the Onemax function. This means that f is easy to optimize by a GA, and most certainly easy for other methods too, since Onemax is easy for hill climbers or global search for instance. When this correlation is close to 1, the fitness function f is misleading, which is the case of fully deceptive functions for instance. When this correlation is close to 0, no indication is really given about f difficulty: f can be a "needle in a haystack" function, or a function with many high peaks located all over the search space.

3 A GA fully deceptive and gradient easy function

We give in this section a first example of a function that is fully deceptive of order 3, and thus hard to optimize for GAs, and unimodal. This function can thus be easily optimized with a gradient search.

Let us consider that the parameter x of this function has been encoded in the binary space $S = \{O, 1\}^l$, with $l = 3$. The function's values are the following:

$$f(000) = 0.2, \ f(001) = 1.4, \ f(010) = 1.6, \ f(011) = 2.9$$

$$f(100) = 3, \ f(101) = 1, \ f(110) = 0.8, \ f(111) = 0.6$$

The optimum of this function is $s^* = 100$, and the deceptive attractor is thus $s^- = 011$. This function is represented on figure 1 and it is obviously unimodal and easy for a gradient search. One can easily show that this function is fully deceptive of order 3. For instance, we have $f(0**) = 1.525$ and $f(1**) = 1.35$, which should mislead the GA when it solves this first order competition. The deception also occurs for the other competitions of order 2 or less, like for instance the following competition:

$$f(00*) = 0.8$$
$$f(01*) = 2.25$$
$$f(10*) = 2$$
$$f(11*) = 0.7$$

where the schema 01* wins.

This function is thus GA-hard. Let us notice that it is also HC-hard, since fully deceptive functions always contain a local optimum in the Hamming space, here the string 011. It would be interesting to generalize this function to any l greater than 3, but we have not succeeded yet in doing so. However, in the next section we show that any fully deceptive function is easy for a global search method.

4 A simple algorithm for GA-easy and GA-hard functions

In section 2, we have seen that generating a string s_1^* by solving independently the order 1 competitions leads either toward:

- s^*, when f is GA-easy,

- $\bar{s}^*$, when f is consistently deceptive,

- any binary string, when f is partially deceptive.

A simple way to extend GS1 is to generate also a string s_2^* equal to the complement $\bar{s}_1^*$, and to output the best of the two strings. Grefenstette has suggested to use this technique with a GA (Grefenstette 1992). Here, we make it simpler and more operational by using a global search algorithm, which is much easier to analyze than a GA. This method is called global search with complement (GSC). When f is consistently deceptive, the search process will be driven toward $s_1^* = \bar{s}^*$ and will output $s_2^* = s^*$. When f is GA-easy, the best string will be $s_1^* = s^*$. It results that GA-hard or GA-easy functions are GSC-easy, a complementary result to Das and Whitley's who had shown that GA-easy functions are GS-easy.

This results in the following global search algorithm with complement (GSC1). In order to perform at least as well as random search, the algorithm may also consider the best string generated in S_1, denoted by s_3^*:

1. Generate a set S_1 of points in $\{0, 1\}^l$,

2. Compute the fitnesses of all the first order schemas,

3. Generate s_1^* by solving the competitions of order 1,

Algo.	# evaluations				
	500	1000	2000	5000	10000
Genitor					53 %
Random	29 %	38 %	33 %	36 %	46 %
	(12)	(11)	(13)	(12)	(13)
GSC1	62 %	73 %	85 %	95 %	99 %
	(16)	(11)	(13)	(6)	(1)

Table 1: Percentages of subproblems solved for Genitor, random search and GSC1 on the 30 bits ugly deceptive function f_1. Numbers in parentheses represent standard deviations. For this function, the best string computed by GSC1 is s_2^*.

4. Evaluate s_1^*,

5. Compute $s_2^* = \bar{s}_1^*$,

6. Evaluate s_2^*,

7. Let s_3^* be the best string in S_1,

8. Output the best string among $\{s_1^*, s_2^*, s_3^*\}$.

This algorithm is very simple and very similar to GS1. However, one major difference is that GSC1 may optimize also GA-hard functions, and this as easily as GA-easy functions. GSC1 solves the competitions independently but could be used in a recursive way also. A t-test could also be used to tell whether a competition is relevant or not.

5 Experiments with GSC1

5.1 Testing methodology

Three functions have been tested. The first function is a classical 30 bits deceptive function studied in (Whitley 1990), denoted by f_1 in the following. This function is obtained by concatenating 10 times the 3 bits fully deceptive function mentioned in section 2.3. The value of the function is the sum of the values of the 3 bits functions. In this function f_1, all competitions of the first and second orders are misleading. Furthermore, this function can be made "ugly" by adding a linkage problem: the 3 bits of a sub-problem are dispatched over the representation and may not be contiguous. However, since GSC1 solves the competitions independently, it is not sensitive to the ordering of bits in the representation, which is not the case of traditional GAs.

Algo.	# evaluations				
	500	1000	2000	5000	10000
Genitor					16 %
Random	26 %	24 %	27 %	29 %	24 %
	(9)	(13)	(12)	(14)	(9)
GSC1	76 %	84 %	92 %	100 %	100 %
	(13)	(12)	(6)	(0)	(0)

Table 2: Percentages of subproblems solved for another deceptive function f_2 defined on 40 bits. GSC1 ouputs also the best string s_2^*.

Algo.	# evaluations					
	500	1000	2000	5000	10000	21240
GA						100 %
Random	46 %	51 %	52 %	55 %	57 %	60 %
	(8)	(11)	(9)	(8)	(9)	(9)
GSC1	88 %	97 %	100 %	100 %	100 %	100 %
	(8)	(4)	(0)	(0)	(0)	(0)

Table 3: Results of GSC1 on the GA-easy function introduced by Wilson, defined here on 30 bits. "GA" denotes Wilson's GA (Wilson 1991). GSC1's best string is s_1^*.

The second function is a 40 bits deceptive function studied also in (Whitley 1990) and denoted by f_2 in the following. This function is build in the same way as the previous function but with a 4 bits fully deceptive function (function f_2 of Whitley's paper),

Finally, the last function is a 30 bits GA-easy function studied in (Wilson 1991) and denoted f_3 in the following. This function is more difficult to optimize than the Onemax function because it contains a local maximum in the Hamming space and is thus hard for steepest ascent hill climbing methods.

We have followed Whitley's testing methodology in order to compare GSC1 with Genitor results. It consists in allowing the tested algorithm to use only a given number of binary string evaluations (10000 for instance). For GSC1, this corresponds to the generation of a set S_1 of 10000 strings. For Genitor, a steady state GA (Whitley 1989), it corresponds to several generations with a population of 200 strings for instance. The quality of the best string found by an algorithm is evaluated by the percentage of sub-problems solved. For instance, in the case of the f_1 function, this percentage corresponds to the number of 3 bits blocks that are equal to 111. All percentages are averaged over 30 runs.

5.2 Results

Tables 1 and 2 give GSC1 and Genitor results for functions f_1 and f_2. Since Genitor was used with several crossover operators, we give here the best results obtained in (Whitley 1990). These comparative tables are given with the only intention of showing the differences between GSC1 and a GA search, and not with the intention of criticizing Genitor which we believe to be an efficient GA. Obviously, GSC1 is well adapted for solving GA-hard functions. With less than 5000 strings, it finds almost the optimum.

Finally, table 3 gives GSC1 results on a GA-easy function. The GA used by Wilson is a generational GA with tournament selection and a one-point crossover. Wilson mentions that for the 30 bits function f_3, with 21240 strings evaluations and ten different runs, the final population always contains the optimum. So Wilson results are not strictly comparable to GSC1, because Wilson's GA could probably find the optimum before the final generation. However, it gives an order of magnitude in the comparison between GSC1 and GAs on GA-easy functions.

One should be careful with the interpretation of these tables. The percentages represent a kind of Hamming similarity measure between the output string and the function global optimum. However, from the Hamming distance point of view, one can be far away from the optimum, but very close to it in performances, especially for GA-hard functions where the global attractor may have a fitness value relatively close to the function optimum.

6 Limitations of GSC1

As mentioned in the introduction, any search method is unlikely to perform well on an arbitrary function, unless P=NP. Thus, GSC1 has of course limitations, and it is useless to say that it does not replace Genitor or more generally any GA. GAs are most certainly more efficient than GSC1 on a broader set of functions than GA-easy or GA-hard functions. Partially deceptive functions which are neither GA-easy nor consistently deceptive are not properly optimized by GSC1, while probably many real world problems are partially deceptive. For instance, if only half of the first order competitions are deceptive, GSC1 does not find the optimum. However, the next section shows how GSC1 could be combined with a GA which would overcome this limitation with a multistrategic search.

Furthermore, some partially deceptive functions can be found that drives directly GSC1 to the worst string of S (Venturini 1995).

7 Conclusion and future work

In this paper, we have shown that GA-hard functions are not challenging problems: they can be easy for other search methods like gradient search, and more generally they can be optimized accurately with a simple global search method with complement (GSC1) that samples points in the space and that finds the optimum by solving the first order schemas competitions. This method performs at least as well as random search on arbitrary functions, and finds directly the optimum of GA-easy and GA-hard functions as shown effectively by the experiments described in this paper. Its efficiency is however limited to such functions.

GSC1 could be extended in at least two ways. From a theoretical point of view, one extension concerns non binary representations (Eshelman and Schaffer 1992) (Radcliffe and Surry 1994). From a practical point of view, GSC1 can be combined with a GA in order to solve more efficiently GA-easy or GA-hard problems as well as partially deceptive problems, and without generating and evaluating additional strings than the GA alone. This cooperation could take place in the following way: GSC1 is used to generate a set S_1 of strings, and outputs the best string is has found, according to the first order schemas. Then, the GA is invoked. A given proportion of the set S_1 is used as the initial population of the GA, which spares many costly evaluations of strings. Then, the remaining points of S_1 can be added during the GA search, which represents a kind of random immigrant operator or hyper-mutation operator (Cobb and Grefenstette 1993). The use of GSC1 would be at virtually no cost for the GA, because all evaluated strings of GSC1 can be used during the GA search. Further more, GSC1 runs in linear time. This method would solve easily GA-easy or GA-hard functions, and would let the GA deal with partially deceptive functions for which GSC1 is not efficient in theory.

Das and Whitley have argued that new extensions of GAs like uniform crossover for instance should be evaluated on deceptive problems only because GA-easy problems can be solved more efficiently with a global search and are thus not challenging. The main conclusion of this paper could be that GAs should not be evaluated either on consistently deceptive functions because such functions are easy to optimize using the global search with complement algorithm introduced here. Challenging problems are the partially deceptive functions only.

All this work is based on the existing theory of GA-easy and GA-hard functions. This theory is however greatly discussed in the community. Some authors have found for instance that GA-easy functions are not always easy (Schaffer et al. 1990) (Mitchell et al. 1991) (Mitchell and Holland 1993). The actual theory of deceptiveness generally does not consider the dynamic properties of the GA and performs only static analysis of deceptiveness (Grefenstette 1992). The stochastic aspects of the GA are however non negligible, as well as the biased sampling which occurs in practice (Schaffer et al. 1990). Studies such as (De Jong et al. 1994) which use Markov chains analysis to capture those dynamic aspects seem promising. There are also other ways of characterizing experimentally GA-easiness like for instance (Davidor 1990) (Davidor and Ben Kiki 1992) (Manela and Campbell 1992).

References

Cobb H.G. and Grefenstette J.J. (1993), Genetic algorithms for tracking changing environments, Proceedings of the Fith International Conference on Genetic Algorithms, 1993, S. Forrest (Ed), Morgan Kaufmann, pp 523-530.

Das R. and Whitley D. (1991), The only challenging problems are deceptive: global search by solving order-1 hyperplane, Proceedings of the Fourth International Conference on Genetic Algorithms, 1991, R.K. Belew and L.B. Booker (Eds), Morgan Kaufmann, pp 166-173.

Davidor Y. (1990), Epistasis variance: a viewpoint on GA-hardness, Proceedings of the first Workshop on Foundations of Genetic Algorithms, 1990, G.J.E. Rawlins (Ed), Morgan Kaufmann, pp 23-35.

Davidor Y. and Ben-Kiki O. (1992), The interplay among the genetic algorithm operators: information theory tools used in a holistic way, Proceedings of the Second Conference on Parallel Problem Solving from Nature 1992, R. Manner and B. Manderick (Eds), Elsevier, pp 75-84.

De Jong K. (1992), Are genetic algorithms function optimizers?, Proceedings of the Second Conference on Parallel Problem Solving from Nature 1992, R. Manner and B. Manderick (Eds), Elsevier, pp 3-13.

De Jong K., Spears W.M. and Gordon D.F. (1994), Using Markov chains to analyze GAFOs, Proceedings of the third Workshop on Foundations of Genetic Algorithms, 1994.

Eshelman L.J. and Schaffer J.D. (1992), Real-coded genetic algorithms and interval schemata, Proceedings of the second Workshop on Foundations of Genetic Algorithms, 1992.

Goldberg D.E. (1987), Simple genetic algorithms and the minimal deceptive problem, Genetic Algorithms and Simulated Annealing, L. Davis (Ed), Morgan Kaufmann, pp 74-88.

Goldberg D.E. (1989). *Genetic Algorithms in Search, Optimization and Machine Learning:* Addison Wesley.

Goldberg D.E., Deb K. and Korb B. (1991), Don't worry, be messy, Proceedings of the Fourth International Conference on Genetic Algorithms, 1991, R.K. Belew and L.B. Booker (Eds), Morgan Kaufmann, pp 24-30.

Grefenstette J.J. (1992), Deception considered harmful, Foundations of Genetic Algorithms 2, 1992.

Hart W.E. and Belew R.K. (1991), Optimizing an arbitrary function is hard for the genetic algorithm, Proceedings of the Fourth International Conference on Genetic Algorithms, 1991, R.K. Belew and L.B. Booker (Eds), Morgan Kaufmann, pp 190-195.

Holland J.H. (1975). *Adaptation in natural and artificial systems.* Ann Arbor: University of Michigan Press.

Jones T. and Forrest S. (1995), Fitness distance correlation as a measure of problem difficulty for genetic algorithms, Proceedings of the Sixth International Conference on Genetic Algorithms, 1995, L.J. Eshelman (Ed), Morgan Kaufmann, pp 184-192.

Manela M. and Campbell J.A. (1992), Harmonic analysis, epistasis and genetic algorithms, Proceedings of the Second Conference on Parallel Problem Solving from Nature 1992, R. Manner and B. Manderick (Eds), Elsevier, pp 57-64.

Mitchell M., Forrest S. and Holland J.H. (1991), The royal road for genetic algorithms: fitness landscapes and GA performance, Proceedings of the first European Conference on Artificial Life 1991, F.J. Varela and P. Bourgine (Eds), MIT press/Bradford Books, pp 245-254.

Mitchell M. and Holland H. (1993), When will a genetic algorithm outperform hill climbing?, Proceedings of the Fith International Conference on Genetic Algorithms, 1993, S. Forrest (Ed), Morgan Kaufmann, pp 647-647.

Muhlenbein H. (1992), How genetic algorithms really work I.Mutation and hillclimbing, Proceedings of the Second Conference on Parallel Problem Solving from Nature 1992, R. Manner and B. Manderick (Eds), Elsevier, pp 15-25.

Radcliffe N.J. and Surry P.D. (1994), Fitness variance of formae and performance prediction, Proceedings of the third Workshop on Foundations of Genetic Algorithms, 1994.

Schaffer J.D., Eshelman L.J. and Offutt D. (1990), Spurious correlation and premature convergence in genetic algorithms, Proceedings of the first Workshop on Foundations of Genetic Algorithms, 1990, G.J.E. Rawlins (Ed), Morgan Kaufmann, pp 102-112.

Spears W.M., De Jong K.A., Baeck T., Fogel D.B. and de Garis H. (1993), An overview of evolutionary computation, Proceedings of the European Conference on Machine Learning 1993, P. Brazdil (Ed.), Lecture notes in artificial intelligence 667, Springer-Verlag, pp 442-459.

Syswerda G. (1989), Uniform crossover in genetic algorithms, Proceedings of the third International Conference on Genetic Algorithms, 1989, J.D. Schaffer (Ed), Morgan Kaufmann, pp 2-10.

Venturini G. (1994), A GA fully deceptive function of order 3 which is also gradient-easy, published in French in the proceedings of Evolution Artificielle 94, Cépaduès Editions.

Venturini G. (1995), GA consistently deceptive functions are not challenging problems (extended version, unpublished).

Whitley D. (1989), The genitor algorithm and selective pressure: why rank-based allocation of reproductive trials is best, Proceedings of the third International Conference on Genetic Algorithms, 1989, J.D. Schaffer (Ed), Morgan Kaufmann, pp 116-124.

Whitley D. (1990), Fundamental principles of deception in genetic search, Proceedings of the first Workshop on Foundations of Genetic Algorithms, 1990, G.J.E. Rawlins (Ed), Morgan Kaufmann, pp 221-241.

Wilson S.W. (1991), GA-easy does not imply steepest-ascent optimizable, Proceedings of the Fourth International Conference on Genetic Algorithms, 1991, R.K. Belew and L.B. Booker (Eds), Morgan Kaufmann, pp 85-89.

GA Techniques

GAS, a Concept on Modeling Species in Genetic Algorithms[*]

Márk Jelasity[1] and József Dombi[2]

[1] Student of József Attila University, Szeged, Hungary
`jelasity@inf.u-szeged.hu`
[2] Department of Applied Informatics, József Attila University, Szeged, Hungary
`dombi@inf.u-szeged.hu`

Abstract. This paper introduces a niching technique called GAS (S stands for species) which dinamically creates a subpopulation structure (taxonomic chart) using a *radius function* instead of a single radius, and a 'cooling' method similar to simulated annealing. GAS offers a solution to the niche radius problem with the help of these techniques. A method based on the *speed* of species is presented for determining the radius function. Speed functions are given for both real and binary domains. We also discuss the sphere packing problem on binary domains using some tools of coding theory to make it possible to evaluate the output of the system. Finally two problems are examined empirically. The first is a difficult test function with unevenly spread local optima. The second is an NP-complete combinatorial optimization task, where a comparison is presented to the traditional genetic algorithm.

1 Introduction

In recent years much work has been done with the aim of extending genetic algorithms (GAs) to make it possible to find more than one local optimum of a function and so to reduce the probability of missing the global optimum. The techniques developed for this purpose are known as *niche techniques*. Besides the greater probability of the success of the algorithm and a significantly better performance on GA-hard problems (see [13]), niche techniques provide the user with more information on the problem, which is very useful in a wide range of applications (decision making, several designing tasks, etc.).

1.1 Best-Known Approaches

Simple iteration runs the simple GA several times to the same problem, and collects the results of the particular runs. **Fitness sharing** has been introduced by Goldberg and Richardson [6]. The fitness of an individual is reduced if there are many other individuals near it and so the GA is forced to maintain diversity in the population. **Subpopulations** can also be maintained in parallel, usually

---

[*] This work was supported by the ESPRIT project BRA 6020 and by the OTKA grant T14228.

with the allowance of some kind of communication between them (see, for example, [9]). The GAS method has developed from this approach. The **sequential niche technique** is described in [13]. The GA (or any other optimizing procedure) is run many times on the same problem, but after every run the optimized function is modified (multiplied by a derating function) so that the optimum just found will not be located again.

1.2 Problems

These techniques yield good results from several viewpoints, but mention sholud be made of some of their drawbacks, which do not arise in the case of our method, GAS.

Simple iteration is unintelligent; if the optima are not of the same value relatively bad local optima are found with low probability, while good optima are located several times which is highly unnecessary. **Fitness sharing** needs $O(n^2)$ distance evaluations in every step, besides the evaluation of the fitness function. It cannot distinguish local optima that are much closer to each other than the *niche radius* (a parameter of the method); in other words, it is assumed that the local optima are approximately evenly spread throughout the search space. This latter problem is known as the *niche radius problem*. The **sequential niche technique** also involves the niche radius problem. The complexity of the optimized function increases after every iteration due to the additional derating functions. Since the function is modified many times, "false" optima too are found. The method seems difficult to use for combinatorial problems or structural optimization tasks, which are the most promising fields of GA applications.

GAS offers a solution to these problems including the niche radius problem, which is the most important drawback of all of the methods mentioned earlier.

1.3 Outline of the paper

In section 2 we give a brief description of GAS that is needed for an understanding of the following part of the paper. The reader who is interested in more details should refer to the Appendix on how to obtain more information or GAS itself.

In section 4 we give a possible solution to the niche radius problem with the help of the GAS system. Both real and binary problem domains are discussed.

In section 5 we present experimental results. Two problems are examined. The first demonstrates how GAS handles the uneven distribution of the local optima of the optimized function. The second is an NP-complete combinatorial problem, where a comparison is presented to the traditional GA.

2 Species and GAS

Using the notations in the Introduction of [10], let D be the problem domain, $f : D \rightarrow \mathbb{R}$ the fitness function and $g : \{0, 1\}^m \rightarrow D$ $(m = 2, 3, \ldots)$ the coding function. (GAS searches for the *maxima* of f!)

Let us assume that a distance function $d : D \times D \to \mathbb{R}$ and term section (section$: D \times D \to P(D)$, where $P(D)$ is the power set of D) are defined.

Example 1.

- $D = [a, b]^m$ $(a, b \in \mathbb{R})$.

$$\text{section}(x, y) = \{z : z = x + t(y - x) \; t \in [0, 1]\}$$

- $D = \{0, 1\}^m$.

$\text{section}(x, y) =$
$\{z : $ if the jth letter of x and y is t, then the jth letter of z is $t\}$

Definition 1. $R : \mathbb{N} \to \mathbb{R}$ is a *radius function* over D if it is monotonous, positive, $R(0) = \max\{ d(e_1, e_2) : e_1, e_2 \in D\}$ and

$$\lim_{n \to \infty} R(n) = 0.$$

Fig. 1a exemplifies these properties.

Let us fix a radius function R.

Definition 2. $s = (o, r, S)$ is a *species* over D, where S is a population over D and the members of S are the *individuals* of s; $o(\in S)$ is the *center* of s, $f(o) = \max f(S)$; $r(\in \mathbb{N})$ is the *radius index* or the *level* of s, and so the *radius* of s is $R(r)$. Recall that in GAs a population is a multiset (or bag) of individuals (e.g. $S =< x_1, x_1, x_2 >$).

Definition 3. $s = (o, r, S)$ is a species. Let $A(s) = \{ a \in D : d(a, o) \le R(r)\}$ be the *attraction* of s.

Fig. 1b illustrates the terms defined above.

Definition 4. $\rho \subseteq \text{SP} \times \text{SP}$; $\text{SP} = \{s : s$ is a species over $D\})$. $s_1 = (o_1, r_1, S_1)$, $s_2 = (o_2, r_2, S_2) \in \text{SP}$, $(s_1, s_2) \in \rho$ if and only if there is no $e \in D$ on the section connecting o_1 and o_2 such that $f(e) < f(o_1), f(o_2)$.

Though ρ is not an equivalence, it is very important from the viewpoint of species creation, as will be shown later.

Definition 5. Let $T = (V, E)$ be a graph with $V(T) \subseteq \text{SP}$. T is a *taxonomic chart* (t.c.) if T is a tree and there is an $s_r = (o_r, 0, S_r)$ root in T, and if $(s_r =)s_0, s_1, \ldots, s_n$ is a path in T, then $r_0 < r_1 < \ldots < r_n$.

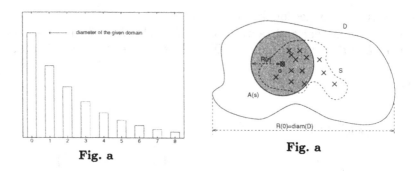

Fig. a

Fig. a

Fig. 1. a: A possible radius function. b: Terms related to species.

```
procedure activity
begin
    while (population size of T_n < maximum allowed) do begin
        choose two parents
        create two offspring
        place the parents and the offspring back in the population
    end
    dying_off
    fusion
end
```

Fig. 2. The basic algorithm that creates T_{n+1} from T_n.

2.1 The Algorithm

Let $V(T_0)$ (T_0 is a t.c.) contain only $s_r = (o_r, 0, S_r)$, where S_r is randomly chosen. The algorithm in Fig. 2 shows how GAS creates a T_{n+1} t.c. from a given T_n t.c.

Before describing the parts of the algorithm, we should make a few remarks.

- It is the flexibility of steady state selection [11] that allows the algorithm to create and manage species, as will be shown later.

- The algorithm can be implemented in parallel on two levels: the level of the while cycle and the level of the **procedure**. (However, our implementation is not parallel.)

Let us now examine the parts of the algorithm.

Population Size. The population size of a given T t.c. is $\displaystyle\sum_{s=(o,r,S)\in V(T)} |S|.$

Choose Two Parents. From a given T, we first choose a vertex with a probability proportional to the number of the elements of the vertices. Then, we choose two parents from this species, using the traditional probability (proportional to the fitnesses of the elements of the species).

Create Two Offspring. From individuals p_1 and p_2, we create p_1' and p_2' by applying onepoint crossover and mutation operators to the parents.

Placing Elements Back in the Population. Since this is the point where new species are created, this is the most important step. We have to decide here whether to separate the given two parents into two different species and we have to find the species of the two newly created offspring. If we decide to separate the parents, we must find a new existing species for them or create new species for them. The placing-back algorithm is shown in Fig. 3. The notations of the algorithm: p_1, p_2 are the parents, p_1', p_2' are the two offspring, e is a random point on the section that connects p_1 and p_2 (note that p_1' and p_2' are not on this section in general), and s_p is the original species of the parents. We always mean $s_x = (o_x, r_x, S_x)$ on s_x for any symbol x.

```
if f(e) < f(p1),f(p2) then
    for p=p1,p2,p1',p2' do
        if (there is a child node s_c of s_p such that p is in A(s_c)) then
            insert(p,s_c)
        else
            create a new child s=(p,max{r_p+1,strict},<p>) for s_p
        { With the restriction that p1 and p2 must not be   }
        { put into the same species and an offspring        }
        { must not create a new species                     }
else begin
    insert(p1,s_p);insert(p2,s_p)
end

for p=(an offspring not put in so far) do begin
    s:=s_p; while (p is not in A(s)) do s:=father node of s
    { if s=s_r then A(s)=D! }
    insert(p,s)
end
```

Fig. 3. The algorithm that places parents and offspring back in the population.

Function insert(p,s) inserts p to S and updates o if necessary. Parameter strict determines the precision of the search and can be varied (typically increased) by the high-level algorithm (see e.g. Fig. 4) that directs the exploration.

It is clear that if $(s_1 = (p_1, r, < p_1 >), s_2 = (p_2, r, < p_2 >)) \notin \rho$, then p_1 and p_2 will not be separated. This means that for a concave or for a unimodal one-dimensional function f GAS will never create a single species.

Dying off. Dying off deletes as many elements from the population of the t.c. as were inserted in the `while` cycle keeping the population size constant. The method used for selecting elements to die is based on the ranking defined by the transformed fitness function $\hat{f}$:

$$\hat{f}(e) := \frac{f(e) - (\text{a global lower bound of } f \text{ on the whole population})}{|S|}$$

where e is in species $s = (o, r, S)$.

This means that species of small size have more chance to survive (and to grow). The precision of the procedure can be varied during the optimization process. In section 5 we discuss how to use this possibility. Dying off has no effect on the species structure (by definition) and does not delete the best individual of a species.

Fusion. The result of `fusion` depends on R and `strict` described earlier. After executing `fusion` for a given T t.c., we get T', for which the following will be true: if s_1, $s_2 \in V(T')$, then $d(o_1, o_2) \geq R(\text{strict})$.

`Fusion` simply unites species of T that are too close to each other, and `strict` tells it what is too close. If s_1 and s_2 are united, the result species is $s = (o, \min\{r_1, r_2\}, S_1 \cup S_2)$, where $f(o) = \max\{ f(o_1), f(o_2)\}$ and o is o_1 or o_2. In view of the tree structure, the species with the lower level absorbs the other. If the species have the same level, either of them may absorb the other.

3 Optimization with GAS

For global optimization with GAS, we suggest the algorithm shown in Fig. 4.

```
create a starting t.c. T_0
for strict=1 to ST_m              { 0 < ST_m(=strict_max) < 8 }
    new species
    evolution                     { evolution is the following macro: }
    stabilize                     {      for i=1 to 10 do activity      }
    iterate evolution until reaching {   immigration                    }
    x_strict function evaluations {      for i=1 to 5  do activity      }
```

Fig. 4. The high-level test algorithm.

The role of ST_m will be examined later. For determination of the vector of evaluation numbers x and the radius function R, we suggest a method in section 4 based on the *speed* of species with a given radius in a given domain.

We now describe the species-level genetic operators, used in the algorithm shown in Fig 4.

Immigration. For every species $s = (o, r, S)$ in a given t.c., $|S|/2$ randomly generated new individuals are inserted from $A(s)$. Immigration refreshes the genetic material of the species and makes their motion faster.

New species. This switch alters the state of the system towards managing species creation. It randomizes dying_off and relaxes competition by decreasing the lower bound of the fitness function, and so decreases the relative differences between individuals. According to some biologists [3], species are born when the competition decreases; our experiments support this opinion.

Stabilize. The effect of this is the contrary of new species. It prohibits the creation of new species and increases competition.

We give here some heuristical arguments that support the subpopulation structure approach and use a radius function instead of a single radius.

- The number of distance calculations grows with the size of the t.c. instead of the size of the population.

- Application of species–level operators (e.g. fusion, immigration) becomes possible.

- Lower-level (closer to root) species manage to create new species in their attraction.

- The advantages of the technique based on the radius function and increasing strict (see Fig. 4) are similar to those of the 'cooling' technique in the simulated annealing method.

Finally, to make our discussion more rigorous, we give the definitions of stability of species and t.c.

Definition 6. $W \subseteq D$. Species s is *stable in* W if $o \in W$, and if $o_1, o_2, \ldots$ is a series of new centers inserted by GAS to s during running, then $\forall i \ o_i \in W$.

Example 2. It is clear that s is stable in $W = \{e \in D : f(e) > f(o)\}$.

Definition 7. $e_0 \in D$, e_0 is a local optimum (with respect to d) of f. s is *stable around* e_0 if, for every $o_1, o_2, \ldots$ series of new centers inserted by GAS to s during running, $o_n \to e_0$ $(n \to \infty)$.

Example 3. $W \subseteq D$, $e_0 \in W$. If s is stable in W and e_0 is the global optimum of f in W (i.e. e_0 is a local optimum of D!) and there are no more optima of f in W, then s is stable around e_0.

Definition 8. T is a t.c. T is *stable* if every species of T is stable around distinct local optima of f.

Definition 9. T is a t.c. T is *complete* if T is stable and there is exactly one stable species around every local optimum of f.

4 Theoretical Results

In this section we discuss the theoretical tools and new twerms that can be used due to the exact definition of the t.c. data structure and GAS algorithm.

4.1 Speed of Species

We do *not* assume that the optima of the fitness function are evenly spread; we create species instead that "live their own lives" and can move in the search space and find the niche on which they are stable. It can be seen that from this point of view determining R depends more upon the *speed* of the species than on the number of spheres (niches) of a given radius that can be packed into the space.

The least value is the base of setting the niche radius parameter of the methods mentioned in the Introduction. However, the solution of the packing problem is useful when evaluating the output of the system since it tells us what percentage of the possible number of optima we have found. In section 4.3 we discuss such packing problems in the case of binary domains.

4.1.1 Real Domains.

First, we define the speed of a species with radius $r \in \mathbb{R}$ over domains from $\mathbb{R}^n$ $(n = 1, 2, \ldots)$.

Definition 10. The *speed* $v(r)$ of a species s with radius r and center o is $(c_1 - o_1)/2$, where c is the center of gravity of the set

$$S_{n,r} = A(s) \cap \{x \in \mathbb{R} : x_1 > o_1\}$$

In other words, let us choose a random element x^* from $A(s)$ with even distribution. Let $\xi = o_1 - x_1^*$ if $o_1 > x_1^*$, and $\xi = 0$ otherwise. Than $M(\xi)$ (the expected value of ξ) is $v(r)$. This means that $v(r)$ is given by the equation

$$v(r) = \frac{1}{2V(S_{n,r})} \int_{S_{n,r}} x_1 \, dx_1 \ldots dx_n \tag{1}$$

where $V(S_{n,r})$ is the volume of $S_{n,r}$. It can be proved that

$$v(r) = \frac{\binom{n}{\frac{n-1}{2}}}{2^{n+1}} r \tag{2}$$

holds. In the general case (if n is even) (2), is defined with the help of the function $\Gamma(t+1)$, the continous extension of $t!$. $\Gamma(t+1) = \int_0^\infty x^t e^{-x} \, dx$, $\Gamma(1/2+1) = \sqrt{\pi}/2$ and $\Gamma(t+2) = (t+1)\Gamma(t+1)$.

4.1.2 Binary Domains.

Let $D = \{0,1\}^n$. We give a definition of speed similar to Definition 10.

Let $e \in D$ such that the number of 0s is equal to or greater by one than the number of 1s. Let

$$S_{n,r} = \{e' \in D \ : \ d(e',e) \leq r\}$$

where d is the Hamming distance (the sum of the bit differences). Let us choose a random e^* from $S_{n,r}$ with even distribution and let $\xi = d(e^*, e)$ if there are more 1s in e^*, and let $\xi = 0$ otherwise.

Definition 11. Let $v(r) = M(\xi)$ be the *speed* of species s in D if the radius of s is r.

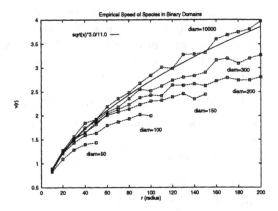

Fig. 5. Speed in binary domains.

We performed experiments to determine $v(r)$ (Fig. 5). It can be seen if $n \gg r$, then the equation

$$v(r) = \frac{3}{11}\sqrt{r} \tag{3}$$

seems to describe the speed. If r approaches n, the growing of the speed becomes slower than (3) would indicate.

4.2 Determining R and x

Let us assume that the evaluation number N, the domain type and the corresponding speed function v are given. We know that

$$\sum_{i=1}^{ST_m} x_i = N \tag{4}$$

We suggest a setting for which the system of equations

$$v(R(i))x_i = C \quad (i = 1, \dots ST_m) \tag{5}$$

holds where C is a constant (independent of i). This simply means that the species of the different levels receive an equal chance to become stabilized. From (4) and (5) it follows that

$$C = \frac{N}{\sum_{i=1}^{ST_m} \frac{1}{v(R(i))}}. \tag{6}$$

We note that C is the distance that a species of level i expectedly crosses during x_i iterations.

In GAS, the upper bound M of the number of species can be set. $M =$ [population size/4] by default. Now we can give the value of C:

$$C = R(0)M\nu \tag{7}$$

Recall that $R(0)$ is the diameter of the domain we examine.

ν is a threshold value. Setting $\nu = 1$ means that every species receives at least sufficient function evaluations for crossing the whole space, which makes the probability of creating a stable t.c. very high. In section [5] we examine the effect of several different settings of ν. Finally let

$$R(i) = R(0)\beta^i \quad (i = 1, \dots ST_m, \ \beta \in (0,1)) \tag{8}$$

Then, R is a valid radius funtion and subproblems defined by the species will be similar in view of the radii.

Using (6), (7) and (8), we can write

$$\frac{N}{R(0)M\nu} = \sum_{i=1}^{ST_m} \frac{1}{v(R(0)\beta^i)} \quad (\beta \in (0,1)) \tag{9}$$

where everything is given except β.

Since v is monotonous, the right side of (9) monotonically decreases if $\beta \in (0,1]$ and so reaches its minimum if $\beta = 1$. Using this fact, the feasibility condition of (9) is

$$\frac{N}{R(0)M\nu} > \frac{ST_m}{v(R(0))} \tag{10}$$

If (10) holds, (9) has exactly one solution. This property allows us to use effective numeric methods for approximating β.

4.3 Evaluating the Output

We based the setting of the parameters of GAS on the speed function. However, it is important to know the maximal possible size of a t.c. for a given radius function R (assuming an arbitrary large evaluation number and population size) since it tells us what percentage of the maximal possible number of optima we have found.

The problem leads to the general sphere packing problem and this has been solved neither for binary nor for real sets in the general case.

Real Case

In n-dimensional real domains Deb's method [5] can be used.

$$p = (\frac{\sqrt{n}}{2r})^n$$

where r is the species radius, the domain is $[0,1]^n$ and p is the number of optima, assuming that they are evenly spread. We note that this is only an approximation.

Binary Case

Results of coding theory can be used to solve the packing problem in binary domains since it is one of the central problems of this field. We will need the definition of binary codes.

Definition 12. $d, n \in \mathbb{N}$, $d \leq n$. $C \subseteq \{0,1\}^n$ is a $(n, |C|, d)$ *binary code* if $\forall c_1, c_2 \in C$: $\mathrm{dist}(c_1, c_2) \leq d$ (The function "dist" is the Hamming distance, the sum of the bit differences.)

Definition 13. $d, n \in \mathbb{N}$. $A(n,d) := \max\{|C| : C$ is a $(n, |C|, d)$ binary code$\}$.

$A(n,d)$ has not yet been calculated in the general case; only lower and upper bounds are known. Such bounds can be found for example in [12], [2] or [1]. One of these is the Plotkin bound:

Theorem (Plotkin bound) *For $d, n \in \mathbb{N}$, we have*

$$A(n,d) \leq \frac{d}{d - \frac{1}{2}n} \quad \text{if } d \geq \frac{1}{2}n$$

Proof. [12].

In a special case, the exact value is also known:

Theorem 14. *For binary codes and $m \in \mathbb{N}$, we have*

$$A(2^{n+1}, 2^n) = 2^{n+2}.$$

In Table 1 we show the Plotkin upper bounds for $l = 32, 128$ and 1024. The values have been calculated according to the following formulas:

$$A(2m, m+a) \leq \frac{m+a}{m+a-2m/2} = \frac{1}{a}(m+a)$$
$$A(2^{n+1}, 2^n) = 2^{n+2}$$
$$A(2m, m-a) \leq 2^{2a+1}2(m-a)$$

r 2n	32	128	1024
n-3	3 328	15 616	130 304
n-2	896	3 968	32 640
n-1	240	1 008	8 176
n	64*	256*	2048*
n+1	17	65	513
n+2	9	33	257
n+6	3	11	86
n+16	2	5	33
n+40	-	2	13

Table 1. Plotkin upper bounds for $A(2n, r)$. The indicated values are exact.

5 Experimental Results

In this section we examine two problems. The first demonstrates how GAS handles the uneven distribution of the local optima of the optimized function. The second is an NP-complete combinatorial problem, where a comparison is presented to the traditional GA.

5.1 Setting of GA and GAS Parameters

In the following experiments, the settings of the traditional GA parameters are P_m (mutation probability) = 0.03 (see e.g. [7]) and P_c (crossover probability) = 1, while the population size = 100. In the while cycle of the basic algorithm (shown in Fig. 2), the maximum allowed population size is 110. For continuous domains, we used Gray coding as suggested in [4].

The settings of the specific GAS parameters are the following:

- R (radius function) and x (evaluation numbers) can be determined using the method described in section 4.

- M (maximal number of species in the t.c.) is set to $M =$ (pop. size)/4. Setting a larger value is not recommended since too many small species could be created.

- N ($\sum_{i=1}^{ST_m} x_i$) depends on the available time and computational resources. We used $N = 10^4$.

- ν (treshold) and ST_m (maximal strict level) are the parameters we tested so we used several values (see the descriptions of the experiments).

For simplicity, we run evolution only once after new species (see Fig. 4) but we note that increasing that number can significantly improve the performance in some cases. The cost of one evolution is 275 evaluations after new species, and 200 after stabilize at the above settings.

5.2 A Function with Unevenly Spread Optimas

The problem domain D is $[0, 10]$. The fitness function $f : D \to \mathbb{R}$.

$$a = \begin{pmatrix} 3.040 \\ 1.098 \\ 0.674 \\ 3.537 \\ 6.173 \\ 8.679 \\ 4.503 \\ 3.328 \\ 6.937 \\ 0.700 \end{pmatrix} \quad k = \begin{pmatrix} 2.983 \\ 2.378 \\ 2.439 \\ 1.168 \\ 2.406 \\ 1.236 \\ 2.868 \\ 1.378 \\ 2.348 \\ 2.268 \end{pmatrix} \quad c = \begin{pmatrix} 0.192 \\ 0.140 \\ 0.127 \\ 0.132 \\ 0.125 \\ 0.189 \\ 0.187 \\ 0.171 \\ 0.188 \\ 0.176 \end{pmatrix} \quad f(x) = \sum_{i=1}^{10} \frac{1}{(k_i(x - a_i))^2 + c_i}$$

f (shown in Fig. 6) is a test function for global optimization procedures suggested in [8].

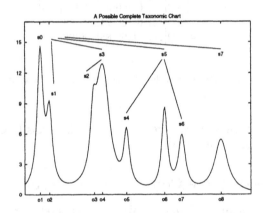

Fig. 6. The function with unevenly spread optima.

We have determined R and x for $\nu = 1/4, 1/2, 3/4$ and 1 (see Table 2). ST_m is 8 in every case. Recall that according to the algorithm in Fig. 4 the elements of x must be divisible by 200 (the cost of evolution after stabilize) and the sum of them must be $10^4 - ST_m \cdot 275$.

We run the corresponding algorithms 100 times. The numbers of stable species that converged to one of the local optima are shown in Table 3. The most important result is that *no* unstable species appeared in the output even for $\nu = 1/4$.

The best results are observed in the case of $\nu = 1/4$. Here, even o_3 was found 2 times in spite of its very small attraction.

Fig. 7 shows the average number of species detected before increasing strict (after stabilizing for the old strict). From these values, we can gain information on the structure of the optima of the fitness function. For example, for radii

	$\nu = 1$		$\nu = 3/4$		$\nu = 1/2$		$\nu = 1/4$	
	R	x	R	x	R	x	R	x
1	6.61	200	6.333	200	5.978	0	5.334	0
2	4.369	200	4.011	200	3.573	200	2.845	0
3	2.888	400	2.54	200	2.136	200	1.517	200
4	1.909	600	1.609	400	1.277	400	0.809	200
5	1.261	800	1.019	800	0.763	600	0.432	600
6	0.834	1200	0.645	1200	0.456	1200	0.23	1000
7	0.551	1800	0.409	1800	0.273	2000	0.123	2000
8	0.364	2800	0.259	3000	0.163	3200	0.066	3600

Table 2. Radius and evaluation numbers for $\nu = 1/4, 1/2, 3/4$ and 1.

	o_1	o_2	o_3	o_4	o_5	o_6	o_7	o_8
$\nu = 1$	100	0	0	100	60	97	48	94
$\nu = 3/4$	100	1	0	100	65	87	72	94
$\nu = 1/2$	100	34	0	100	74	99	58	98
$\nu = 1/4$	100	25	2	100	85	100	90	100

Table 3. Number of stable species around the local optima.

greater than 3, very few species were created, which means that the optima are probably closer to each other than 3.

5.3 An NP–complete Combinatorial Problem

We study the subset sum problem here. We are given a set $W = \{w_1, w_2, \ldots, w_n\}$ of n integers and a large integer C. We would like to find an $S \subseteq W$ such that the sum of the elements in S is closest to, without exceeding, C. This problem is NP-complete.

We used the same coding and fitness function as suggested in [14]: $D = \{0, 1\}^{128}$. If $e \in D$ ($e = (e_1, e_2, \ldots, e_{128})$), then let $P(e) = \sum_{i=1}^{128} e_i w_i$, and then

$$-f(e) = a(C - P(e)) + (1 - a)P(e)$$

where $a = 1$ when e is feasible ($C - P(e)) \geq 0$), and $a = 0$ otherwise.

Here, $\forall w \in W \ 1 \leq w \leq 1000$ and C is the sum of a randomly chosen subset of W (every element is chosen with a probability 0.5).

We do not need a coding function here since D is the code itself.

We tested several values of ST_m. Table 4 shows R and x for $ST_m = 1, 2, \ldots, 6$. The value 8 is not feasible and 7 is also very close to that bound. $\nu = 1$ in every case. We run the corresponding algorithms 50 times.

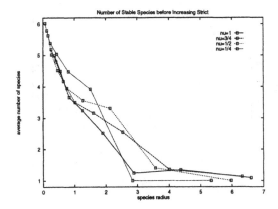

Fig. 7. The species number increasing history (average of 100 runs).

$ST_m = 1$		$ST_m = 2$		$ST_m = 3$		$ST_m = 4$		$ST_m = 5$		$ST_m = 6$	
R	x	R	x	R	x	R	x	R	x	R	x
1	2	9600	20	2600	47	1800	72	1400	93	1200	109	1200
2			3	6800	17	2800	41	1800	67	1400	92	1200
3					6	4600	23	2400	49	1600	79	1400
4							13	3200	35	2000	67	1400
5									26	2400	57	1600
6											48	1600

Table 4. Radii and evaluation numbers for $ST_m = 1, 2, \ldots, 6$.

For comparison, in experiments on the same problem with two times more (i.e. $2 \cdot 10^4$ instead of 10^4) evaluation numbers in [14], 0.93 optimal solutions were found per run. Here, this value is at least one for every ST_m, and for $ST_m = 2$ it is 2.62 (see Table 5).

Besides this, many near-optimal solutions were found (as shown in Fig. 8) so we received much more information with only 10^4 function evaluations.

6 Summary

In this paper we have introduced a method called GAS for multimodal function optimization (or multimodal heuristic search). GAS dynamically creates a sub-population structure called a taxonomic chart, using a radius function instead of a single radius, and a 'cooling' method similar to simulated annealing.

We based setting of the parameters of the method on the speed of species instead of their relateive size to the search space and we gave speed functions

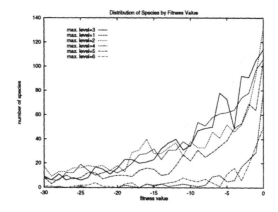

Fig. 8. Number of near-optimal solutions found during the 50 runs.

ST_m	opt. found/ run	avg. fitness of all spec.	number of species
1	1.56	-3.194	201
2	2.62	-15.616	1250
3	2.1	-10.866	1250
4	2.12	-12.568	1238
5	2.08	-12.84	846
6	1.0	-6.943	211

Table 5. Result of the experiment (50 runs).

for both real and binary domains.

We performed experiments for a difficult test function with unevenly spread local optima and for an NP-complete combinatorial problem.

In both cases our results are encouraging though much work will have to be done to examine the effects of the parameters of the method more thoroughly.

Appendix

GAS and a detailed description of the system is available via anonymous ftp at the following URL:

```
ftp://ftp.jate.u-szeged.hu/pub/math/optimization/GAS
```

This is a directory. Please read the file readme.

The authors would highly appritiate it if you informed them about any problems regarding GAS (compilling, using, etc.).

References

[1] F.J. MacWilliams, N.J.A. Sloan (1977) *The Theory of Error-correcting Codes*. North Holland.

[2] R.J. McEliece (1977) *The Theory of Information and Coding.*, Addison-Wesley.

[3] V. Csányi (1982) *General Theory of Evolution*. Publ. House of the Hung. Acad. Sci. Budapest.

[4] R.A. Caruana, J.D. Schaffer (1988), Representation and hidden bias: Gray vs. binary coding for genetic algorithms, in *Proceedings of the fifth international conference on machine learning* (editor: John Laird), pp153-161.

[5] K. Deb (1989) *Genetic algorithms in multimodal function optimization*. Masters thesis, TCGA Report No. 89002. The University of Alabama, Dep. of Engineering mechanics.

[6] K. Deb, D.E. Goldberg (1989) An investegation of niche and species formation in genetic function optimization. In *Proceedings of the Third ICGA* (pp42-50). Morgan Kaufmann.

[7] D. E. Goldberg (1989), *Genetic algorithms in search, optimization and machine learning*, Addison-Wesley, ISBN 0-201-15767-5.

[8] A. Törn, A. Žilinskas (1989), *Global Optimization*, in Lecture Notes in Computational Science 350. Springer-Verlag.

[9] Y. Davidor (1991) A naturally occuring niche and species phenomeon: the model and first results. In *Proceedings of the Fourth ICGA* (pp257-263). Morgan Kaufmann.

[10] G.J.E. Rawlins (ed.) (1991) *Foundations of Genetic Algorithms.*, Morgan Kaufmann.

[11] G. Syswerda (1991) A Study of Reproduction in Generational and Steady State Genetic Algorithms. in [10].

[12] J.H. van Lint (1992) *Introduction to Coding Theory.*, Springer-Verlag.

[13] D. Beasley, D.R. Bull, R.R. Martin (1993) A Sequential Niche Technique for Multimodal Function Optimization. *Evolutionary Computation* 1(2), pp101-125, MIT Press.

[14] S. Khuri, T. Bäck, J. Heitkötter (1993), An Evolutionary Approach to Combinatorial Optimization Problems, in *The Proceedings of CSC'94*.

[15] M. Jelasity, J. Dombi (1995) GAS, an Approach to a Solution of the Niche Radius Problem. In *The proceedings of GALESIA'95*.

A Genetic Algorithm with Parallel Steady-State Reproduction

Cyril Godart[1] and Martin Krüger[2]

[1]INRIA, 2004 route des Lucioles, B.P.93, 06902 Sophia Antipolis, France
[2]UPM, ETSI de Telecomunicación, Ciudad Universitaria s/n, 28040 Madrid, Spain
<cyril.godart@inria.fr>, <krueger@die.upm.es>

Abstract. During the last years, heterogeneous computing became more and more popular. Its advantage is availability, since any network of workstations can be used as a parallel virtual machine. On the other hand, programming these machines efficiently is even more difficult than programming a parallel dedicated machine. Each machine in the network may dynamically change performance, so optimal load balancing is the main challenge. One also has to deal with reliability problems on the network. With these two points in mind, we designed a parallel genetic algorithm featuring a new reproduction operator, called *parallel steady-state reproduction*. This operator adapts to a heterogeneous and unpredictable environment by allowing some "chaotic" behaviour. This parallel genetic algorithm revealed to be robust and efficient, and is particularly easy to implement.

1 Introduction

For parallel function optimization, genetic algorithms (GA's) are very popular for several reasons. They can be used as robust and efficient function optimizers, which has been shown in a wide variety of applications in fields like control, optimization and artificial intelligence (see [4, 7] for a summary). They also have a large amount of inherent parallelism that is relatively easy to exploit to accelerate execution (see for instance [9, 10]).

Most of these parallel implementations are designed for use on dedicated parallel computers, like transputer networks. These are *homogeneous* parallel computers, where processing power of all units is constant over time. In the last years however, *heterogeneous* architectures are getting more and more popular, so there is a growing need of powerful, robust and efficient parallel optimization algorithms specially suited to this type of architecture.

Our work was focused around the optimization of control parameters for various types of stochastic optimization algorithms (see [6]). This type of optimization problem demands a very high amount of computation in the evaluation phase of the genetic algorithm. On a VAX 8550, one control parameter optimization often took a week or more. To accelerate execution, we decided to implement a parallel version on a large network of workstations.

In this paper we present a genetic algorithm with a new reproduction operator, *parallel steady-state reproduction*, enabling the genetic algorithm to make optimal use of a network of workstations. The algorithm is as easy to implement as a standard genetic algorithm with steady-state selection, and there are no additional control-parameters.

The outline of the paper is as follows. After an introduction to heterogeneous computing in Sect. 2, Section 3 introduces some representative parallel implementations of genetic algorithms and shows why they are not suitable for use on a heterogeneous machine. Section 4 presents the parallel steady-state reproduction operator. It is validated by means of simulation in Sect. 5, using a specially designed network simulator that models random execution times occurring on a real network. In Sect. 6 we show speedup figures obtained on a real network.

2 Heterogeneous computing

What is meant by heterogeneous computing is to combine the computing power of locally available machines. This can be workstations or large scale computers, but also desktop computers. In order to do heterogeneous computing, all computers have to be interconnected in a suitable way so they can communicate with each other (an example is a Local Area Network connected through Ethernet, using TCP/IP protocol).

The addition of the performance ratings of all available machines can lead to quite impressive MIPS figures. Unfortunately, this computing power is particularly difficult to exploit efficiently:

1. in most cases, the network consists of workstations with different processors and memory configurations, leading to a **static distribution** of the available computing power;

2. other users and background processes temporarily reduce the available computing power on some workstations. This leads to a **dynamic distribution** of available computing power;

3. any machine in the network may crash and subsequently reset, leading to the lost of all work done so far on this machine. During several hours of computing on a large network the probability of isolated crashes may be quite high.

As a consequence, an efficient algorithm for a heterogeneous architecture must have elements that allow to **adapt dynamically** to any given network configuration. Moreover, to be reliable, the algorithm has to be **fault-tolerant**, too. In particular, solutions with a global control process running on one machine should be avoided.

3 Parallel Implementations of Genetic Algorithms

The following is a short discussion of the most common parallel implementations of genetic algorithms and their suitability for our application.

One of the particularities of a genetic algorithm is that instead of working with a unique solution, a genetic algorithm it works with a whole population of solutions simultaneously. Several operators are applied to this population simulating natural evolution as described by Darwin [2].

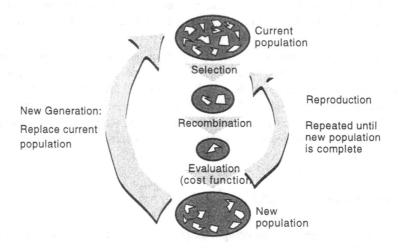

Fig. 1. Operators of a standard genetic algorithm: the reproduction operator is composed of selection, recombination and evaluation. The cost function is used to evaluate a new solution. Reproduction is applied until the new population is of the same size as the current population.

The first operator is *selection*. The probability for any solution to be chosen is proportional to its quality. This quality is determined by the cost function. For instance, if we look for the minimum of a function, a good-quality solution is one with low cost. The second operator to be applied is *recombination*. This operator takes the parent solutions and builds a new solution using parts of each of the parent solutions. The recombination operator usually consists of crossover and mutation. The new solution has to be assigned a quality, which is done by the *evaluation* operator and will be inserted into a new population.

The application of the three operators selection, recombination and evaluation is referred to as *reproduction*, which will be repeated until the new population contains the same number of solutions as the current population. The current population is then replaced by the new population, and a new generation begins. New generations will be created until a given criterion is satisfied, for example a maximum number of generations. Figure 1 sums up the operators of a standard genetic algorithm.

3.1 The worker-model

In the genetic algorithm, each reproduction is an independent process. Since it is executed as many times as there are solutions in the population, this allows a very straightforward parallelization of the genetic algorithm. This *worker-model* is the most widely used parallel model of a genetic algorithm.

This model is also known as "agenda parallelism" [1], where each worker in a group of identical workers grabs a task descriptor and computes the results (i.e. the new solution), then puts it back. In case of the genetic algorithm, each task consists of applying reproduction once in order to compute and evaluate a new solution. The data to be visible for all workers is the current population as well as the new population. If the number of workers equals the size of the population N, this model allows to divide the genetic algorithm into N parallel tasks, thus speeding up the execution by a factor of up to N in the ideal case (with no time lost in communication and synchronisation).

This speedup is only valid for identical workers. On a network of workstations, load balance will be very uneven. Different processors will have different execution times for a given process, and unpredictable load by other users can slow down the execution of one or several machines, whereas other machines have already finished and are idling. The worst case would be a machine that is blocked — although all other tasks are done, the algorithm waits eternally for the termination of the last one. In fact, execution time would be governed by the slowest machine and speedup may get smaller than 1, meaning that execution on the network may be slower than on the fastest machine alone.

One remedy would be to have a scheduler checking for idling workers, trying to even load balance. An idling worker would be loaded with one of the tasks another worker is still computing. The first to finish inserts its solution in the new population. Such a strategy would be able to avoid dramatic decrease of speedup. But a considerable amount of redundant computing would take place, since identical tasks are computed twice or more.

3.2 The neighbourhood-model

This model is based on *neighbourhood selection*. Each solution in the population is allocated to one processor, and a neighbourhood is defined for each processor. When a new solution has been computed, it tries to find a partner in its neighbourhood. This model is easily scalable for large numbers of processors, since communication does not increase locally. As for the worker-model, the speedup of the neighbourhood-model is roughly proportional to the number of processors in the network as long as most of the time is spent on reproduction. This model is widely used on message-passing machines like Transputer networks, as shown for instance in [13], where a neighbourhood is implicitly defined by the architecture of the network.

If we applied the neighbourhood-model to a heterogeneous system (for instance a Transputer network composed of different types of Transputers), synchronisation between neighbours will lead to the same problems as for the worker-model. Nevertheless, it is possible to implement the neighbourhood-model in an asynchronous way (see for instance [5, 12]). In that case, an individual ready for mating searches in its neighbourhood for other individuals. If no one is ready, it adopts a strategy that may consist in extending its neighbourhood or waiting for the first to get ready.

The neighbourhood-model is working well if the communication between processors is done locally. In that case, the physical layout of the local communication defines the neighbourhood. In case of our network of workstations, all communication usually go over the same physical medium (the Ethernet cable). Therefore the neighbourhood-model is likely to create bottleneck situations, since its total communication bandwidth is significantly higher than that of the worker-model.

3.3 The island-model

This model is well suited if the genetic algorithm is running with a large population. The individuals are grouped in subpopulations that evolve independently. At some point, individuals are exchanged between subpopulations, which is referred to as *migration*. This model imitates effects like sharing and cataclysms without any explicit control.

The model is easily parallelized by putting each subpopulation on a different processor, and migration takes place between these subpopulations [9]. On a heterogeneous parallel machine, different evolution speeds between subpopulations (islands) would not lead to any synchronization problems, since migration can take place at any time.

Nevertheless, the island model is more complicated to set up, since additional parameters controlling migration have to be adjusted. Furthermore, a large population may not always be efficient or feasible. In our application, where evaluation is a time consuming process, the number of individuals should be small (at most 50 to 100) to make the genetic algorithm evolve in a reasonable amount of time. If they are spread over several subpopulations, the remaining genetic material in each subpopulation would become insufficient, and problems like premature convergence will compromise the genetic algorithm.

4 Parallel steady-state reproduction

To overcome the drawbacks of the parallel models in respect to our application, we designed a new reproduction scheme, parallel steady-state reproduction. It is based on *steady-state reproduction*, a very well known (sequential) reproduction operator introduced by Syswerda [11]. To avoid confusion, we will call the latter *sequential steady-state reproduction*.

Syswerda's operator simplifies the genetic algorithm by reducing the number of necessary control-parameters and often behaves better than a generational genetic algorithm. Steady-state reproduction does a continuous replacement of individuals in the current population rather than a complete replacement of the current population by a new population (see Fig. 2). This operator allows parents and children to coexist in the same population, which is closer to the natural evolution scheme.

4.1 Principle

Syswerda's operator is a sequential process. The principle of parallel steady-state reproduction is to apply **several** sequential steady-state reproductions **concurrently**, as shown in Fig. 3. Each machine in the network is running the sequential steady-state reproduction process, and all these processes work on the same unique population, independently from each other.

A genetic algorithm using this parallel steady-state reproduction models another feature of natural evolution: the fact that reproduction takes place in parallel without

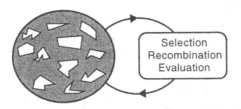

Fig. 2. Principle of sequential steady-state reproduction. After selection, recombination and evaluation the new solution is inserted into the current population. At each insertion another solution must be removed from the current population. This is done by reverse selection, i.e. the worst solution is eliminated with the highest probability.

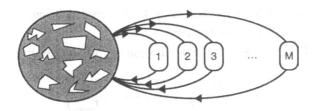

Fig. 3. Principle of parallel steady-state reproduction, where each numbered block stands for one reproduction process. If M machines are available in the network, M such processes may take place concurrently, all working on the same current population.

synchronization. In other words, in a population several individuals may be "pregnant" at the same time, and give birth at any moment.

Parallel steady-state reproduction allows each machine to contribute with all its available computing power to the evolution of the population. It does so without any need of a control process. The fastest processor will do most of the work, but even the slowest processor contributes.

4.2 Speedup estimations

The algorithm we describe focuses on an optimization problem that demands a very high amount of computation in the evaluation phase of the genetic algorithm. Therefore, in the following sections we assume that reproduction time is much larger than the time needed for all other computations. This may seem somewhat coarse, but as we will see it is sufficient for our type of application, and many real-world optimization problems (for instance optimization of neural networks) share this characteristic.

Relative speedup of the worker-model

Given the number of machines M (M should be a divider of the population size N), the *relative speedup* S can be calculated as follows:

$$S = \frac{\text{time on sequential machine}}{\text{time on parallel machine}} = \frac{T_{\text{sync}} + NT_{\text{repro}}}{T_{\text{sync}} + \frac{N}{M}T_{\text{repro}}} = 1 + \frac{M-1}{1 + \frac{M}{N}\frac{T_{\text{sync}}}{T_{\text{repro}}}} . \quad (1)$$

T_{repro} is the time necessary to do all the reproduction (selection, recombination and evaluation), whereas T_{sync} is the time necessary for all the other computations. T_{sync} is the sum of the times spent for the following computations:

1. communication between processes;
2. replacement of the current population by the new population;
3. computation of statistics, halt condition etc.;

If $T_{\text{repro}} \gg T_{\text{sync}}$, the above equation for S can be approximated in first order by

$$S = M - \frac{M(M-1)}{N}\frac{T_{\text{sync}}}{T_{\text{repro}}} + O\left(\frac{T_{\text{sync}}}{T_{\text{repro}}}\right) , \quad (2)$$

and thus

$$S \cong M . \quad (3)$$

We can see that speedup is proportional to M when most of the time is spent on reproduction.

Relative speedup of parallel steady-state reproduction

We estimate the relative speedup for a heterogeneous network, where the performance of each machine may be different but is constant over time. Let us first consider the case of two different types of machines in the network, where:

1. M_1: is the number of fastest machines, with execution time T^1_{repro}
2. M_2: is the number of second fastest machines, with execution time T^2_{repro}
3. T_{sync} is the time used to communicate with the population file.

We define the following ratio c:

$$c = \frac{T^1_{\text{repro}}}{T^2_{\text{repro}}} < 1 \quad (4)$$

The relative speedup S is then given by the expression

$$S = \frac{T_{sync} + N T^1_{repro}}{T_{sync} + \dfrac{N}{M_1 + cM_2} T^1_{repro}} , \tag{5}$$

where cM_2 is the equivalent number of fast machines. Again, if $T_{repro} \gg T_{sync}$ the above equation can be approximated in first order by

$$S = (M_1 + cM_2) - \frac{(M_1 + cM_2)(M_1 + cM_2 - 1)}{N} \left(\frac{T^1_{repro}}{T_{sync}}\right) + O\left(\frac{T^1_{repro}}{T_{sync}}\right) . \tag{6}$$

and thus

$$S \cong cM_2 + M_1 . \tag{7}$$

which is linear with the number of machines.

Let us now consider the case of several types of machines in the network, where:

1. M_1: is the number of fastest machines, with execution time T^1_{repro}
2. M_2: is the number of second fastest machines, with coefficient $c_2 = T^1_{repro} / T^2_{repro}$
3. ...
4. M_p: is the number of slowest machines, with coefficient $c_p = T^1_{repro} / T^p_{repro}$

In that case the relative speedup S becomes

$$S \cong M_1 + \sum_{i=2}^{p} c_i M_i . \tag{8}$$

which still is linear with the number of (equivalent) machines, taking into account their performance coefficients. Since these performance coefficients reflect the raw computing power of the machines, S indicates the maximum possible speedup on the real network.

5 Simulating the "Hibernatus Effect"

A genetic algorithm using parallel steady-state reproduction operator introduces a new stochastic element: the **order** in which new solutions are inserted into the current population is unpredictable. We call this the chaotic behaviour of the algorithm since the order of insertion depends on events external to the algorithm (machine load, network traffic etc.).

In the extreme case, if one machine is much slower than average, we will have the "Hibernatus Effect" — named after the novel of Jean-Bernard Luc [Pierre Horay Editor, 1969]. A man of the past was frozen in an iceberg, discovered and brought to life again in our era.

For the genetic algorithm, this means that a solution computed by a very slow machine will be inserted into a population that has evolved a lot. Such an "ancestor" may have a positive effect on the evolution, for example by re-introducing some forgotten genetic material into the genepool. But it may also disappear very quickly, meaning that computing resources have been wasted as the individual may have been useful at its time, but not later.

To ensure that the possible speedup is not ruined by a significant loss of efficiency of the original algorithm, it is necessary to validate parallel steady-state reproduction. Current genetic algorithm theory is not sufficient to treat such a problem analytically. It applies only to isolated operators, like selection and recombination. The only way to validate parallel steady-state reproduction is to compare it to existing genetic algorithms by applying them on a series of test problems.

Averaging over several experiments to get reliable results would take too much time on a real network. Therefore we needed a simulator that models the random executions times encountered on a real network on only one machine. This allows to determine the order in which each process is accessing the population. The duration in between these accesses is of no importance to the evolution.

5.1 Simulation of a network of workstations

The crucial part of this simulator is to model the random execution times so they correspond to what happens on a real network. In order to accurately simulate the behaviour of a real machine, we studied the distribution of execution times of an existing network.

It is intuitively clear that execution time of a given process lies between a nominal time t_N and infinity. The nominal time is composed of cpu-time plus the time used by the system to run the process. In the worst case, execution time can be infinite, for instance when the machine is blocked or has been rebooted.

To know how a real distribution might look like, we monitored execution times of a given process on two types of workstations (Sparc1+ and Sparc10). Figure 4 shows the resulting distribution of execution times. The distribution looks like a sawtooth which decreases rapidly for higher execution times. The first peak corresponds to t_N. The next peak occurs at approximately $2t_N$. The reason why peaks show up at multiples of t_N lies in the time-sharing system, where resources are equally allocated to each process. If two processes with same level of priority are present, the execution time is about twice the nominal time. In our case, it seems that the distribution is governed mainly by the number of processes present on the machine and much less by user-activities like editing, scrolling etc.

For simulation we used two pseudo random number generators, one for each machine type. They have been biased with the two probability distribution of Fig. 4 so we could be sure to obtain realistically distributed random execution times.

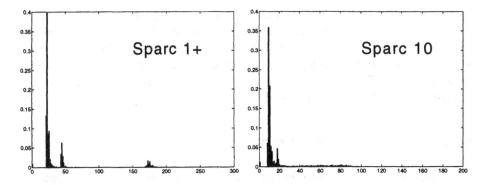

Fig. 4. Distribution of execution times on the real network. The vertical axis gives the probability of a given process to have the execution time (in seconds) on the horizontal axis. Both distributions are averages over all available stations of this type in the network, tested every 20 minutes during one week.

5.2 Results from simulation

We tested the parallel steady-state reproduction on the five standard test functions of de Jong [3]. Table 1 shows the definition of two of them, F_4 and F_5. The results for these two functions are shown in Table 2. We tested generational and sequential steady-state reproduction against three parallel configurations:

1. only Sparc10's (5);
2. only Sparc1's (20);
3. mixed (5 Sparc10's, 15 Sparc1's).

Population size is set to 50. The other control parameters have been tuned automatically with a method described in [6].

In comparison to generational reproduction, sequential steady-state reproduction showed slightly better results, especially on F_4. This agrees with already published results in [14]. Results obtained with parallel steady-state reproduction are close to those of sequential steady-state reproduction, with a slight advantage for the parallel

Table 1. The definitions of the two functions of de Jongs test-suite used for comparison.

Test function	Equation	Application interval	Optimum
F_4: quartic with noise	$y = \sum_{i=1}^{30} i x_i^4 + \text{GAUSS}(0, 1)$	$-1{,}28 \leq x_i \leq 1{,}28$	$y = 1248{,}2$
F_5: Shekel's foxholes	$y^{-1} = 0{,}002 + \sum_{j=1}^{25} \dfrac{1}{j + \sum_{i=1}^{2}(x_i - a_{ij})^6}$	$-65{,}536 \leq x_i \leq 65{,}536$	$y = 500$

operator in case of F_5. This is quite interesting, since it suggests that parallel steady-state reproduction may be in some cases more efficient than its sequential version. In other words, the evolution may benefit from the parallel reproduction and the unpredictable order in which individuals are inserted into the population. This point certainly deserves further study. However, our goal here was to show that parallel steady-state does not introduce a significant loss of efficiency.

6 Application on a real network

In this section we show speedup results obtained on a real network. We concentrate only on the speedup of the parallel implementation. During all tests, the parallel genetic algorithm consistently converged around the same solutions for our optimization problems than the sequential version.

We tested the parallel steady-state reproduction on a relatively large local network of Unix-workstations, composed of 7 Sparc10 and 13 Sparc1+ with different memory and disc configurations. Other local networks can be reached via gateways, thus allowing up to 48 workstations to be used simultaneously, forming a distributed heterogeneous system.

6.1 Implementation details

We implemented the algorithm in C-language, running under Unix system 5. All volumes are visible through the NFS protocol, which means that remote files can be accessed by all machines. For sake of simplicity we decided to put the current

Table 2. Results of performance measures on the network simulator. Parallel steady-state reproduction (PSSR, with M_1 the number of Sparc1's and M_2 number of Sparc10's) is compared to sequential steady-state reproduction (SSR) and generational reproduction (standard GA). The measures are off-line performance according to [3], averaged over 100 experiments, and displayed as percentage of the optimal solution quality (see Table 1).

Test function	Evaluations	standard GA	SSR	PSSR $M_1 = 5$	PSSR $M_2 = 20$	PSSR $M_1 = 5$ $M_2 = 15$
F_4	1000	98,84	99,49	99,47	99,38	99,04
F_4	2000	99,67	99,82	99,83	99,80	99,73
F_4	5000	99,94	99,98	99,97	99,95	99,95
F_5	1000	96,83	95,46	95,80	96,32	96,42
F_5	2000	97,67	97,36	97,46	97,52	97,57
F_5	5000	97,93	97,89	97,90	97,91	97,92

population on a file where all machines have access to it. To avoid concurrent writing, a lock mechanism had to be implemented.

Each machine runs the same identical process. Each process consists of the following operations:

1. Open the population file for reading. Apply selection operator and choose two solutions according to how fit they are. Close population file.
2. Apply recombination operators (crossover and mutation) to create one new solution.
3. Evaluate new solution (calculate cost function).
4. Open and lock population file for writing. Apply inverse selection operator and choose the solution to be replaced by the new solution. Replace solution. Unlock and close population file. Continue at 1.

It is possible that one process must wait before applying the lock if the file is locked by another process. Whether or not this occurs frequently depends on the ratio T_{repro} / T_{sync}, where T_{sync} is the time necessary for steps (1) and (4). It also depends on the distribution of execution times. After start-up, when all machines try to take their first task, a lot of wait phases will occur. Later, the communication phases will be more uniformly distributed.

6.2 Speedup results

Table 3 shows the theoretical as well as the experimental speedups for different configurations of the parallel algorithm, running until 2000 new solutions have been evaluated (including the initial population). Experimental speedup has been calculated dividing the CPU-time of the sequential algorithm by the execution-time of the parallel algorithm. Since the execution time depends on the load of the network, the obtained speedup also depends on it. Execution times are relatively long, and load heavily the network, therefore only one experiment was done with each configuration.

The column containing the experimental speedup shows that its approximately linear to the number of (equivalent) machines. This indicates that the model chosen to calculate the theoretical speedup is sufficient. The last row of Table 2 seems to invalidate this result. But the additional 12 machines have to be reached through gateways, so communication time is significantly longer. The purpose of this experiment was mainly to achieve a speedup of more than 10 ! The last column shows the quotient of theoretical and experimental speedups, which can be interpreted as the total usage coefficient of the network.

The parallel steady-state GA has a very high fault tolerance. Each process is able to supervise any other process in the network, and can abort and restart a faulty process. Even if several machines are rebooted, execution of the algorithm is not disturbed and normal operation is re-established very quickly.

The major problem with this algorithm, and with heterogeneous computing in general, is the heavy load of the network. Unix is a time-sharing system, and therefore not very well suited for supporting long-time jobs in background. Even when working with lowest priorities, other users often complain about saturation of their machines.

Partly, this is a psychological problem: many users, when they see their performance meters at 100% all the time, overreact even if it does not affect their work a lot. However, the problem is real especially for slower machines and when other processes are already running on them. Possible solution are to let the processes check for interactive user activity, or simply to stop computation periodically.

7 Conclusions

For an application like ours, i.e. long evaluation time, and implementation on a network of workstations, parallel steady state reproduction is best suited, both in terms of simplicity and efficiency.

Our parallel algorithm is easy to implement and makes efficient use of the available computing power. A genetic algorithm using this operator has proven to be at least as efficient than a sequential (generational or steady-state) genetic algorithm. It would be interesting to study its behaviour more closely, as some results suggest that using parallel steady-state reproduction on a single machine might be better than using a purely sequential version.

On a network of workstations, the algorithm leads to a relative speedup approximately linear with the number of machines. This indicates almost optimal use of each machine in the network. Furthermore it has a very high fault tolerance, which is important for this kind of systems.

Those results are encouraging, and we think this parallel genetic algorithm is a powerful and very useful instrument for optimization of complex problems that require a lot of computation.

Table 3. Experimental speedup with parallel steady-state reproduction. $S_{exp} = T_{seq(CPU)} / T_{exec}$, where $T_{seq(CPU)}$ is the CPU time on a Sparc10. The configurations are given by M_1 (number of Sparc 10's) and M_2 (number of Sparc 1's). In the last configuration, "12" stands for the number of Sparc stations of another department, reachable via gateways (performance rating of these Sparcs: about half of Sparc10).

Configuration	T_{exec} (min : sec)	S_{exp}	S_{max}	$\dfrac{S_{exp}}{S_{max}}$
$M_1 = 1, M_2 = 0$	28 : 57	0,76	1	0,76
$M_1 = 5, M_2 = 0$	7 : 16	4,2	5,0	0,84
$M_1 = 0, M_2 = 10$	8 : 53	2,5	3,3	0,76
$M_1 = 5, M_2 = 10$	3 : 11	6,9	8,3	0,83
$M_1 = 7, M_2 = 13$	2 : 43	8,1	11,3	0,72
$M_1 = 7, M_2 = 13 + 12$	2 : 04	10,6	17,3	0,62

8 References

1. N. Carriero & D. Gelernter, 1989. How to Write Parallel Programs: a Guide to the Perplexed. ACM Computing Surveys, Vol. 21, No. 3.
2. Ch. Darwin, 1859. The Origin of Species by Means of Natural Selection. Penguin Classics, London.
3. K. A. de Jong, 1975, Analysis of the Behaviour of a Class of Genetic Adaptive Systems. Ph.D. dissertation, University of Michigan.
4. D. E. Goldberg, 1989. Genetic Algorithms in Search, Optimization and Machine Learning. Addison-Wesley.
5. M. Georges-Schleuter, 1992. ASPARAGOS: An Asynchronous Parallel Genetic Optimization Strategy. Proc. 3rd ICGA. Morgan Kaufman Publishers, pp. 422–427.
6. M. Krüger, 1993. Using Genetic Algorithms to improve Stochastic Optimization Methods. Eighth Int. Symposium on Computer and Information Sciences, pp. 237–244, Istanbul, Turkey.
7. E. Lutton, 1994. Etat de l'art des Algorithmes Génétiques. Rapport d'Expertise SGDN, INRIA-Rocquencourt, B.P. 105, 78153 Le Chesnay Cedex, France.
8. H. Mühlenbein, M. Schomisch & J. Born, 1991. The Parallel Genetic Algorithm as Function Optimizer. Parallel Computing 17, pp. 619–632. North-Holland.
9. C. B. Petty, M. R. Leuze & J. J. Grefenstette, 1987. A Parallel Genetic Algorithm. Proc. 2nd ICGA. Lawrence Erlbaum Associates, pp. 155–161.
10. P. Spiessens & B. Manderick, 1991. A Massively Parallel Genetic Algorithm – Implementation and First Analysis. Proc. 4th ICGA. Morgan Kaufmann Publishers, pp. 279–286.
11. G. Syswerda, 1991. A Study of Reproduction in Generational and Steady-State Genetic Algorithms. Foundations of Genetic Algorithms, Morgan Kaufmann Publishers, pp. 94–101.
12. E-G Talbi, P. Bessière, 1991. A Paralleled Genetic Algorithm for the graph partitioning problem, ACM Int. Conf on Supercomputing, Cologne, Germany.
13. H. Tamaki & Y. Nishikawa, 1992. A Paralleled Genetic Algorithm based on a Neighbourhood-Model and its Application to the Job-Shop Scheduling. Parallel Problem Solving from Nature, 2. North-Holland, pp. 573–582.
14. D. Whitley, 1989. The Genitor Algorithm and Selective Pressure: Why Rank-Based Allocation of Reproductive Trials is Best. Proc. 3rd ICGA. Morgan Kaufmann Publishers, pp. 116–121.

Induction-based Control of Genetic Algorithms

Caroline Ravisé[1,3], Michèle Sebag[1,3], Marc Schoenauer[2]

(1) : LMS, CNRS-URA 317, Ecole Polytechnique, F-91128 Palaiseau

(2) : CMAP, CNRS-URA 756, Ecole Polytechnique, F-91128 Palaiseau

(3) : LRI, CNRS-URA 410, Université Paris-Sud, F-91405 Orsay

{Caroline.Ravise,Michele.Sebag,Marc.Schoenauer}@polytechnique.fr

Abstract

This paper presents a Machine Learning approach to control genetic algorithms. From examples gathered through spying evolution or experimenting on populations, induction extracts a rule-based characterization of *which evolutionary events* are good or bad for evolution. Such rule base allows for further generations to escape most disruptive or unproductive changes, according to a *civilized* rather than *Darwinian* evolution scheme. An evolutionary event is described as mutating a chromosome (at given bit—string positions) or crossing over two chromosomes (with given crossing points), and labeled by comparing the fitness of the offspring with that of its parents. Knowledge induced from such events allows to predict the effects of further operators, thereby filtering further undesirable events. Experimentations on some artificial problems are discussed.

1   Introduction

One of the main difficulty when using Genetic Algorithms (GAs) is the tuning of their numerous parameters (e.g. population size, crossover type, crossover and mutation rates). Different attempts have been proposed to automatically perform such tuning. Davis [Dav91] cites four known approaches, namely, hand optimization, optimization by a meta–GA, brute–force search, and adaptive parameterization, to which should be added now statistical methods [Gre95]. These approaches are concerned with the basic GA parameters, e.g. mutation and crossover rate [Gre86, Dav89, LT93]. Similar techniques have been employed to adjust the structure of the operators, eg. crossover masks [SM87], crossover types [Spe91, Spe95] and mutation amplitude [Sch81, FFAF92]; in all cases the cited parameters are encoded in the representation of the individual and thereby optimized 'for free' by the evolutionary algorithm at run–time.

This paper proposes to control the choice of GA operators by using an *external* machine learning algorithm, which observes the effect of crossover and mutation, constructs rules defining the good and bad operators and ultimately rejects those operators the rule base diagnoses as bad. This rule–based approach can accommodate both global rules, and rules that describe appropriate masks in terms of individual schemas. This paper extends previous work devoted to the global rule–based control of crossover only [SS94].

The plan of this paper is the following: Section 2 is concerned with extracting knowledge by observing evolution and experimenting on a population. Section

3 discusses how such knowledge could be used in order to guide the next evolution steps, by decreasing the probability of disruptive or unproductive changes. Section 4 presents some comparative results with reference methods on well known problems. Lastly, section 5 discusses the advantages and limitations of learning-based control with respect to related works [SM87, Spe95].

2 Learning from Evolution

Our goal is to achieve some coupling between ML and GAs, in order for ML to take in charge part of the GA control. As far as we know, this goal is original: all works dealing with GAs and ML we are aware of are concerned with using GAs to reach ML goals in the line of Holland's pioneering work [Hol86]; but inductive learning has not yet been applied to support genetic search.

This section introduces the basic concepts of Inductive Learning that will be used further, and defines more precisely the 'genetic events' characterizing the behavior of the GAs to be learned.

2.1 Requirements for Induction

Of course, we do not aim here at presenting an exhaustive state of the art in ML (see [Mic83, Qui86, MDR94]), but rather present the basic ML goal and terminology with respect to the classical GA terminology [Gol89a]. Only inductive supervised learning will be considered throughout this paper. In this frame, examples are points of the search space which have been classified or labeled (e.g. by an expert). The goal of induction is to extract rules from training examples; a rule can be viewed as a schema of the search space associated to a given label.

Induction attempts to optimize a quality function involving several features: (a) *Generality*, i.e. order of the schema in the rule; (b) *Accuracy*, i.e. ratio between the number of examples the rule generalizes and the number of examples it covers; and (c) *Significance*, i.e. number of examples the rule generalizes.

Table 1 shows some examples in $\{0,1\}^6$ representing classes *good* and *bad*, together with a rule R. Rule R covers E_3, E_4 and E_5, generalizing only examples E_3 and E_4 (the label of E_3 and E_4 can be deduced from R).

Induction proceeds by exploring the training examples either in a top-down or in a bottom-up way. In the top-down approach, illustrated by the ID3 family [Qui86], one builds a decision tree by repeatedly selecting the most discriminant attribute, i.e. the gene whose value gives a maximal information regarding the label of the examples. In the bottom-up approach, illustrated by the Star algorithms [Mic83], one starts from a given example and finds out the rules that cover this example and maximize some user-supplied quality function. Then the examples covered by these rules are removed from the training set, and another example is considered. The learning algorithm used in this paper is a

E_1	1	1	1	0	0	1	*good*
E_2	0	0	0	1	1	1	*good*
E_3	1	1	0	0	1	1	*bad*
E_4	1	0	0	0	1	1	*bad*
E_5	0	0	0	0	1	1	*good*
R	$\star$	$\star$	$\star$	0	1	$\star$	*bad*

Table 1: *Training examples and a rule*

star-like algorithm termed *Constraint Based Induction* (CBI), that determines all rules maximally general with a given prescribed (user-supplied) accuracy; a constraint-based formalism allows to build such rules with a polynomial complexity [Seb94].

2.2 Examples about Evolution

The central idea is interleaving evolution and induction, so that what is learned can immediately be used to control further evolution. In order to do so, examples relevant to the course of evolution are needed; moreover, such examples must be easy to describe and easy to gather.

The possibility investigated in this paper is to take as examples the elementary events of evolution, namely the crossing-over of two chromosomes and the mutation of a chromosome.

A crossover event is defined by a pair of parents and the crossover mask applied on these parents. Following Syswerda [Sys89] a crossover c can be represented by a binary mask $(c_1, \ldots c_N)$, $c_i \in \{0, 1\}$:

$$\left. \begin{array}{l} x_1 \ldots x_N \\ y_1 \ldots y_N \end{array} \right\} \rightarrow \left\{ \begin{array}{l} x'_1 \ldots x'_N \\ y'_1 \ldots y'_N \end{array} \right.$$

$$\text{with } \left\{ \begin{array}{ll} x'_i = x_i \quad y'_i = y_i & \text{if } c_i = 1 \\ x'_i = y_i \quad y'_i = x_i & \text{otherwise} \end{array} \right.$$

Likewise, a mutation event is defined by a parent and the mutation applied on this parent. A mutation can also be represented through a binary mask $m = (m_1, .. m_N)$, such that $x_1 \ldots x_N \rightarrow x'_1 \ldots x'_N$ with :

$$x'_i = \left\{ \begin{array}{ll} 1 - x_i & \text{if } m_i = 1 \\ x_i & \text{otherwise} \end{array} \right.$$

Hence, both kinds of events can be represented through the parent chromosome(s) and the operator mask. In this paper, the description of a genetic event consists of the operator mask, together with the description of the chromosome the operator applies on (the fittest parent in the crossover case).

2.3 Gathering and Labeling Genetic Events

Next step is to define classes of events: examples must be classified (labeled) in order to permit supervised learning. It seems natural, as far as learning intends to serve control, to classify events as to whether they contribute to the current optimization task. The choice made in this paper is the following: the class of an event depends on the way the fitness of offspring compares to the fitness of parent(s): the event falls within class

- *good* if the best offspring resulting from a crossover has higher fitness than both its parents or the offspring resulting from a mutation has higher fitness that its single parent,

- *bad* if the (best) offspring has lower fitness than the (best) parent

- *unproductive* if the (best) offspring and the (best) parent have the same fitness.

Examples of genetic events are gathered by experimenting on the population at a given stage of evolution:
1. An operator mask is randomly generated according to the parameters of the GA (e.g. mutation rate, n-point crossover or uniform crossover); inactive operators (e.g. mutation mask 00...0) are a priori discarded;
2. One or two chromosomes (depending on the operator, mutation or crossover) are randomly selected in the given population, after the selection step;
3. The operator is performed according to the mask and parent(s) selected. The fitness of the offspring is computed and compared to that of the parent(s). This comparison determines[1] the label of the event, *good, bad* or *unproductive*;
4. The example composed of the operator mask, the (fittest) parent description, and the associated label, is stored in the example base.

This approach involves a computational overhead due to the induction cost itself on one hand, and to the computation of the fitness of the experimental offspring on the other hand[2]. The number of additional fitness evaluations is the number of training examples, set to the size of the population in all following experiments.

---

[1] Note that it could happen that crossing over $parent_1$ with $parent_2$ according to a given crossover mask is *good*, while crossing over $parent_1$ with $parent_3$ according to the same crossover mask is *bad*. Then, since only one parent (say $parent_1$) is considered in the event description, one gets two examples with same description and distinct labels, i.e. examples are inconsistent. Fortunately the learning algorithm we used can deal with inconsistencies, so this is not a real limitation.

[2] This extra cost could be avoided if examples were gathered by "spying" the current generation instead of performing experimentations.

	Chromosome						Mask						Label
E_1	1	1	1	0	0	0	1	1	1	0	0	1	*good*
E_2	1	1	1	0	0	0	0	0	0	1	1	1	*good*
E_3	1	1	1	0	0	0	1	0	1	1	1	1	*bad*
E_4	1	1	1	0	0	0	0	0	1	1	1	1	*bad*
E_5	0	1	0	0	0	0	1	0	1	1	1	1	*good*
R	1	1	1	$\star$	$\star$	$\star$	$\star$	0	1	$\star$	$\star$	$\star$	*bad*

Table 2: *Induction from genetic events*

3   Using Knowledge to Control Evolution

This section discusses how to use rules induced from genetic events, and makes precise the interaction between evolution and induction.

3.1   What is Learned and How to Use it

Table 2 shows examples of 2-point crossovers together with a rule induced from these examples.

Rule R says that : Crossing over parent(s) belonging to schema $111 \star \star\star$ according to a crossover mask in schema $\star 01 \star \star\star$, will end in a *bad* result, i.e. the offspring will be less fit than the parents. This can be interpreted as: if the parent belongs to schema $111 \star \star\star$, don't cross-over between bits 2 and 3.

So, induction delivers general statements as to which operator masks are disruptive (*bad*), or beneficial (*good*) or without effects (*unproductive*) for a chromosome in a given schema. Induction can also process the information relative to the operator masks only, without considering the parent description[3]: it then states which kinds of operator masks are generally beneficial, or disruptive, or unproductive, independently from the chromosome(s) they apply on.

In both cases, the rules strongly reflect the population considered to build the examples. To see this, assume schema $111 \star \star\star$ has crowded the population; then rule R will not be learned any more, for breaking apart bits 2 and 3 has no visible effects if both parents belong to this very schema.

Actually, rules can be thought of as an estimate of the effects of operators: they predict that applying a given crossover or mutation mask on a given chromosome will be *good, bad* or *unproductive*, i.e. will lead to offspring that have higher, lower or equal fitness than that of the parent(s). This estimate allows to control evolution in several possible ways: given a chromosome (or pair of chromosomes), one can

- Select operator masks classified *good* only. However, this filter would strongly break the balance between the exploration and the exploitation

---

[3]The chances of inconsistency are thereby increased, but are still affordable for the learning algorithm at hand.

tasks devolved to evolution [Gol89a], as it would chiefly allow for exploitation.

- Reject operator masks classified *bad*. The resulting control strategy is termed *classical*. When a mask is classified *bad*, it can be interpreted as disrupting a promising schema. It follows that, in the case of mutation, this mask will 'always' be disruptive; in the case of crossover, this mask will be disruptive until the promising schema has crowded the population, and thereafter it will be unproductive. So, rejecting *bad* masks would not really limit the exploration of the search space, but rather prevent to explore again regions that are known (from induction) to be uninteresting.

- Reject operator masks classified *unproductive*. The resulting control strategy is termed *modern*. When the population tends to be converged, most crossovers are unproductive. It then appears that, while disruptive crossovers still ensure some exploration of the search space, unproductive crossovers do nothing at all. In that case, rejecting them will ensure a really efficient exploration.

This paper investigates a mixed strategy: when the percentage of unproductive examples is less than a threshold I (typically 70%), rules are used to reject operator masks classified *bad*. This classical control strategy is primarily concerned with avoiding disruption. When the number of unproductive examples is greater than threshold I, rules are used to reject operator masks classified *unproductive*. This modern control aims at preventing evolution to "fall asleep". Finally, when too many operators belong to the same class (more than a prescribed percentage F), the induction is considered useless, and further control is disabled: such a period of evolution is termed *Darwinian*.

3.2 Global Coupling

The induction-based control presented so far obviously does not pretend to universality. It is only too easy to devise a fitness landscape fooling up the above control strategy. For instance, the strict avoidance of disruptions is misleading if one is to find a peak situated in a well. Similarly, rejecting unproductive operators is misleading if the peak is on a large plateau.

However this approach does not suffer from the classical limitations of deterministic heuristics, stressed by Goldberg [Gol89b]: it inductively captures and exploits the main trends only about disruption or un-productivity, as nonsignificant rules generalizing less than a given number of examples are not considered. Nevertheless, two possible causes of failure can occur:

When induction finds no rule significant enough, rule-based control is disabled: the cost of induction as well as the extra fitness computations are a waste.

Even worse, the control derived from the induced rule-base may mislead evolution, either because it applies on a population much different from the one used to generate the rules, or because the fitness function itself evolves along time. Under the hypothesis of a fixed fitness function, this case of failure can be avoided by a periodical update of the rule-base: This updating is at present achieved by performing a new learning stage from scratch every M generations.

Finally, the integrated scheme of a GA controlled by induction is as follows:

1. INIT. A traditional genetic evolution is performed during the first M generations. This period expectedly allows for the first promising low-order schemas to appear in the population, in order for significant rules to be learned: the first case of failure discussed above is prevented.

2. KNOWLEDGE EXTRACTION. Let P be the size of the population ; P examples of genetic events are gathered by randomly generating operator masks and experimenting them on the current population. Rules are induced [Seb94] from these examples, by processing either the complete description of the examples (including both the chromosome and the operator mask), or the operator mask only.

3. KNOWLEDGE-GUIDED EVOLUTION. In the next generations, when crossover or mutation are applied, the masks are selected according to the rules. The rule-based control is either *classical* (masks classified bad are rejected) or *modern* (masks classified unproductive are rejected), depending on whether the ratio of unproductive events observed in step 2 is less or greater than a given threshold I.

4. KNOWLEDGE UPDATING: goto step 2. After M generations, the population has evolved; re-learning guarantees that rule-based control will adaptively follow the evolution of the population.

An important heuristic was found beneficial: when almost all operators used to generate the example base of genetic events belong to the same class, the next M generations are *Darwinian*, i.e. no control is performed. The Darwinian threshold F is used to trigger this feature.

The proposed integrated process involves three parameters (besides the GA parameters): (a) The number M of generations between two learning stages; (b) The threshold I of unproductive events, used to switch from classical to modern control; (c) The Darwinian threshold F, indicating control should be temporarily abandoned because more than $F\%$ among the operators belong to the same class.

3.3 Controlling Crossover or Mutation?

The representations for both crossover and mutation operators are based on boolean masks[4] (section 2.2). Hence, from a formal point of view, the rules

---

[4]except that mutation masks have very few 1s

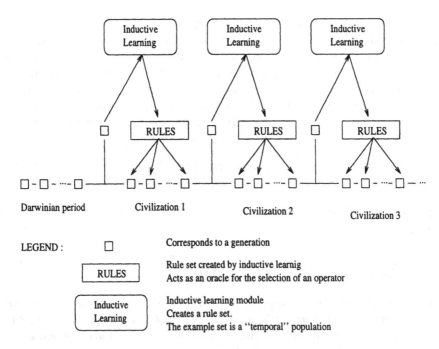

Figure 1: Coupling a genetic algorithm and inductive learning

induced form one operator can be used to predict the behavior of the other operator. The following four possibilities will be discussed here:

- *Crossover-based learning controlling crossover.* Since the seminal work of Holland [Hol75], crossover operators have been thought of as the major operator of evolution. Moreover, the mutation rate is usually small compared to the crossover rate, and many more crossovers than mutations are thus being applied during a GA run. So, the firsts studies [SS94] were done on using rules induced from crossover examples to control crossover.

- *Mutation-based learning controlling mutations.* More and more studies [HM90, Bäc93] stress on the so far unnoticed importance of mutations in the GA framework. Even more, Fogel and Stayton [FS94] and Jones [Jon95] claim that crossovers could simply be macro-mutations and could be replaced by high mutation rates. So the second stage of this approach was to learn and control mutations.

- *Mutation-based learning controlling of both.* As stated above mutations rules can be used to control crossovers too: consider for instance the rule R_m (*11***⇒ bad). From the mutation point of view, this rule means *do not mutate simultaneously bits 4 and 5*. From the crossover point of view, it means *avoid to separate bits 4 and 5*. In that respect, a mutation can be viewed as a specialized crossover: mutating x is equivalent to crossing-over x with $\neg x$ ($\neg x$ is the opposite of x). Hence, if a mask is disruptive for mutations, a reasonable hypothesis is that it is disruptive for crossovers too. But an important point is that a mutation is much more "severe" than a crossover: during a crossover between x and y, y is likely to be closer to x than $\neg x$. Moreover, a mutation has only one offspring, while crossovers have two offspring: for these two reasons, mutations rules should be too "severe" when applied to control crossovers.

- *Crossover-based learning controlling both.* The reverse argument indicates that crossovers rules should be too "loose" when controlling mutations.

The above discussion will be enlightened by the experimental results of next section.

4 Experimental Validation

The method of control of the genetic operators described in the above sections is experimented on different artificial problems: the GA-easy *Twin Peaks* problem, the GA-difficult *Royal Road* problem, the GA-deceptive *Ugly* problem and the combinatorial NP-complete *Multiple Knapsack* problem.

4.1 Reference works

The proposed induction-based control is compared to the **canonical** GA [Gol89a], and to two other methods of crossover control: the *à la* Schaffer and Morishima control [SM87] and the *à la* Spears control [Spe91, Spe95]:

- Schaffer and Morishima add to the bitstring representation of an individual another bitstring of the same length encoding the loci for crossover. In their results, they notice that the rate of "productive" crossovers (when parents and offspring fitnesses differ) is almost constant during the evolution, even when the population tends to converge.

- Spears only adds one bit to each bitstring indicating the type of crossover (two-point or uniform) to apply to the bitstring. The analyze stresses that this approach does not work well with small populations. Moreover, at the end of the evolution, the bit-position of control tends to get uniform as the population gets uniform [Spe91]. But, further experiments seem to indicate that the simple fact of simultaneously using two crossovers is beneficial [Spe95].

Four kinds of induction-based control are considered, as discussed in section 3.3: crossover-based learning controlling crossovers (**X-X**) or controlling both operators (**X-XM**) ; mutation-based learning controlling mutations (**M-M**) learning on crossovers and controlling crossovers and mutations, or controlling both operators (**M-MX**).

4.2 Protocol

The GA parameters are the following: Population size $P = 25$ for all problems. Selective pressure sp is set to 1.2 and 2.0. Crossover probability p_c is 0.6 and mutation probability p_m is 0.005.

The coupling parameter, defined in section 3.2, are the following: The length of a civilization (number of generations between two inductive steps) is $M = 3$. After systematic study of possible values for thresholds I and F, values of $I = 67\%$ and $F = 95\%$ were found to be experimentally robust (note that if I is greater than F, there can never be any modern control), and all following experiments use these values.

All experiments involve 15 independent runs. For each problem, the following plots and tables are given:

- The dynamics of evolution is illustrated by the plot of the average best fitness vs the number of fitness evaluations (including the ones needed to generate the training set for the inductive steps described in section 3.2). Two of such plots are provided for the sake of readability: the first one involves the reference techniques (standard GA, Spears' and Schaffer and Morishima's) plus the best of the induction-based controls; the second one compares the four different induction-based methods.

- The rates of success (percentage of hits of the – known – maximum value of the fitness) for all algorithms are provided in the same table.

- In order to gain some insight on the actual effectiveness of the proposed control, the rates of Classical, Modern and Darwinian periods (section 3.1, together with the global percentage of operator masks rejected according to the rules, are given in a last table.

4.3 The Twin Peaks Problem

This problem consists in two equal maximal peaks [ES93]. The length of the individuals is here $n = 100$. Let $N_1(x)$ be the number of '1' of individual. The twin peaks fitness is defined by

$$f(x) = \begin{cases} N_1(x) & \text{if } N_1(x) > 50 \\ n - N_1(x) & \text{if } N_1(x) < 50 \\ 0 & \text{else } (N_1(x) = 50) \end{cases}$$

In this problem, for all individuals, most mutations induce a modification of fitness. In the neighborhood of both optima, mutations is highly disruptive

sp	Canonical	Sc-Mo	Spears	X-X	X-XM	M-M	M-MX
1.2	86.67	93.33	93.33	73.33	66.67	100	100
	(11.07)	(9.96)	(10.12)	(28.23)	(28.75)	(20.89)	(19.58)
2.0	100	100	100	100	93.33	100	100
	(3.03)	(2.37)	(2.69)	(8.43)	(14.41)	(7.63)	(8.12)

Table 3: *Success rate (thousands of evaluations) for the Twin Peaks problem.*

while there are more and more good mutations as the Hamming distance to one optimum increases (all mutations are good when the number of '1's is exactly 50).

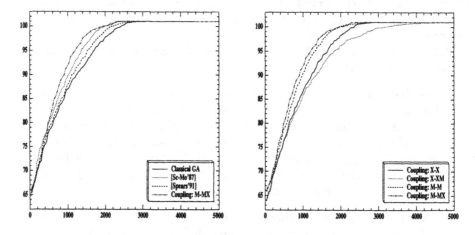

Figure 2: *Dynamics of evolution for the Twin Peaks problem. The maximum fitness is 101. The parameters are $P = 25, sp = 2.0, F = 95\%, I = 67\%$.*

Method	Classical	Modern	Darwinian	Rejected masks
X-X	15.62	83.21	1.17	90.52
X-XM	17.93	80.58	1.49	90.97
M-M	96.97	2.59	0.44	97.14
M-MX	96.61	2.95	0.44	94.80

Table 4: *Effectiveness of control on the Twin Peaks problem.*

On this GA-easy problem, all methods succeed with $sp = 2.0$ though GAs with induction-based mutation control (M-M and M-MX) are slightly faster (Figure 2). With $sp = 1.2$ induction-based control is outperformed by all other methods (Table 3).

The statistics about the type of control in Table 4 lead to the following remarks:

- crossover-based control leads to a majority of modern periods (during which inactive masks are rejected). Together with the very small number of Darwinian period, this is a sign of the effectiveness of crossover.

- as all mutations modify the fitness, mutation-based control remains classical (rejecting bad masks), and very effective (many masks do get rejected): the GA resembles here a local hill-climber.

- in both cases, control is highly applied (more than 90% masks are rejected by the rule base), and is sufficient (very few Darwinian periods are observed). Note that if the Darwinian threshold F is lowered down to 67%, the ratio of Darwinian generations gets close to 70%, as almost all masks rapidly become *bad* when climbing either peak.

4.4 The Royal Road Problem

This problem was conceived by Mitchell, Holland and Forrest [MFH93, FM93] to study the combined features that seemed most adapted to GAs. It was later simplified [Jon94] with minimal order schemas length of 2 bits. Let $\Omega = \{0,1\}^{64}$ be the search space and $H_{i,j}$ be the schema beginning at the bit j composed of sequence of i '1'. This defines the schemas $H_{2,1}, H_{2,3}, \ldots, H_{32,1}, H_{32,33}, H_{64,1}$. There are 6 orders of schema (number of defined bits) and a schema of order 2^{k+1} is the intersection of two schemas of order 2^k. The fitness of an individual is computed as follows:

$$f(x) = \sum_{k=1}^{6} 2^k + (NS(x,k) * 0.2)$$

where $NS(x,k)$ is the number of schema of order 2^k appearing in x.

This problem was designed to illustrate the *building block* phenomenon and the usefulness of crossover. However, a close look at mutations shows that there are very few good mutations (building up a new schema by completing a block of 1s), opposed to inactive mutations (mutations acting on bits in incomplete schema) or destructive mutations (mutations disrupting some already complete schema). Together with the *hitch-hiking* phenomenon (once a schema of some order is built up, it takes with it all other incomplete schema of the same order), this explains the difficulty encountered by standard GAs on the Royal Road problem [MFH93, MH93].

It is worth noting that according to these results, the *à la Schaffer and Morishima* control of crossovers offers no improvement over classical GA, while the *à la Spears* control of crossover outperforms all other methods.

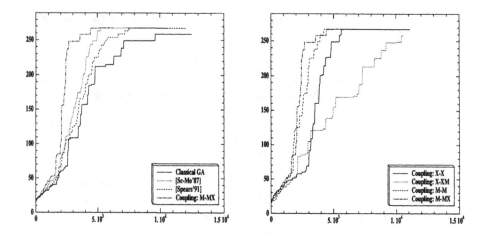

Figure 3: *Dynamics of evolution for the Royal Road problem. The maximum fitness is 267.6. The parameters are $P = 25, sp = 2.0, F = 95\%, I = 67\%$.*

sp	Canonical	Sc-Mo	Spears	X-X	X-XM	M-M	M-MX
1.2	80	73.33	83.33	53.33	46.67	100	86.67
	(10.03)	(10.16)	(10.47)	(23.31)	(26.16)	(22.40)	(21.21)
2.0	93.33	100	100	100	86.67	100	100
	(7.16)	(6.19)	(6.21)	(19.67)	(24.14)	(11.75)	(10.73)

Table 5: *Success rate (thousands of evaluations) for the Royal Road problem*

On this problem, crossover-based control turns out to be quite inefficient while mutation-based control allows to catch up with other methods, slightly outperforming them (Figure 3 and Table 5). Moreover, the a priori considerations on mutations are confirmed by the figures of Table 6: in the beginning of evolution, most mutations are inactive; but as higher order schema appear, they are likely to be disrupted by mutations, and more and more mutations are classified bad, ending in absolutely no modern period at all when using the mutation-based control (Table 6).

4.5 The Ugly Problem

This problem was designed by Whitley [Whi91]. A 30-bit individual is the concatenation of ten instances of the following 3-bit deceptive problem:

$$f(x) = f'(x_1) + \ldots + f'(x_{10}) \text{ with } x_i \text{ as a 3-bit gene and}$$

Method	Classical	Modern	Darwinian	Rejected masks
X-X	25.72	67.07	7.20	88.62
X-XM	21.60	75.59	2.81	90.63
M-M	99.31	0.00	0.69	96.20
M-MX	99.40	0.00	0.60	94.29

Table 6: *Effectiveness of control on the Royal Road problem.*

$$f'(x_i) = \begin{cases} 3 & \text{if } x_i = 111 \\ 2 & \text{if } x_i \in H_0 = 0 * * \\ 0 & \text{else} \end{cases}$$

The first bit of a schema is very mutation-sensitive: when it is turned into '0' the mutation is good, but when it is turned into '1', this corresponds generally to a bad mutation, except when the remaining bits are '11'. Hence, half of the mutations are inactive, and the majority of other mutations is bad.

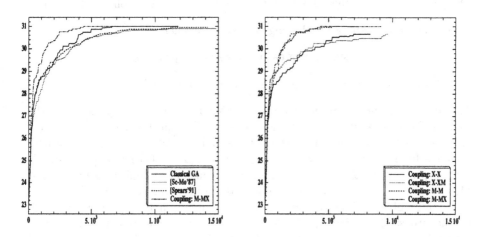

Figure 4: *Dynamics of evolution for the Ugly problem. The maximum fitness is 31. The parameters are $P = 25, sp = 2.0, F = 95\%, I = 67\%$.*

Mutation-based control methods are the clear winners of this set of experiments (Table 7 and Figure 4). But here again, almost no modern period is observed (Table 8). Moreover, crossover-based control is rather inefficient, and when the best control is not modern (which reflects that most crossovers are inactive), the best strategy is a Darwinian (canonical) strategy (Table 8): crossover seems of poor interest on this problem.

sp	Canonical	Sc-Mo	Spears	X-X	X-XM	M-M	M-MX
1.2	80	86.67	83.33	66.67	13.33	80	86.67
	(7.45)	(7.90)	(8.75)	(21.54)	(20.54)	(16.55)	(16.93)
2.0	93.33	86.67	90	53.33	60	100	100
	(5.49)	(8.07)	(7.76)	(20.03)	(18.63)	(12.78)	(13.27)

Table 7: *Success rate (thousands of evaluations) for the Ugly problem*

Method	Classical	Modern	Darwinian	Rejected masks
X-X	1.59	41.80	56.61	88.00
X-XM	2.31	70.93	26.76	91.96
M-M	99.55	0.00	0.45	93.56
M-MX	99.42	0.13	0.45	91.82

Table 8: *Effectiveness of control on the Ugly problem.*

4.6 The Knapsack Problem

The general multiple knapsack problem can be described as follows [KBH94]:

- There are m knapsacks having each one a capacity c_j.

- There are n objects having each one having a profit p_i.

- Each object i has a defined weight for each knapsack j: $w_{i,j}$.

The aim is to find a subset of objects which maximizes the profit, i.e. the sum of the profits of objects in this subset. This problem is a constrained optimization problem since the capacity of each knapsack must not be exceeded. Hence, a feasible solution is a vector $x = (x_1, \ldots, x_n)$ where $x_i \in \{0, 1\}$ stands for the presence of the object in the subset considered, and such that: $\forall j \in \{1, \ldots, m\} \sum_{i=1}^{n} w_{i,j} x_i \leq c_j$. For these feasible individuals, $f(x) = \sum_{i=1}^{n} p_i x_i$. The fitness of infeasible individuals involves some penalty term: what would be the fitness is divided by 2 and is reduced with respect to the number of constraints satisfied m_{ok} in order to distinguish between different levels of non-feasibility, giving the following formal description:

$$f(x) = \begin{cases} \sum_{i=1}^{n} p_i x_i & \text{if } x_i \text{ is a feasible solution (f.s.)} \\ (\sum_{i=1}^{n} p_i x_i) * m_{ok}(x)/2m & \text{else} \end{cases}$$

For these problems, almost all mutations are effective (good or bad), with roughly the same number of good ones and bad ones.

Three of the multiple knapsack problems originally due to Petersen [Pet67] have been used, demonstrating similar results. Only the case $n = 20$ and $m = 10$ is presented here.

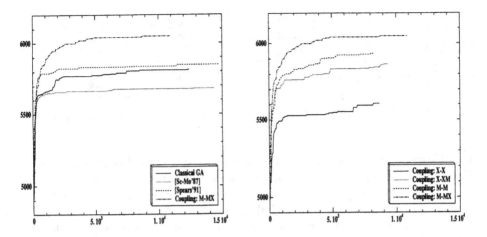

Figure 5: *Dynamics of evolution for the Knapsack Petersen 4 problem. The maximum fitness is 6120. The parameters are $P = 25, sp = 2.0, F = 95\%, I = 67\%$.*

sp	Canonical	Sc-Mo	Spears	X-X	X-XM	M-M	M-MX
1.2	13.33	0	20	13.33	26.67	40	26.67
	(5.20)		(3.73)	(30.48)	(14.72)	(15.45)	(22.62)
2.0	0	0	0	0	6.67	13.33	6.67
					(7.21)	(15.16)	(27.86)

Table 9: *Success rate (thousands of evaluations) for the Petersen 4 Knapsack problem*

For this problem, no method proves very effective (Table 9), but the mutation-based control again outperforms other methods. Note that crossover-based control gives good results *when controlling also mutations*, even giving results more rapidly than mutation-based control applied to both operators. Very few *inactive* mutation masks are expected, ending again in no modern period.

5 Discussion

The first general remark on these results is that mutation-based control is more efficient than crossover-based control – the latter being sometimes totally inefficient. Moreover, except on the knapsack problem with selective pressure 1.2, mutation-based control performs better when applied to mutation only.

A possible explanation, as quoted in Section 3.3, is that mutation-based rules are more severe than crossover-based ones. Furthermore, there are much more

Method	Classical	Modern	Darwinian	Rejected masks
X-X	3.46	43.80	52.74	89.30
X-XM	4.49	66.32	29.19	91.07
M-M	99.56	0.00	0.44	91.16
M-MX	99.56	0.00	0.44	88.08

Table 10: *Effectiveness of control on the Petersen 4 Knapsack problem.*

possible combinations of "Parent1-Parent2-Crossover" than "Parent-Mutation": Crossover rules might be in general less relevant than mutation rules.

Forbidding inactive crossover might speed-up evolution, but it does not seem to help convergence toward the global maximum. On the opposite, forbidding mutations diagnosed bad from past mutation examples increases the success rate and compares favorably to the canonical GA and both other GA-based control methods.

Except for the knapsack problems, better results are obtained with a selective pressure of 2.0 than with a selective pressure of 1.2. Probably, when there is less diversity in the population, as with a selective pressure of 2.0 (the examples are generated *after* the selection step), the rules more likely apply to most individuals.

An important hidden effect of the mutation-based control mechanism is to raise the actual mutation rate: As the prescribed mutation rate is usually small, the example masks feeding the learning process contain very few '1' bits. Rejecting mutation masks from the rules induced from such examples hence amounts to choose mutation masks in regions of $\{I\!R^n\}$ containing more '1's than prescribed by the initial mutation rate. The a posteriori experimental mutation rate on the Twin Peaks problem was found to be 0.01 with a prescribed mutation rate of 0.005.

Regarding the different periods of control (classic, modern or Darwinian), the results are similar for all problems: mutation-based control leads to classical control only (bad masks are rejected). The disruptive effects of mutations can only be avoided by controlling the choice of masks. On the opposite, Achilles heel of crossover lies in its ineffectiveness when convergence has started, ending up in either modern control, or even the need to go back to Darwinian evolution (trying to control would be worse than doing nothing).

6 Conclusion

This paper has introduced an hybrid scheme for GAs, interleaving the GA evolution and inductive learning. This work extends previous work devoted to the control of crossover operators to mutation operators. Despite the small mutation rates generally used in the GA community, the high disruptiveness of mutations when approaching the global optimum of the fitness function is worth counterbalancing: A control based on examples of mutation operators, and further rejecting mutations considered as bad (*modern mutation-based control*) is clearly the best choice among the different possibilities. Avoiding to repeat past errors is beneficial.

The anthropomorphic analogy to the explicit memory implemented by the rule set is the invention of writing.

Many points remain to be deepened. The most important deals with the overall benefit of the hybridization: the cost of induction has not been taken into account in the presented experiments, and is far from negligible. On the other hand, the cost of fitness computations is very small on the artificial problems used in section 4. Large scale problems will be considered, as well as some simpler – and hence cheaper – learners.

Future work will be concerned with applying such hybrid scheme to continuous problems: in that context, when linear recombination crossover operators are considered [Mic92], crossover masks can be represented by real vectors, hence allowing to apply the same type of inductive learner.

Another field of application is Genetic Programming [Koz94b, Koz94a], and more generally variable length order independent representations (e.g. neural networks): The underlying inductive problem is amenable to first-order logic.

Acknowledgments

The readability and clarity of this paper was greatly improved by fruitful comments of E. Lutton and E. Ronald.

References

[Bäc93]   T. Bäck. Optimal Mutation Rate in Genetic Search. In S. Forrest, editor, *Proceedings of the 5th International Conference on Genetic Algorithms*, pages 2–8, 1993.

[Dav89]   L. Davis. Adapting operator probabilities in genetic algorithms. In J. D. Schaffer, editor, *Proceedings of the 3rd International Conference on Genetic Algorithms*, pages 61–69, 1989.

[Dav91]   L. Davis. *Handbook of Genetic Algorithms*. Van Nostram Reinhold, New York, 1991.

[ES93]   L.J. Eshelman and J.D. Schaffer. Crossover's niche. In S. Forrest, editor, *Proceedings of the 5th International Conference on Genetic Algorithms*, pages 9–14. Morgan Kaufmann, 1993.

[FFAF92]   D. B. Fogel, L. J. Fogel, W. Atmar, and G. B. Fogel. Hierarchic methods of evolutionary programming. In L. J. Fogel and W. Atmar, editors, *Proceedings of the 1st Annual Conference on Evolutionary Programming*, pages 175–182, La Jolla, CA, 1992. Evolutionary Programming Society.

[FM93]   S. Forrest and M. Mitchell. What makes a problem hard for a genetic algorithms : Some anomalous results and their explanation. *Machine Learning*, pages 285–319, 1993.

[FS94]   D.B. Fogel and L.C. Stayton. On the effectiveness of crossover in simulated evolutionary optimization. *BioSystems*, 32:171–182, 1994.

[Gol89a]   D. E. Goldberg. *Genetic algorithms in search, optimization and machine learning*. Addison Wesley, 1989.

[Gol89b]   D. E. Goldberg. Zen and the art of genetic algorithms. In J. D. Schaffer, editor, *Proceedings of the 3rd International Conference on Genetic Algorithms*, pages 80–85, 1989.

[Gre86]   J. J. Grefenstette. Optimization of control parameters for genetic algorithms. *IEEE Trans. on Systems, Man and Cybernetics*, SMC-16, 1986.

[Gre95]   J. J. Grefenstette. Virtual genetic algorithms: First results. Technical Report AIC-95-013, Navy Center for Applied Research in Artificial Intelligence, February 1995.

[HM90]   J. Hesser and R. Männer. Toward an Optimal Mutation Probability for Genetic Algorithms. In Hans-Paul Schwefel and Reinhard Männer, editors, *Proceedings of the 1st Parallel Problem Solving from Nature*, pages 23–32. Springer Verlag, 1990.

[Hol75]   J. Holland. *Adaptation in natural and artificial systems*. University of Michigan Press, Ann Arbor, 1975.

[Hol86]   J. Holland. Escaping brittleness : The possibilities of general purpose learning algorithms applied to parallel rule-based systems. In R.S Michalski, J.G. Carbonell, and T.M. Mitchell, editors, *Machine Learning : an artificial intelligence approach*, volume 2, pages 593–623. Morgan Kaufmann, 1986.

[Jon94]   T. Jones. **A description of Holland's Royal Road Function**, 1994.

[Jon95]   T. Jones. Crossover, macromutation and population-based search. In L. J. Eshelman, editor, *Proceedings of the 6th International Conference on Genetic Algorithms*, pages 73–80. Morgan Kaufmann, 1995.

[KBH94]   S. Khuri, T. Bäck, and J. Heitkötter. The zero/one multiple kmapsack problem and genetic algorithms. In Proceedings of the ACM Symposium of Applied Computation (SAC'94), 1994.

[Koz94a]   J. R. Koza. *Genetic Programming II: Automatic Discovery of Reusable Programs*. MIT Press, Massachussetts, 1994.

[Koz94b]   J. R. Koza. *Genetic Programming: On the Prorgamming of Computers by means of Natural Evolution*. MIT Press, Massachussetts, 1994.

[LT93]   M.A. Lee and H. Takagi. Dynamic control of genetic algorithms using fuzzy logic techniques. In S. Forrest, editor, *Proceedings of the 5th International Conference on Genetic Algorithms*, pages 76–83, 1993.

[MDR94] S. Muggleton and L. De Raedt. Inductive logic programming: Theory and methods. *Journal of Logic Programming*, 19:629–679, 1994.

[MFH93] M. Mitchell, S. Forrest, and J.H. Holland. The Royal Road for genetic algorithms : Fitness landscapes and GA performance. In F. J. Valera and P. Bourgine, editors, *Proceedings of the First European Conference on Artificial Life-93*, pages 245–254. MIT Press/Bradford Books, 1993.

[MH93] M. Mitchell and J.H. Holland. When will a genetic algorithm outperform hill-climbing ? In S. Forrest, editor, *Proceedings of the 5th International Conference on Genetic Algorithms*, page 647, 1993.

[Mic83] R.S. Michalski. A theory and methodology of inductive learning. In R.S Michalski, J.G. Carbonell, and T.M. Mitchell, editors, *Machine Learning : an artificial intelligence approach*, volume 1. Morgan Kaufmann, 1983.

[Mic92] Z. Michalewicz. *Genetic Algorithms+Data Structures=Evolution Programs*. Springer Verlag, 1992.

[Pet67] C.C. Petersen. Computational experience with variants of the balas algorithm applied to the selection of r & d projects. *Management Science*, 13:736–750, 1967.

[Qui86] J. R. Quinlan. Induction of decision trees. *Machine Learning*, 1:81–106, 1986.

[Sch81] H.-P. Schwefel. *Numerical Optimization of Computer Models*. John Wiley & Sons, New-York, 1981.

[Seb94] M. Sebag. Using constraints to building version spaces. In L. De Raedt and F. Bergadano, editors, *Proceedings of ECML-94, European Conference on Machine Learning*. Springer Verlag, April 1994.

[SM87] J.D. Schaffer and A. Morishima. An adaptive crossover distribution mechanism for genetic algorithms. In J. J. Grefenstette, editor, *Proceedings of the 2nd International Conference on Genetic Algorithms*, pages 36–40. Morgan Kaufmann, 1987.

[Spe91] W. M. Spears. Adapting crossover in a genetic algorithm. In R. K. Belew and L. B. Booker, editors, *Proceedings of the 4th International Conference on Genetic Algorithms*. Morgan Kaufmann, 1991.

[Spe95] W. M. Spears. Adapting crossover in evolutionary algorithms. In J. R. McDonnell, R. G. Reynolds, and D. B. Fogel, editors, *Proceedings of the 4th Annual Conference on Evolutionary Programming*, pages 367–384. MIT Press, March 1995.

[SS94] M. Sebag and M. Schoenauer. Controlling crossover through inductive learning. In Y. Davidor, H.-P. Schwefel, and R. Manner, editors, *Proceedings of the 3rd Conference on Parallel Problems Solving from Nature*. Springer-Verlag, LNCS 866, 1994.

[Sys89] G. Syswerda. Uniform crossover in genetic algorithms. In J. D. Schaffer, editor, *Proceedings of the 3rd International Conference on Genetic Algorithms*, pages 2–9. Morgan Kaufmann, 1989.

[Whi91] D. Whitley. Fundamental principles of deception in genetic search. In G. J. E. Rawlins, editor, *Foundations of Genetic Algorithms*. Morgan Kaufmann, 1991.

New Multicriteria Optimization Method Based on the Use of a Diploid Genetic Algorithm: Example of an Industrial Problem

Rémy Viennet[1], Christian Fonteix[2] and Ivan Marc[1]

[1] Laboratoire des Sciences du Génie Chimique - U.P.R. C.N.R.S. 6811
E.N.S.I.C., 1 rue Grandville, B.P. 451, 54001 NANCY Cédex, France
[2] Laboratoire des Sciences du Génie Chimique - U.P.R. C.N.R.S. 6811
E.N.S.A.I.A., 2 Av. de la forêt de Haye, B.P. 172, 54505 VANDOEUVRE Cédex,
France

Abstract. The design of a new product or of its manufacturing process consists in reconciling multiple objectives with each other to take into account their different features. In this paper, a new multicriteria optimization algorithm is presented. This method is based on the use of (*i*) a genetic algorithm (GA) which optimizes each system response and (*ii*) a selection algorithm which sorts Pareto-efficient points. This technique presents the great advantage of being of wide use. There is no particular mathematical condition about functions that are simultaneously optimized and, unlike the other multicriteria optimization methods which depend on the user's choice, our algorithm permits to obtain an optimal surface in which the user will be able to pick up his own working conditions. Efficiency of this new method is here illustrated with one mathematical example and with an industrial application.

1 INTRODUCTION

In a lot of domains, processing or product formulation depend on several criteria and consequently industrials are often confronted with multicriteria decision problems. Food processes are a good illustration for this case. For example, it can be necessary to optimize different parameters such as texture, flavour, and so ever, in order to formulate a new product or else, before to use bacteria or yeasts, it is very interesting to maximize yields, production and product quality with minimal investment whereas these four criteria are not optimal for the same working conditions. To help to solve such problems, traditionnally, objectives were whether all combined to form a scalar objective, through a linear combination of multiple attributes [2] or a geometric average of these ones [3]. In an other classical method, a function, called primary response function, was optimized whereas the others, called secondary response functions, were turned into constraints [8]. These two techniques depending on the user's choice, for weights or constraints determination, they are consequently not adapted to solve multiple objective problems found in food industry where it will be more interesting to find a compromise solution.

Methods incorporating a domination criterion are more interesting because they are of more general use and more accurate. For example, an interactive algorithm permits a reference direction approach [9]. The decision maker has to choose his solution in a set of solutions which are not dominated. A solution is nondominated or Pareto-efficient if no other solution exists that is equally good for every objective and better for at least one of them [10], [7]. This algorithm generates Pareto-optimal alternatives without defining the whole Pareto area and the solution depends on the user's choices. In this case, a set of points that are the nondominated solutions, is defined. The domination criterion of Pareto is widely used to perform multicriteria optimization [6], [11] and [5]. The definition of the zone of Pareto is easy but it is more difficult to determine it pratically.

A method based on a diploid GA previously elaborated [1], [4] and a nondom-inated solutions selection procedure has been developed [12] to define a set of Pareto-efficient points. This algorithm performs the simultaneous minimization of p functions (the maximization of f_i being equivalent to the minimization of $-f_i$). As a consequence, the search of working conditions concerning a culture medium which maximize quality and production and minimize investment no more represents any problem. The number of real-valued decision variables and the constraints can be different for each function.

This new method is presented through the intermediary of the two algo-rithms it uses. It is then illustrated with one mathematical example. In the last section, this technique is compared with two classical methods on an industrial application.

2 DESCRIPTION OF THE ALGORITHM

2.1 Genetic algorithm

Fonteix et al.[4] proposed a GA which modelizes the genetic of diploid individ-uals. This GA has been compared with a haploid one [1] and its performances were found to be better that is why we used the diploid version.

Each individual (which can be a possible solution of the problem) is described by a four-tuple (a_j, a'_j, D_j, x_j). a_j and a'_j represent the two alleles of one gene. The genotype is composed of two chromosomes and phenotype, x_j, is the result of the combination of the respective alleles, a_j and a'_j, with the respect to the dominance, D_j, of one allele over the other. D_j is a value randomly chosen in $[0, 1]$ interval; it is a specific value for the both alleles of each gene. An initial population is created by generating a set of N points (or individuals) from the search space. Each point is tested and evaluated. If this population is not the solution, then probabilistic rules are used to make it evolve. Only the better individuals will survive (elitist selection) and participate to the creation of a new generation. The reproduction of the individuals in the diploid model is caracterized by a multi-crossover on the two chromosomes of each parent, and by the mutation and the homozygosity [4]. When a child is created, it is tested with respect to all constraints which must be satisfied; if not, it is rejected and another child is created. The new population is evaluated and if it is not a solution, a new population is generated and so on.

This GA can be described by the six following parameters:

N: Population size. If N is very small then the algorithm does not perform an efficient search and if it is very large then the rate of convergence becomes slower.

L: Number of genes.

M: Mutation rate. it is a secondary operator which prevents from premature convergence to suboptimal solutions.

G: Generation patrimory. It determines the population ratio which remains unchanged at each generation. The objective function value of the last individual kept to be a parent is noted f_c. When a child is created, it survives if and only if its objective function value is better than f_c.

H: Homozygosity rate. It permits to control "genetic material" evolution.

S: stop criterion of the algorithm. In our case, it is the absolute clustering of the individuals; $f_{max} - f_{min} \leq S$ where f_{min} and f_{max} are respectively the minimal and the maximal objective function values in the current population.

2.2 Generation of intermediate populations

This GA is used in the first two steps of this multicriteria optimization method. In these ones, intermediate populations of possible solutions are created. In the third and last step, individuals of these populations wich are dominated, are eliminated.

Firstly, p functions are separately minimized with GA. A function f_i has its minimum for the point x_i^*. Then, each function is calculated for the minimun values of the $p - 1$ other functions and the upper value is noted F_i: $F_i = Sup\{f_i(x_j^*) \quad \forall j \neq i\}$. Secondly, for each function, one zone Ω_i defined by $f_i(x) \leq F_i$ is searched.

In the present step, the running of GA can be separated in two parts: a mortality stage and a birth stage. So:

Firstly, the stop criterion of GA is modified and becomes: $f_{ci} \leq F_i$ for the current function (f_{ci} is the objective function i value of the last individual kept to be a parent). In each last obtained population, where f_{imax} is the maximal value of the objective function i in the population i, if elitist selection was applied as previously, then the number of survivors would be $N * G$. In this case, the obtained zone, delimited by the contour line defined by $f_i(x) = f_{ci}$, would belong to Ω_i but would not be Ω_i searched. Actually, f_{ci} is generally lower than F_i, consequently, the points in the surface comprised between the two contour lines respectively defined by: $f_i(x) = f_{ci}$ and $f_i(x) = F_i$, which could be Pareto-efficient, would be lost. In order to avoid this loss, the selection stage in each last population has been modified and it becomes the following: all individuals with a function i adaptation value lower than F_i are kept to generate children. The number of survivors (all individuals belonging to Ω_i) becomes, in these conditions and for each function f_i, N_i.

Secondly, the birth step of the GA is performed with survivors previously defined. The maximal adptation function i value of children is F_i.

At the end of these two steps, one population P_i has been defined for each function f_i. In each P_i, adaptation function value of the worst individual is F_i. In fact, P_i is the set of points comprised between F_i and the minimum value of f_i $(f_i(x_i^*))$.

2.3 Final selection

In this last step, all populations P_i, previously found, are combined to form P. The multicriteria optimization is performed in this step by application of the Pareto domination criterion which is applied by using the following procedure: the first individual x of P is taken as reference and all other points x' are compared with it. An intermediate function h_i is introduced for these comparisons. The value of h_i is 0, 1 or 2 when $f_i(x)$ is respectively lower, egal or greater than $f_i(x')$.

If $h_1 h_2 \ldots h_p > 1$, the individual x is better than x' which is consequently eliminated. The following not eliminated point becomes the reference and so ever. All individuals which have not been eliminated define the optimal Pareto surface. The more the number of remaining points is important, the more the Pareto surface is accurate. Thus, it is possible to adapt the accuracy by adjusting the population size of GA.

3 RESULTS

This multicriteria optimization technique is applied to solve one mathematical example: simultaneous minimization of three functions presenting two real-valued decision variables. Then, this method is compared with two classical multicriteria optimization techniques for one industrial application.

3.1 Mathematical example

The three functions are:

$$
\begin{aligned}
f_1 &= \frac{(x_1-2)^2}{2} + \frac{(x_2+1)^2}{13} + 3 \quad \text{minimum: } f_1(2,-1) = 3 \\
f_2 &= \frac{(x_1+x_2-3)^2}{36} + \frac{(-x_1+x_2+2)^2}{8} - 17 \quad \text{minimum: } f_2(2.5, 0.5) = -17 \quad (1) \\
f_3 &= \frac{(3x_1-2x_2+4)^2}{8} + \frac{(x_1-x_2+1)^2}{27} + 15 \quad \text{minimum: } f_3(-2,-1) = 15
\end{aligned}
$$

with $x = (x_1, x_2) \in [-4, 4]^2$. Figure 1 shows optimal zone obtained for the simultaneous minimization of f_1, f_2 and f_3, and these defined for the simultaneous minimization of f_1 and f_2, f_1 and f_3 and f_2 and f_3.

The superimposition of the zones A, B and C gives the optimal Pareto zone D obtained for the minimization of the three functions f_1, f_2 and f_3.

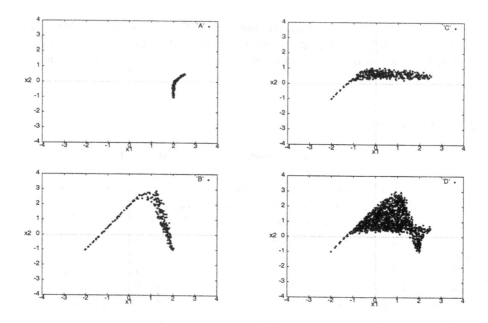

Fig. 1. Optimal Pareto zones obtained for the minimization of f_1 and f_2 (A), f_1 and f_3 (B), f_2 and f_3 (C), f_1, f_2 and f_3 (D)

3.2 Industrial application: food's granulation

The goal is the optimization of working conditions of a press used to make animal food [2]. A pulverulent product is converted into granules due to the conjugated effects of heat, moisture and pressure. The industrial would like to simultaneously minimize moisture and friability of his product and the energetic consumption of process (Figure 2). Factors having an influence on this process are feed rate, speed of rotation for mixing, flour temperature, speed of rotation and drawplate profile. For this study, only two factors are taken into account, flour temperature comprised between 35 and 75°C and drawplate profile between 2 and 6 cm.

In figure 3, the optimal zone of Pareto is represented by a scatter of black points. Acceptable working space defined by the constraints dictated by the user is illustrated with hachures in the same figure (friability index < 2.5, moisture < 14 % and energetic consumption < 15 kwh). The optimal working conditions obtained by minimizing a linear combination of criteria [2], square in figure 3, is placed on the bordline of Pareto zone. Consequently, any variation in the flour temperature could entail a degradation of the quality of the product (the conditions will no more belong to the Pareto zone). The Pareto set knowledge is very interesting for the decision maker who can pick up his optimal working conditions to take into account constraints and a specific choice. This example shows how a decision maker can choose the best working parameters to take into account variabilities of raw materials and of process.

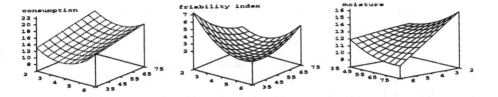

Fig. 2. Evolutions of energetic consumption (from 0 to 30 kwh), friability index (from 0 to 8) and moisture (from 10 to 17 %) of granules vs temperature (from 35 to 75°C) and drawplate profile (from 2 to 6 cm)

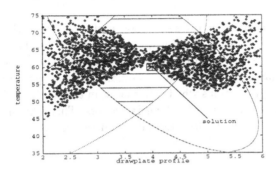

Fig. 3. Multicriteria optimization of animal granulated food. The optimal Pareto zone is represented by a scatter of black points. The constraints dictated by the industrial are drawn with bold lines. The working space delimited by these constraints is the stippled zone. The square (4 cm, 60°C) represents the minimum of a linear combination of the three considered criteria

4 CONCLUSION

In most multicriteria optimization techniques, the decision maker has to choose weights for each criterion or constraints, and it is very difficult to take into account maximal variabilities of raw materials and working conditions which assure optimal quality criteria of the product in accordance with Pareto domination criterion. If the decision maker has a good knowledge of the Pareto zone, then his specific choices can be performed after having used the multicriteria optimization technique instead of before it with other methods. So it is now possible to obtain a better robustness for the quality of the product.

The algorithm proposed in this paper is an efficient tool able to solve multicriteria optimization problems industrials can be confronted with. This method permits to obtain an optimal zone, in accordance with Pareto. The user can choose his working conditions in this zone taking into account his constraints. The algorithm is based on a diploid GA and a function that eliminates Pareto-inefficient points. Unlike existing algorithm [5], the pressure of Pareto is not applied in the selection stage of the GA, but in an original selection procedure of Pareto-efficient points.

A comparison between the two techniques has been carried out [12]. These ones give similar results but our algorithm needs less generations: 27 compared with 200. A more interesting comparison, would consist in quantifying the number of births, therefore informations given by Horn et al. do not permit this comparison.

The proposed method has been tested with one mathematical example and results show that Pareto zone can have complex shape even though functions are simple. The size of population of GA permits to determine the accuracy of the Pareto zone.

The second example illustrates the good performances of this method compared with the minimization of a linear combination of the three criteria. Intersection between Pareto set and the zone delimited by some constraints dictated by the decision maker, defines a working domain. This one permits the definition of the best running conditions which will maintain the stability of the process to take into account variabilities of influencial factors. Thus, the proposed method is an efficient tool for the multicriteria decision taking in an environment with incertainty.

References

1. Bicking F., Fonteix C., Corriou J.P., Marc I.: Global Optimization by Artificial Life: a new technique using Genetic Population Evolution. RAIRO-Operation Research, 1, 28 (1994) 23–36
2. Courcoux P., Qannari E.M., Melcion J.P., Morat J.L.: Optimisation multiréponse : application à un procédé de granulation d'aliments. Récents Progrès en Génie des Procédés : Statégie expérimentale et procédés biotechnologiques, 36, 9 (1995) 41–47
3. Derringer G., Suich R.: Simultaneous optimization of several response variables. Journal of quality Technology 12 (1980) 214–219
4. Fonteix C., Bicking F., Perrin E., Marc I.: Haploid and diploid algorithms, a new approach for global optimization: compared performances. Int. J. of Systems Science (to appear)
5. Horn J., Nafpliotis N., Goldberg D. E.: A niched Pareto genetic algorithm for multiobjective optimization. Proceedings of the first Conference on Evolutionary Computation, IEEE World Congress on Computational Intelligence, 1 (1994) 82–87
6. Korhonen P., Laakso J.: A visual interactive method for solving the multiple criteria problem. European Journal of Operational Research 24 (1986) 277–287
7. Kouada I.: Sur la propriété de domination et l'existence de points Pareto-efficaces en optimisation vectorielle convexe. RAIRO-Operation Research 1, 28 (1994) 77–84
8. Logothetis N., Haigh A.: Characterizing and optimizing Multi-response Processes by the Taguchi Method. Quality and Realibility Engineering International 4 (1988) 159–169

9. Narula S.C., Kirilov L., Vassilev V.: Reference Direction Approach for solving Multiple Objective Nonlinear Programming Problems. IEEE Transactions on Systems Man and Cybernetics **5**, **24** (1994) 804–806

10. Steuer R.E.: Multiple criteria optimization: theory, computation and application. J. Wiley & Sons ed. (1986)

11. Tarvainen K: Generating Pareto optimal alternatives by non feasible hierarchical method. Journal of Optimization Theory and Applications **1**, **80** (1994) 181–185

12. Viennet R., Fonteix c., Marc I.: Multicriteria optimization using a genetic algorirhm for determining a set of Pareto. accepted for publication in Journal of Systems Sciences.

9. [faded, illegible]
10. [faded, illegible]
11. [faded, illegible]
12. [faded, illegible]

Coevolution

Lotka Volterra Coevolution at the Edge of Chaos

Paul Bourgine[1] and Dominique Snyers[2]

[1] CREA, Ecole Polytechnique, 1 Rue Descartes, Paris - France
[2] LIASC, Télécom Bretagne, B.P. 832, F-29285 Brest Cédex, France
email: Dominique.Snyers@enst-bretagne.fr

Abstract. In this paper, we study the coevolution of species by combining a theoretical approach with a computer simulation in order to show how a discrete distribution of viable species emerges. Coevolution is modelled as a replicator system which, with an additional diffusion term representing the mutation, leads to a Schrödinger equation. This system dynamics can be interpreted as a survival race between species on a multimodal sinking and drifting landscape whose modes correspond to the eigen modes of the Schrödinger equation. This coevolution dynamics is further illustrated by a simulation based on a continuous phenotypic model due to Kaneko in which the interactions between species are interpreted through a Lotka-Volterra model. This simulated coevolution is seen to converge to viable species associated with a dynamics at the edge of chaos (i.e with a null Lyapounov exponent). The transition from such a viable species to another results from some kind of tunnel effects characteristic of the punctuated equilibrium classically observed in biology in which rapid changes in the species distribution follow long plateaus of stable distribution.

1 Introduction

More than 100 years ago, Darwin revolutionarized the way science thought about evolution with his concept of *natural selection*. Since then, modern genetics have added new grounds to his hypothesis with the discovery of the underlying coding mechanism for the genetic information of living beings. Evolution is now widely regarded as a combination of natural selection and mutation at the gene level. But even if some of the driving forces behind evolution have started to be understood, mathematical model of them are still in their infancy. Lotka and Volterra pioneered the field in 1931 [14],[12] with a model for the predation of one species by another explaining the oscillatory levels of certain fish catches in the Adriatic sea, but coevolution in more complex situations remains an open problem.

Computer simulations could bring new insights to this intriguing problem. Kauffman, for example, studied the coevolution of species with a computer model of genotypic fitness landscape: the NK-landscape model [9]. He showed how this model leads to self-organized criticality as introduced by Per Bak for the sand piles [2], [3], with a power law distribution on the avalanche size. Bak, Flyvberg and Lautrup proposed a variant of this NK-landscape based on random graphs which leads to an even more robust self-organized critical state [7] [4]. Kaneko presented a model of phenotypic fitness landscape based on an imitation game [8]. Punctuated equilibrium emerged from this simulation with the species on the plateaus being associated with a behavior at the edge of chaos.

In this paper, Kaneko's landscape is interpreted in terms of a continuous Lotka-Volterra coevolution model which is seen as a special case of replicator systems [1]. Following Ebeling [5], we present this coevolutionary process as a survival race between species on a multimodal sinking landscape whose modes correspond to the eigen modes of an associated Schrödinger equation.

Section 2 introduces the notion of genotype and phenotype fitness landscapes and section 3 presents Ebeling's view on coevolution as a hill climbing on a sinking landscape. Kaneko's imitation game is then detailed in section 4, it leads to the continuous unidimensional phenotype landscape whose simulation results are given in section 5.

2 Evolution on a Fitness Landscape

Modern biology tells us that the genetic information of all living beings is stored in a linear macro-molecule (DNA or RNA) as a long chain of A, C, G and T (DNA) or A, C, G and U (RNA) amino acids. Each individual with a genotype g_k is seen as an element of an abstract space, the *Genotype Space G* with

$$g_k = \left(a_1{}^k,...,a_r{}^k\right)$$

where $a_i{}^k$ is the allele (i.e. A, C, G in the DNA case) associated with the ith gene of individual k (r is the dimension of G). The genotype space is also a metric space with, for example, the generalized hamming distance [5]

$$d(g_k,g_l) = \min_j \sum_{i=1}^{r-j} \delta^\star \left(a_i{}^k,a_{i+j}{}^l\right)$$

(with $\delta^\star$ being the dual Kronnecker function $\delta^\star(a, b) = 1$ only if $a \neq b$). This distance represents the minimal number of non-coincidences between the two sequences g_k,g_l which have to be shifted before comparison since the beginning of the amino-acid sequence is usually unknown. With this metric, G can then be represented as a multi-dimensional lattice on which the mutation of one allele defines immediate neighborhood.

Natural selection makes the fittest individuals among the population to reproduce more often. This individual fitness is specified by the environment and we define such an evolutionary fitness landscape, called here the *Genotype Fitness Landscape*, as a smooth homomorphism from G to $\Re$, the set of real numbers. Evolution can therefore be seen as a hill climbing process on such an evolutionary landscape where mutant species associated with a better fitness reproduce more often and therefore push the whole population to climb on the phenotype landscape. In the case of coevolution, the environment is not only composed of the physical environment but also of the other species: the fitness depends on the species interactions.

Morphogenesis transforms the genotypes into the phenotypes which describe the physical properties of the associated individuals. This paper however, ignores this morphogenesis step to focus on the phenotypes only. Each phenotype is described by a set q of n properties represented by real numbers $q_1,...q_n$, and the associated vector space of dimension n is the *Phenotype Space Q*:

$$q = (q_1,...q_n) \in Q.$$

This phenotype space Q is also a metric space with a distance associated with the classical scalar product and the corresponding *Phenotype Fitness Landscape* is defined as another smooth homomorphism from Q to $\Re$. As Ebeling points out [5], the postulate of smoothness for the homomorphism between G, Q and $\Re$ in the strict mathematical sense is probably too strong for biological purposes. What is really required is a conservation of neighborhoods in most cases.

The fitness landscape can become very complex. It usually combines several values such as the replication rate, the mutation rate, the death rate, etc. , and in many cases it is simply unknown. This is the reason why many of the studies use random landscapes (i.e NK-landscape [9], random graphs [4], random field models [5]) to draw some conclusions on the evolution dynamics.

3   The Replicator System

In order to take advantage of the power of differential equations for describing the system dynamics, Ebeling interpolates continuous values between the phenotype lattice points and considers a continuous phenotype space [5] where $x(q, t)$ is the density of species with phenotype q at time t in the population. Those species evolves as a replicator system:

$$\frac{dx(q,t)}{dt} = x(q,t)\, w(q;\{x\})$$

where $w(q;\{x\})$ is the replication rate depending on the phenotype q and on the population distribution $\{x\} = \{x\left(q^1,t\right), x\left(q^2,t\right),...\}$. Two classical examples of such a replicator system are:

1. **The Fisher-Eigen model:**

$$w(q;\{x\}) = E(q) - \frac{\int_Q E(q')x(q',t)dq'}{\int_Q x(q',t)dq'} = E(q) - \langle E(t)\rangle$$

where $E(q)$ is the fitness landscape, $\langle E(t)\rangle$ the population mean fitness value at time t and Q the phenotype domain. A special case of such a dynamic is represented by genetic algorithms which preferentially select individuals with an above average fitness.

2. The Lotka-Volterra model:

$$w(q;\{x\}) = a(q) - \int_Q b(q,q')x(q',t)dq'$$

where $a(q)$ is the internal dissipation parameter (e.g. birth rate minus dead rate) and $b(q,q')$ is the species to species interactions. Figure 1, computed from Kaneko's imitation game, gives an example of such a $b(q,q')$ function.

Mutation is added to this replicator system as a diffusion term:

$$\frac{dx(q,t)}{dt} = x(q,t)w(q;\{x\}) + D\,\nabla_q^2 x(q,t) \tag{1}$$

where D is the diffusion parameter (we suppose here that $D(q) = D, \forall q$). This can lead to a Schrödinger-type of equation inducing a discreteness of the phenotypic states emerging from evolution.

3.1 Eigen-Fisher Model

In the case of Eigen-Fisher, we transform the density variable $x(q,t)$ into a new variable $y(q,t)$:

$$y(q,t) = x(q,t)\exp\left(\int_0^t \langle E(t')\rangle\, dt'\right).$$

This leads to the following linear equation [5]:

$$\frac{dy(q,t)}{dt} = y(q,t)[w(q;\{x\}) + \langle E(t)\rangle] + D\,\nabla_q^2 y(q,t) = y(q,t)E(q) + D\,\nabla_q^2 y(q,t)$$

which is equivalent to a Schrödinger equation with a potential $-E(q)$.

The eigen functions of the associated operator for different eigen value ω_k form a basis of the Hilbert Space in which $y(q,t)$ can be expressed as follows:

$$y(q,t) = \sum_k c_k\,\phi_k(q)\psi_k(t) = \sum_k c_k\,\phi_k(q)\,\exp(\omega_k t)$$

(the sum should be replaced by an integral for the continuous part of the spectrum of eigen values, but Ebeling showed that this continuous part of the spectrum disappears through evolution [5]). By dividing the Schrödinger equation by the eigen function $\phi_k(q)\psi_k(t)$ indeed, we get for the kth eigen solution:

$$\frac{1}{\psi_k(t)}\frac{d\psi_k(t)}{dt} = E(q) + D\,\frac{1}{\phi_k(q)}\nabla_q^2\phi_k(q).$$

The two sides of this equation depend on different variables and must therefore equal a constant ω_k :

$$\begin{cases} \frac{d\psi(t)}{dt} = \omega_k\psi(t) \\ \left[E(q) + D\nabla_q^2\right]\phi_k(q) = \omega_k\,\phi_k(q), \end{cases}$$

where ω_k corresponds to the eigen values of the system $H\phi_k(q) = \omega_k$ in which H is the operator $H = \left[E(q) + D\,\nabla_q^2\right]$ and $\phi_k(q)$ is its eigen function associated with ω_k.

3.2   Lotka-Volterra Model

In the case of the Lotka-Volterra model, the replication rate can be expressed by an integral operator $K\,x(q,q') = \int_Q b(q,q')x(q',t)dq'$ where $b(q,q')$ is the kernel :

$$w(p;\{x\}) = a(q) + Kx\,(q',t).$$

This integral operator is defined on $L^2(\Delta) = \{f \mid \int_\Delta \mid f(y) \mid^2 dy < \infty\}$ with Δ an interval in $\Re$ $(x(q,t) \in L^2(\Delta)$ since $x(q,t) \in [0,1]$, $\forall q \in Q$ et $\forall t)$. In the case of an *Hermitian Kernel* (i.e $K(x,y) = \overline{K}(y,x)$ where $\overline{K}(y,x)$ is the complex conjugate of $K(y,x)$), the integral operator is a self-adjoin Hilbert-Schmidt operator and $b(q,q')$ can be expanded to a bilinear form in the following way (cfr. [10]) :

$$b(q,q') = \sum_{k=0}^{\infty} \mu_k \Phi_k(q)\overline{\Phi}_k(q') \tag{2}$$

where functions $\Phi_k(q)$ are the eigen functions of the associated integral operator :

$$\int b(q,q')\Phi_k(q')dq' = \mu_k \Phi_k(q);$$

The replication rate can then be rewritten as follows:

$$w(q,t) = a(q) - \sum_{k=0}^{m} \mu_k \Phi_k(q) \int_Q \overline{\Phi}_k(q')x(p',t)dq',$$

$$w(q,t) = a(q) - \sum_{k=0}^{m} \mu_k \Phi_k(q)\, z_k(t). \tag{3}$$

where $z_k(t) = \int_Q \overline{\Phi}_k(q')x(q',t)dq'$ and where ∞ has been replaced by m following the property that an Hilbert-Schmidt operator is also a compact operator and corresponds, therefore, to the limit of a serie of operators of finite rank (cf. [10]). m is the number of ecological niches. Each niche i is associated with a prey-predator strategy defined by the function pairs $(\Phi_i(p), \overline{\Phi}_i(p'))$. Under the hypothesis that these niches have a disjoint compact support, equation (3) can be approximated on every niche k by:

$$w(q,t) \approx a(q) - b_k(t),$$
$$b_k(t) = \Phi_k \mu_k z_k(t),$$
$$\Phi_k = max\Phi_k(q).$$

In this case, each niche has its own replicator system k that follows equation (1) :

$$\frac{dx(q,t)}{dt} = [a(t) - b_k(t)]x(q,t) + D\,\nabla_p^2 x(q,t). \tag{4}$$

The same analysis than the one performed in the Eigen-Fisher model applies by deriving the Schrödinger equation by the following variable change:

$$y(q,t) = x(q,t)\exp\left(\int_0^t b_k(t')\,dt'\right).$$

Under the hypothesis of disjoint compact support for the niches, the evolution dynamics of a niche is similar to the one associated to the Eigen-Fisher Model.

3.3 The Fisher Law

It follows from the general solution that the phenotypic distribution evolves as a superposition of a discrete number of eigen modes. Mutation is supposed to be small enough to ensure a separation between these eigen modes in such a way that an eigen mode k is concentrated around a barycenters q^k defined as the phenotypic mean value in the associated domain Q_k:

$$q^k = \frac{\int_{Q_k} q\, x(q,t)dq}{\int_{Q_k} x(q,t)dq} = \langle q \rangle_k.$$

The evolutionary motion of these barycenters q^k is given by the following time derivative:

$$\frac{dq^k(t)}{dt} = \frac{\int_{Q_k} q\, \frac{\partial x(q,t)}{\partial t} dq}{\int_{Q_k} x(q,t)dq} - \frac{\int_{Q_k} qx(q,t)dq \int_{Q_k} \frac{\partial x(q,t)}{\partial t} dq}{\left(\int_{Q_k} x(q,t)dq\right)^2}$$

$$= \frac{\int_{Q_k} (q-q^k(t))\, \frac{\partial x(q,t)}{\partial t} dq}{\int_{Q_k} x(q,t)dq}$$

$$= \frac{\int_{Q_k} (q-q^k(t))\, w(q,\{x\})\, x(q,t)dq}{\int_{Q_k} x(q,t)dq}$$

For small clusters, $w(q;\{x\})$ is approximated by

$$w(q;\{x\}) = w(q^k;\{x\}) + \nabla_q w(q^k;\{x\}).(q - q^k(t))$$

and this leads to :

$$\frac{dq^k(t)}{dt} = w(q^k;\{x\})\frac{\int_{Q_k} (q-q^k(t))\, x(q,t)dq}{\int_{Q_k} x(q,t)dq} + \frac{\int_{Q_k} (q-q^k(t))\, \left[\nabla_q w(q^k;\{x\}).(q-q^k(t))\right] x(q,t)\, dq}{\int_{Q_k} x(q,t)dq}.$$

The first term is null by definition of barycenters and this equation simplifies as follows:

$$\frac{dq(t)}{dt} = \frac{\int_{Q_k} (q-q(t))\, \left[\nabla_q w(q;\{x\}).(q-q(t))\right] x(q,t)\, dq}{\int_{Q_k} x(q,t)dq}.$$

This indeed tells us that the small clusters climb on the growth rate landscape since:

$$\nabla_q w(q;\{x\}) \cdot \frac{dq(t)}{dt} = \frac{\int_{Q_k} \left[\nabla_q w(q;\{x\}).(q-q(t))\right]^2 x(q,t)\, dq}{\int_{Q_k} x(q,t)dq} \geq 0.$$

This Fisher law specifies the hill climbing nature of evolution which causes the formation of an "iceberg" structure in the fitness landscape corresponding to the viable species. Viable species are defined by a non negative replication rate so that we can picture evolution as an ice breaker floating on the ocean of null replication rate. Natural selection increases the mean fitness value and pushes the drifting icebergs into the water, hence forcing the species to mutate and climb for survival. Those icebergs may sink totally or break into pieces leading to species extinction or creation.

This hill climbing dynamic on a sinking landscape for which discontinuity emerged from a continuous phenotypic model may serves as an indication that the stepwise character of evolution observed in nature (the so-called *punctuated equilibrium*) is a general feature of selection-mutation processes instead of just a consequence of the discreteness of mutations. This is exactly the behavior observed on the simulation based on Kaneko's bird landscape with the interaction between species corresponding to a Lotka-Volterra model.

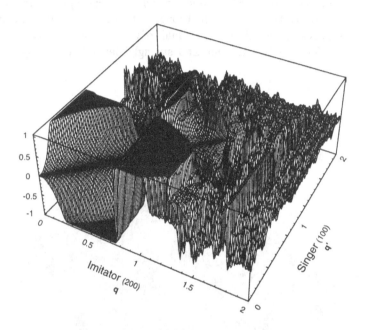

Fig. 1.: Mean gain $b(q, q')$ over 100 games between two species q and q' starting with different initial values $s_0{}^q$ and $s_0{}^{q'}$. The x (resp. y) axis gives the phenotype values associated with the imitator (resp. singer) . The phenotypic values have been taken here by 0.01 intervals between 0 and 2. The ridge around $q = 1.75$ corresponds to a robust imitator strategy winning against a large range of singer strategies.

4 The Bird Landscape

4.1 Bird Phenotype

The bird landscape was introduced by Kaneko [8] as a one-dimensional phenotype landscape. A species corresponds here to some kind of "bird" whose song is defined by the following time series:

$$s_{n+1} = 1 - q.s_n{}^2 = f(s_n)$$

where s_i takes values in $[0,1]$. The parameter q (taking real value in $[0,2]$) is the associated bird phenotype. Depending on the value taken by this parameter, the time series is known to go from periodic to chaotic dynamics (see for example [11] or [13]). Chaotic dynamic is characterized by a positive /it Lyapounov exponent, the average exponential divergence between closely adjacent points during one iteration [13].

Coevolution is defined by a distribution of individuals through the phenotypic space as a function $x(q,t)$, the density of individuals with phenotype q at time t in the population. The fitness value associated with a phenotype q is dynamically evaluated relatively to the population through a tournament between a typical member of the species q and all the individuals of the population. The phenotype landscape, therefore, evolves in time and strongly depends on the population distribution.

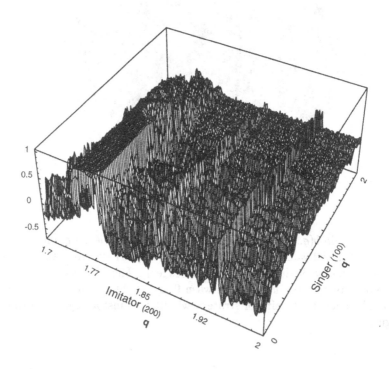

Fig. 2.: This is a zoom of Figure 1 showing the mean gain $b(q, q')$ over 100 games between two species q and q'. The imitator phenotypic values, this time, have been taken with a better resolution on 0.0015 intervals between 1.7 and 2 and new ridges appear around $q = 1.86$ and $q = 1.94$. The ridge around $q = 1.94$ is thinner than the one around $q = 1.75$ but seems to be more robust against singer strategies above $q = 1.6$ values.

4.2 Tournament Fitness

The tournament is based on a one-to-one imitation game. One bird α sings by generating a time series from an initial random value s_0^α. We shall refer to this bird as the "singer". The other bird β, called the "imitator", starts from an initial random value s_0^β and is given a time $t_{learning}$ to learn its opponent song, i.e. to reduce the distance between their two time series. The learning strategy is a gradient descent:

$$s_{n+1}{}^\beta = f\left((1-\epsilon)s_n{}^\beta + \epsilon s_n{}^\alpha\right).$$

After this learning phase, the euclidian distance between the two bird songs is computed:

$$D(\alpha,\beta) = \sqrt{\sum_{\tau=t_{learning}}^{t_{learning}+t_{game}} (s_\tau{}^\alpha - s_\tau{}^\beta)^2}.$$

The imitator and singer rôles are then switched and a similar $D(\beta,\alpha)$ distance computed. If $D(\alpha,\beta) < D(\beta,\alpha)$ (resp. $D(\alpha,\beta) > D(\beta,\alpha)$), α is declared as the winner (resp. looser) and it gains 1 point (resp. -1 point) while β, the looser (resp. winner), receives -1 point (resp. 1 point). In case $D(\alpha,\beta) = D(\beta,\alpha)$ the game is declared a draw and none of the two birds receives any point. (Let us note that in our simulation equality was defined up to the third decimal).

Figure 1 shows the mean gains between two bird species corresponding to the $b(q, q')$ interaction term of the Lotka Volterra model.

Figure 2 shows the same $b(q, q')$ interaction function with a better resolution exhibiting a new ridge around $q = 1.94$.

The fitness of a phenotype q at time t is computed during a tournament against all other members q' of the current population by summing up the $b(q, q')$ gains of the one-to-one games. With a selection mechanism based on this fitness landscape, the replication rate corresponds to the Lotka Volterra model without dissipative term:

$$w(q,\{x\}) = \int b(q,q')x(q',t)dq'.$$

5 Simulated Evolution on the Bird Landscape

The simulation starts on a population of 400 Kaneko's birds uniformly distributed on the phenotype space. At each time step a bird is chosen at random and is replaced by another bird selected from the population with a probability proportional to its fitness value $w(q,\{x\})$. Before replacement this selected bird mutates, i.e. a random value from a normal distribution is added to its phenotype. The standard deviation associated with this normal distribution is the mutation rate σ_μ.

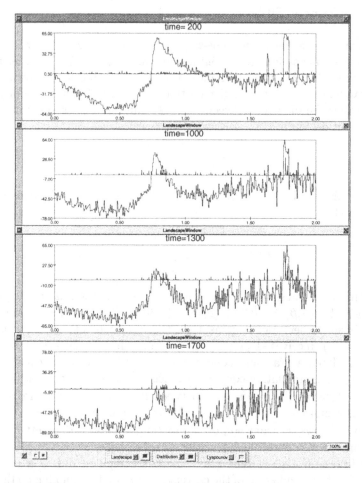

Fig. 3.: This figure shows the species distribution (the curve with small peaks proportional to the number of individuals of the corresponding species above the X-axis) and its sinking replication rate landscape (the rugged curve with positive and negative values) at time 200, 1000, 1300 and 1700 with a mutation rate $\sigma_{mut} = 0.0001$. The phenotypes takes values between 0 and 2 on the X-axis, $t_{learning} = 100$, $t_{game} = 50$ and $\epsilon = 0.7$. From time 200 to time 1700 we clearly observe an aggregation of individuals around a restricted number of species under the pressure of a positive replication rate. This is especially visible around the two replication landscape peaks at $q = 0.8$ and $q = 1.75$. But at the same time, these aggregations are associated with a decreasing replication rate and the peaks sink.

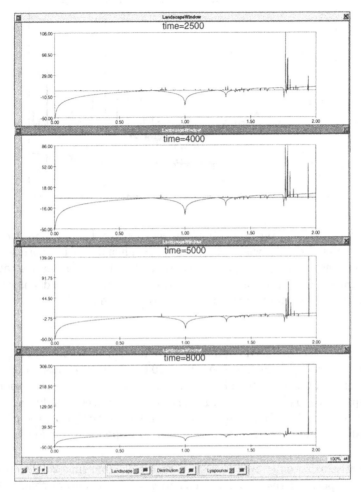

Fig. 4.: This shows the same species distribution (peaks above the X-axis) and the Lyapounov exponents associated with the time series for each parameter q value (the curve with positive values corresponding to chaotic dynamics after $q = 1.4$ only) at time 2500, 4000, 5000 and 8000 with a mutation rate $\sigma_{mut} = 0.0001$. The species aggregation seen on Figure 3 continues and we can also clearly see the shift of the species in majority from values around $q = 1.75$ to values around $q = 1.94$. This is the punctuated equilibrium with the viable species on the plateaus corresponding to dynamics at the edge of chaos characterized with a null Lyapounov exponent. This correlation is further pointed out on Figure 5.

5.1 Punctuated Equilibrium on a Sinking Landscape

With small mutation rates, the coevolution aggregates the species around discrete phenotype values as can be seen on Figure 3. Starting from a uniform distribution, the population concentrates around fewer and fewer species under

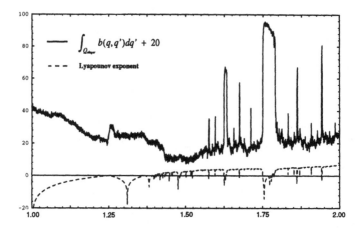

Fig. 5.: The maxima of the replication rate $w(q,\{x\}) = \int_0^2 b(q,q')x(q',t)dq'$ (dark curve) computed on the whole range of singer phenotypes are correlated with the zeros of the Lyapounov exponents (dashed curve). The replication rate curve has been raised for clarity reason by addition of a bias of 20 and was computed for 4000 imitator phenotype values between $q = 1$ and $q = 2$.

the pressure of the fitness landscape. Figure 3 indeed, shows the sinking icebergs dynamics announced in section 3. At time 200 the replication rate is high for phenotype between 0.7 and 0.9 and many birds are attracted to that region of the phenotype space. This replication rate however decreases rapidly and becomes negative at time 1700. This pushes the population to the right of the phenotypic space around value 1.77 of phenotype which corresponds to the longest ridge of Figure 1.

Discreteness becomes even clearer on Figure 4. From time 2500 to time 8000 the largest species derives from $q = 1.77$ to $q = 1.94$ by abrupt changes characteristic of the punctuated equilibrium discussed in section 3.

5.2 Coevolution at the Edge of Chaos

Furthermore, most of the surviving species correspond to a null Lyapounov exponent (see the lower curve on Figure 4). And indeed, there is a strong correlation between replication rate and Lyapounov exponent as can be seen from Figure 5 which represents the integration of Figure 1 on the singer axis. Kaneko's birds evolve thus to the edge of chaos (i.e null Lyapounov exponent) where their song dynamics is a compromise between periodic songs which are too easily imitated and chaotic ones which are usually not very successful at imitating others. The edge of chaos is also associated with a robust behavior corresponding to a ridge on the $b(q, q')$ landscape which extends on the whole phenotype range as can be seen on Figure 2.

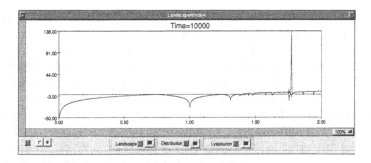

Fig. 6.: Species distribution and Lyapounov exponent for mutation rate $\sigma_\mu = 0.001$ at time 10000. With this mutation rate the coevolution could not aggregate on the thin ridge around $q = 1.94$ and coevolution gets stuck at $q = 1.75$.

5.3 Critical Mutation Rate

With a mutation rate $\sigma_\mu = 0.0001$, the coevolution system converges to phenotype 1.94 (see Figure 4) corresponding to the narrow but long ridge on the far right of Figure 2. With a larger mutation rate (e.g. $\sigma_\mu = 0.001$) the system is not able to stay on this narrow ridge and the system gets stuck on the wider one around 1.76 as shown on Figure 6. This mutation rate therefore is critical for convergence. We could decrease its value during evolution in analogy with the temperature of the simulated annealing algorithm or we could look at the mutation rate as a phenotypic property and let it evolve itself as in some evolutionary algorithms[6]. This remains to be investigated.

6 Conclusion

We have studied the coevolutionary phenomenon through a simulation based on a continuous fitness landscape model defined at the phenotype level. In this simulated coevolution the continuous phenotype space has rapidly shrank to a discrete set of viable species with the evolving population exhibiting a punctuated equilibrium: a majority of its individuals concentrates around a succession of such viable species. The viable phenotypic states have also been shown to be correlated with an intrinsic dynamical behavior at the edge of chaos associated with a null Lyapounov exponent.

These simulation results have been interpreted as a replicator system with a Lotka-Volterra model of species interaction. Coevolution in a continuous phenotype space was described by differential equations and punctuated equilibrium was shown to result from the combination of natural selection and mutation. We pictured coevolution as a climbing race on a sinking landscape and showed that the viable species correspond to robust phenotypes, i.e. those performing well against the whole range of phenotypes. Robustness is preferred to local performance, but mutation rate is critical to reach these robust states and stabilize on them. Viable species behaviors at the edge of chaos have also been interpreted as a compromise between the easy to learn periodic behaviors and the chaotic behaviors with a poor imitation capability.

Coevolution is a very complicated and poorly understood process. This paper is yet another attempt to grasp some of its intrinsic features. The originality of this work lies the connection it makes between a theoretical approach and a simulation, each one of them shedding new lights on the other one. We are currently investigating further the theoretical model and we are also pursuing the simulation with, for example, a variable mutation rate. This study could also offer new directions to some optimization problems in which several criteria compete and especially when the fitness landscape is dynamic and depends on the population distribution (i.e. in this case on the way the criteria are met).

References

1. J.P. Aubin. *Viability Theory*. Birkhauser, 1991.
2. P. Bak. Self-organized criticality. *Physica A*, 1990.
3. P. Bak and K. Chen. Self-organized criticality. *Scientific American*, pages 26–33, January 1991.
4. P. Bak, H. Flyvbjerg, and B. Lautrup. Coevolution in a rugged fitness landscape. *Phys. rev A*, 46:6724–6730, November 1992.
5. R. Feistel and W. Ebeling. *Evolution of Complex Systems: Selforganization, Entropy and Development*. Kluwer Academic,Dordrecht, 1989.
6. F.Hoffmeister and T. Báck. Genetic algorithms and evolution strategies: Similarities and differences. Technical report, University of Dortmund, 1992.
7. H. Flyvbjerg and B. Lautrup. Evolution in a rugged fitness landscape. *Phys. rev A*, 46:6714–6723, November 1992.
8. K. Kaneko and J. Suzuki. Evolution to the edge of chaos in an imitation game. In *Artificial Life III*, 1994.
9. S. A. Kauffman and S. Johnsen. Co-evolution to the edge of chaos : coupled fitness landscape, poised states and co-evolutionary avalanches. In J. D. Farmer C. G. Langton, C. Taylor and S. Rasmussen, editors, *Artificial Life II*, Santa Fe, New Mexico, 1991. Addison Wesley.
10. S.G. Mikhlin. *Integral equations*. Pergamon Press, 1957.
11. H.-O. Peitgen, H. Jürgens, and D. Saupe. *Chaos and Fractals: New Frontiers of Science*. Springer-Verlag, New York, 1992.
12. M. Peschel and W. Mende. *The Predator-Prey Model*. Springer-Verlag, Wien, 1986.
13. H. G. Schuster. *Deterministic Chaos: An Introduction*. VCH Verlagsgesellschaft, 1989.
14. V. Volterra. *Leçons sur la théorie mathématique de la lutte pour la vie*. Herrmann & Cie, Paris, 1931.

Evolution Through Cooperation: The Symbiotic Algorithm

Renaud Dumeur

Département Informatique
Institut d'Intelligence Artificielle
Université de Paris-8
2, rue de la Liberté
93256 St Denis cedex 02
renaud@uparis8.univ-paris8.fr

C^2V
82, boulevard Haussmann
75008 Paris
renaud@ccv.fr

Abstract. This article describes a new problem solving paradigm: the Symbiotic Algorithm. Problem solutions are considered as embedded organisms whose genetic materials are expressed as evaluable phenotypes. This organic hierarchy can be viewed as a biosphere containing the whole set of creatures manipulated by the algorithm. Being immortal, organisms do not replicate. The only changes occuring in the biosphere are the creation or destruction of symbiotic relationships between organisms. To let these relationships appear and evolve, we provide a fitness function which decides the outcome of organic interplay. The higher an organism's fitness is, the more likely it is to remain unchanged and the more likely it is to invade weaker participants.

Some experiments, addressing the optimization of sinusoidal functions, show that the organic hierarchy adopts configurations in which appear substructures corresponding to optimal solutions. Moreover, the use of multimodal fitness functions induce phenotype distributions matching the fitness function peaks.

1 Introduction

In her book "Symbiosis in Cell Evolution", Lynn Margulis [6] presents the existence of mithocondrial DNA as a strong argument in favor of the symbiotic origin of eukaryotic cells. If endosymbiosis between prokaryotes has produced sophisticated cells, able to solve the difficult problem of survival in harsh environments, it is legitimate to think that such process can be useful as a computation paradigm.

This kind of technology transfer from nature to computer science is not new. It has already been used to establish other computation techniques, such as genetic algorithms [4, 2, 1, 8] and neural networks [7, 9, 5].

Following the example of genetic algorithms, our symbiotic algorithm tackles problem solving by processing a set of artificial organisms – which represent potential solutions to the stated problem – endowed with a genetic material. However, in contrast with traditional GAs:

- these organisms are immortal and unable to replicate.
- the notion of population is only implicit. Any organism can be viewed as a population of interacting symbionts.
- an expression mechanism is employed to convert the genetic material into an evaluable structure: a phenotype.

2   Definitions

The symbiotic algorithm uses:

a structure used to represent solutions as organisms.

an evaluation function which quantifies an organism performance (solution fitness).

an interaction law ruling the creation of endosymbiotic relationships between organisms. Following this law, a valuable organism (with a high fitness value) tends to keep its symbionts while a weak one is likely to lose them or to be invaded.

3   The organisms

Each organism owns a structure which can be generated by the following BNF-like grammar:

```
<organism>  : <gene> <symbionts>
<symbionts> :
            | <organism>
            | <symbionts> <organism>
```

Each organism contains an elementary gene and may also contain other organisms (its symbionts). The entire genetic material, of both the organism and its symbionts, determines the organism's fitness. This strategy models the fact that, in endosymbiosis, the invading organism alters the invaded's fitness. If the fitness is increased, then the relation will last and will be observed as symbiosis[1].

An holobiont survives as long as it preserves its structure. Symbiont gain or loss results in a structural change of the organism. For instance, if a weak (low fitness) organism hosts strong symbionts (with high fitness) it will tend to lose them. On the contrary, if a recently invaded host becomes stronger than its invader(s), it will succeed to confine them. The same (and recursive) way, the trapped symbionts are put in interplay, the weak ones being likely to be invaded by stronger ones.

---

[1] Of course, as biological holobionts are able to replicate, they manage to trigger symbiont replications at the right time. Eventually, after numerous generations, host and symbionts genetics materials may fuse and form an single organism which may no longer be described as an association.

4 Evaluation

The evaluation function must process a genetic hierarchy to yield a scalar value. This process must overcome the following problems:

lacunas in the genetic material contained in the hierarchy: if the genetic information is incomplete, it is not possible to create an evaluable phenotype.

conflicts between elementary genes which code incompatible phenotype features.

redundancies in the genetic material: two or more elementary genes can code the same phenotype features.

An expression mechanisms allows us to solve all of these problems. It converts the hierarchical, messy [3] and recursive genetic structure into a more traditional one: a feature vector. To do so, we consider an elementary gene as a (feature, position) pair. The probability for such a gene to be expressed – to find a feature at a given position in the phenotype – can be determined on the basis of the number of occurrences of this elementary gene in the expressed genetic hierarchy: the greater the number of occurrences, the more probable the expression. The probability for a feature f_i to be expressed at position p in a phenotype ϕ can be computed from the number occurrences $\mathcal{N}(f_i, p)$ in the genome with the following formula:

$$P(\phi[p] \to f_i) = \frac{\mathcal{N}(f_i, p)}{\sum_j^{N_{values}} \mathcal{N}(f_j, p)}$$

where N_{values} is the number of possible features values.

When, at a given phenotype position, the genome does not specify a feature value, we have:

$$\sum_j^{N_{values}} \mathcal{N}(f_j, p) = 0$$

and we consider that all feature values have the same probability of expression :

$$P(\phi[p] \to f_i) = \frac{1}{N_{values}}$$

At the end of the expression phase, the feature vector fitness value is computed and assigned to the organism. This value will be used to determine the symbiosis relationships.

The figure 1 illustrates the different stages of the evaluation process.

It is interesting to note that such a coding makes the genome act like a random sampler whose sampling region is probabilistically constrained by the genetic information. Thus, if good solutions are independent of certain phenotype positions, the corresponding good organisms can harmlessly miss the elementary genes coding for these positions.

We can also consider our genetic hierarchies as implementations of J. Holland schemata[4], with an additional piece of information: the recursive structure which encodes schema robustness.

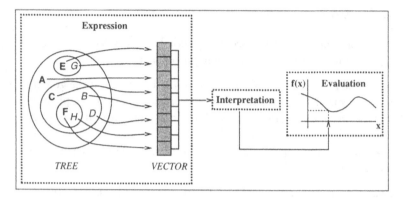

Fig. 1. Expressing genetic hierarchies.

5 The Interaction Law

The interaction rule governs the behavior of the interacting organisms: it favors valuable organisms by protecting their genetic material from intruders. This rule also applies when the holobiont interacts with its symbionts. As long as a holobiont is more valuable than its symbionts, the symbionts will not be able to escape. Such a rule will favor the emergence of good organisms formed by successive embedding of less fit organism[2].

Consider, for instance, two organisms A and B contained in the same host B. Assume that the result of an interaction between A and B is a function of their respective states. These states are :

open: denoted $(X_{][})$: X is ready to be invaded.
closed: denoted $(X_{[]})$: X can resist invasion.

We denote by $\langle\rangle$ the probability for an organism to adopt a given state at interaction time. Thus, $\langle A_{[]} \rangle$ represents the probability for $A[]$ to stay open while interacting.

We compute the state probability of organism A by using its fitness value $\mathcal{A}$. As a bad organism is likely to be found open during interaction (while a good one tends to be closed), we compute $\langle A_{[]} \rangle$ by normalizing $\mathcal{A}$:

$$\langle A_{[]} \rangle = \frac{\mathcal{A} - \min \mathcal{O}}{\max \mathcal{O} - \min \mathcal{O}}$$

where $\max \mathcal{O}$ and $\min \mathcal{O}$ respectively denote the maximal and minimal fitness values of all the organisms which represent the interaction context of A. This interaction context is a set of organisms comprising A's parent (O) and the symbionts it contains.

---

[2] such a process has been inspired from the Serial Endosymbiosis Theory (SET) presented by L. Margulis [6]

Consequently, the probability for a single organism A to be open is :

$$\langle A_{][} \rangle = 1 - \langle A_{[]} \rangle$$

For a binary system composed of organisms A and B, we can find three possible interactions :

$$A + B \rightsquigarrow \begin{vmatrix} A.[B] & B \text{ invades } A \\ B.[A] & A \text{ invades } B \\ A + B & A \text{ and } B \text{ do not change} \end{vmatrix}$$

From A and B respective states, we then define the probability for each interaction to occur with:

$$\langle A_{[]} + B_{][} \rightsquigarrow B.[A] \rangle = \langle A_{[]} \rangle \langle B_{][} \rangle + \frac{\langle A_{][} \rangle \langle B_{][} \rangle}{2}$$

$$\langle A_{][} + B_{[]} \rightsquigarrow A.[B] \rangle = \langle A_{][} \rangle \langle B_{[]} \rangle + \frac{\langle A_{][} \rangle \langle B_{][} \rangle}{2}$$

$$\langle A_{[]} + B_{[]} \rightsquigarrow A + B \rangle = \langle A_{[]} \rangle \langle B_{[]} \rangle$$

Applying these formuli, the probability for an interaction to yield a structural change is

$$\left\langle A + B \rightsquigarrow \begin{vmatrix} A.[B] \\ B.[A] \end{vmatrix} \right\rangle$$

as illustrated by figure 2. The complementary probability (no structural change)

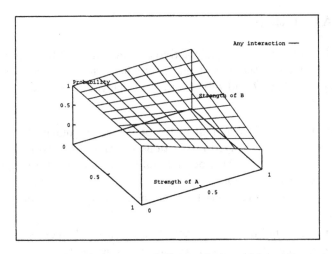

Fig. 2. Probability of structural change during a binary ineraction.

is shown in figure 3.

We will note that structural changes are likely to happen when two interacting organisms are weak. Conversely, when two strong organisms are in competitive interaction, no change is anticipated.

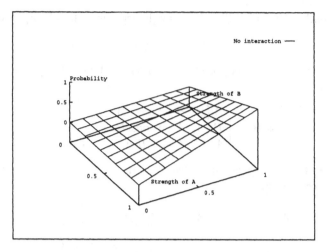

Fig. 3. Structural conservation probability during binary interaction.

The function we use to model interactions probabilities also respect the following logical property:

$$\sum_{i}^{\{0,1\}} \sum_{j}^{\{0,1\}} (\langle A_i \rangle \langle B_j \rangle) = 1$$

6 The Algorithm

In our simulated biosphere, organic activity is implemented through a recursive function which starts at the root at the genetic herarchy. This function (**Evolve**) processes an organism O and returns a list of delayed interactions. It proceeds as follows:

1. apply **Evolve** to each symbiont of the processed organisms. Each recursive calls returns a set of delayed interactions which are merged into a set I.
2. add to I the $(N * (N - 1))/2$ possible interactions between the organisms in interplay: O and its symbionts.
3. eliminate all possible conflicts between interactions in I by constructing two sets:
   - a set of isolated binary interaction: I_{binary} involving O's symbionts.
   - a set of delayed binary interactions involving O: I_{delayed}.
4. execute each interaction of I_{binary} by modifying the genetic hierarchy in agreement with the outcome of the interaction: invasion or evasion of a symbiont.
5. if O's genetic structure has been modified (it has been invaded, or has lost some of its symbionts) then rebuild and evaluate its phenotype.
6. return I_{delayed}.

Step 5 is a lazy evaluation phase which reduces the number of calls to the phenotype evaluation procedure.

Evolve can easily be implemented to take advantage of a parallel architecture. Indeed, for a given recursion level, recursive calls on symbionts can be modeled by parallel processes synchronized at step 2.

7 Experiments on sinusoidal functions

We have used the symbiotic algorithm to optimize sinusoidal functions of the type: $y = 1 + \sin 2\pi nx$. We use n to arbitrary change the number of peaks of the evaluation function.

The other parameters of the system are:

growth rate of the biosphere. This is the probability for a new elementary organism (a single gene) to appear at the root of the genetic hierarchy (the biosphere). We use a value of 0.1 to allow time for the biosphere to structure itself under the effect of the algorithm. The resulting structure reduces the number of possible interactions between organisms and thus reduces also the associated (quadratic) computational load.

maximum number of organisms contained in the biosphere. Once this limit has been reached, the biosphere stops growing. For our experiments, this value has been set to 256.

precision (in bits) used to encode the parameter x. We use 12 bits in our tests.

number of cycles : the number of execution of the Evolve algorithm. This parameter is set to 2560.

In phenotypic distribution graphs presented below, the horizontal axis represents the possible values for x and the height of vertical bars gives the number of organisms whose phenotype is x. The tested function is also represented (and scaled 10 times) and allows us to see the matches between the phenotype distribution and the fitness peaks.

In the following example ($n = 1$) the larger phenotype class – the number of occurrence of a specific solution – resides under the single peak of the evaluation function (cf. figure 4).

When doubling the frequency ($n = 2$), two major phenotypes classes appear, each of them centered on a fitness peak (cf. figure 5).

With $n = 3$, again, the number of major phenotype classes matches the number of fitness peaks. The total number of organisms being constant, the density peaks are lower than in the previous examples (cf. figure 6).

When we raise the frequency to 4, we still obtain the correct number of distribution peaks (cf. figures 7). But, for the same reasons as in the previous examples, these peaks are smaller.

We can observe – by plotting the distribution of organisms cardinalities for different sinus frequencies[3] – that complex organisms exist, which have been formed through successive integration of symbionts (cf. figure 8).

---

[3] by cardinality, we mean the number of elementary genes contained in the organism.

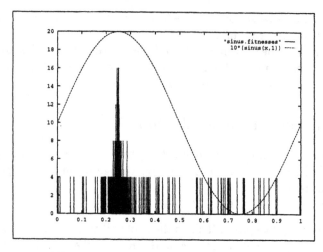

Fig. 4. Phenotype distribution with $n = 1$.

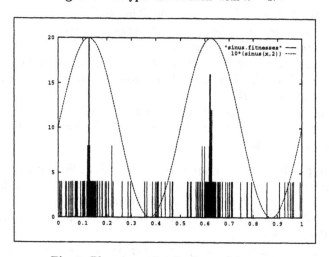

Fig. 5. Phenotype distributions with $n = 2$.

When there is a large number of atomic symbionts, we nevertheless observe some organisms formed of a dozen symbionts. These correspond to specific solutions.

By analyzing the biosphere contents with $n = 3$, we can identify some valuable organisms. The graph given in figure 9 places each organism – represented by a point – using its (phenotype, fitness) as (x, y) coordinates.

When the vertical axis is used to represent the total number of symbionts contained in each organism, we obtain the graph 10. One notices that the three highest vertical dot sequences correspond to optimal structures containing a large number of symbionts.

Considering organism lifetimes, we find three peaks which correspond to peaks in the fitness function (cf. figure 11). The lifetime is the number of cycles

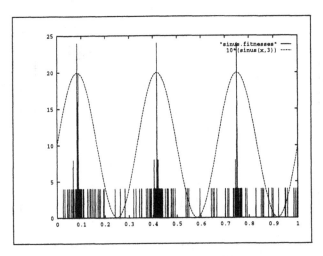

Fig. 6. Phenotype distribution with $n = 3$.

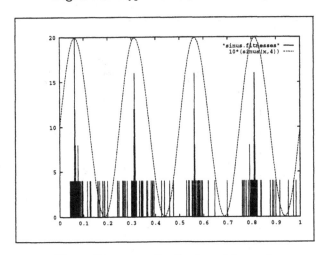

Fig. 7. Phenotype distribution with $n = 4$.

during which the organism's total genetic material has not been modified – this is true even if some of the organism's symbionts have been changed, being invaded by other symbionts. This confirms to us that good composite organisms tend to preserve their genetic content by resisting to the invasion or loss of symbionts.

Let's take an optimal biosphere organism (illustrated by figure 12) whose expressed phenotype is 101111111111. In base ten, this value is decoded as 3071 which, once divided by $2^{12} - 1$, yields .7499389. For such value, the fitness formula $(1 + \sin 2\pi n x$, with $n = 3)$ delivers – after computer approximations – the optimal value 2.

One can see that a conflict exists in the coding of lower bits – $(0, 0)$ and $(0, 1)$ genes – and that no value is specified for positions 4 and 11.

To understand this representation better, consider a phenotype produced from

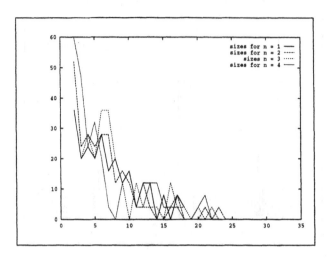

Fig. 8. Size distributions with $n = 1, 2, 3, 4$.

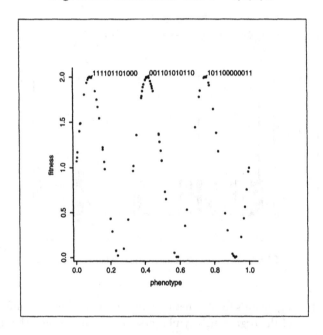

Fig. 9. Organism fitnesses.

the same genome with 0 expressed at position 4 (101111101111). The corresponding fitness value is 1.98229 which is near the optimum. We conclude that the feature value contained in position 4 is not very important for the phenotype fitness. By contrast, if we put a 1 at position 11 we obtain a phenotype with all bits to set 1: 111111111111. The decoded value being 1, it yields a low fitness value: 1. Thus, chance has played an important role in generating the good phenotype 101111111111 from the described genetic tree. The expression probability

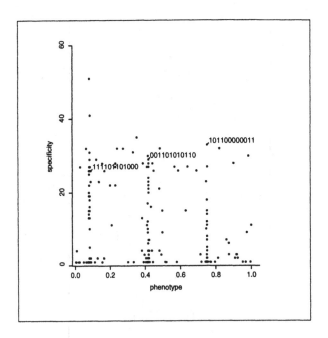

Fig. 10. Number of symbionts per phenotype.

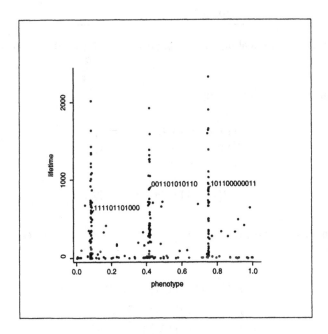

Fig. 11. Organism lifetime vs. phenotype.

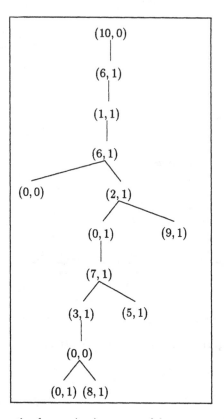

Fig. 12. An optimal organism's genome (phenotype 101111111111).

of which can be computed exactly:

$$P(1**********) = 1/2$$
$$P(*******1****) = 1/2$$
$$P(**********1*) = 2/4$$
$$\text{then}$$
$$P(101111111111) = 1/8$$

But, considering the fitnesses of other phenotype alternatives generated by flipping bits 1 and 4:

$$f(101111101101) = 1.99647$$
$$f(101111101111) = 1.9972$$
$$f(101111111101) = 1.99995,$$

we can then aproximate the expression probability of a good phenotype (whose fitness is near 2) as the probability of bit 11 to be set: 1/2 only!

8 Conclusion

We have implemented endosymbiosis between artificial organisms evolving in interplay within a simulated biosphere.

We have introduced a representation of genetic information and an interpretation mechanics allowing artificial organisms (similar to Holland schemata) to evolve through endosymbiosis. An interaction law, valid in any point of the biosphere, rules the evolution of the organisms.

By optimizing a sinusoidal function, we have shown that the Symbiotic Algorithm is able to produce good solutions. Analysis of the generated solution distributions following their size, fitness and lifetime, shows that the biosphere favors the production of high-fitness, resistant structures.

We plan to develop the present approach and combine it with proven techniques, to:

- try new codings and new expression mechanics for genetic hierarchies (i.e. $\mathcal{L}$-systems) to tackle more complex problem classes (i.e. simulation of multicellular development, the TSP problem). The automatic clustering of production rules though symbiotic transformations can also be envisioned.
- endow the artificial organisms with replication. This should allow the diffusion and the reuse of robust genetic structures in different places (organisms) of the biosphere.
- search for new interaction laws from which replication could emerge.
- extend the symbiotic approach to structures other than trees. For instance, it is possible to imagine cooperation rules allowing a self organizing directed graph to evolve. Graph points would be individual organisms and connections would represent cooperation links conveying information. Through the action of an interaction rule, stable subgraphs (i.e. highly fit closed loops) could appear, cooperate with other subgraphs and even migrate by changing their connections. What kind of mechanics could model the graph geometry dynamics?
- to create a development tool suite allowing easy control of the biosphere activity. An interactive graphical representation of the biosphere has proven essential during the debugging of the test programs.

To conclude, it seems that modeling the biosphere activity is difficult because of the following conceptual loop:

- the biosphere geometry determines potential organism interactions.
- the result of these interactions is a change of the biosphere geometry.

which requires to be theoretically studied.

Many Thanks

to Dr. Howard Gutowitz for having so kindly reread and improved this article.

References

1. Lawrence Davis. *Genetic Algorithms and Simulated Annealing.* Research Notes in Artificial Intelligence. Morgan Kauffmann Publishers, Inc., Los Altos, California, 1987.
2. David E. Goldberg. *Genetic Algorithms in Search, Optimization, and Machine Learning.* Addison-Wesley, Reading, Mass., 1989.
3. David E. Goldberg, Kalyanmoy Deb, and Bradley Korb. Don't worry, be messy. In Richard K. Belew and Lashon B. Booker, editors, *Proceedings of the Fourth International Conference on Genetic Algorithms*, pages 24–30. University of California, San Diego, Morgan Kaufmann Publishers, 1991.
4. J. H. Holland. *Adaptation in Natural and Artificial Systems.* University of Michigan Press, Ann Arbor, 1975.
5. J.J. Hopfield. Neural networks and physical systems with emergent collective computational abilities. *Proc. Nat. Acad. Sci.*, 79:2554–2558, 1982.
6. Lynn Margulis. *Symbiosis in Cell Evolution (second edition).* Freeman, 1993.
7. W.S. McCulloch and W. Pitts. A logical calculus of the idea immanent in nervous activity. *Bulletin of Mathematical Biophysics*, 5:115–133, 1943.
8. Zbigniew Michalewicz. *Genetic Algorithms + Data Structures = Evolution Programs.* Springer Verlag, 1992.
9. M. Minsky and S. Papert. *Perceptrons: An Introduction to Computational Geometry.* MIT Press, Cambridge, 1969.

The Artificial Evolution of Cooperation

Nicolas Meuleau[1] and Claude Lattaud[2]

[1] CEMAGREF, Artificial Intelligence lab,
Parc Tourvoie, BP 121, F-92185 Antony Cedex, France.
e-mail : nicolas.meuleau@elan.cemagref.fr
[2] Université Paris 5, LIAP 5,
45, rue des Saints Pères, F-75006 Paris, France.
e-mail : latc@math-info.univ-paris5.fr

Abstract. We propose here a new approach to study co-evolution and we apply it to the well-known iterated prisoner's dilemma. The originality of our work is that it uses a simplified version of the game, and thus, restrict the search space of evolutionary dynamics. This allows to have a look at the totality of the search space in permanence, and so, a complete understanding of the phenomenon of co-evolution in process. The paper includes a little game-theoretic introduction to iterated prisoner's dilemma, a survey of previous works on evolution in this game and the exposition of the questions that were still asked to us. We describe then our special approach to the problem, using populations larger than the search space, or even infinite. The experimental results that we present complete the actual knowledge of iterated prisoner's dilemma.

1 Introduction

In a symmetric two-players game, each one secretly chooses a decision among the N available. The $N \times N$-matrix $\nu = [\nu_{ij}]$ that defines the game, gives the utility earned by each player consequently to these decisions : ν_{ij} is the result of the player who chooses decision i while its opponent chooses decision j.

In the Prisoner's Dilemma (PD), there are two possible decisions (i.e. $N_{PD} = 2$), number 1 being interpreted by "Defect" (D) and number 2 by "Cooperate"(C). The utility matrix is given by the following table :

player 1 \ player 2	D	C
D	P / P	T / S
C	S / T	R / R

(result of player 1 / result of player 2)

i.e. :

$$\nu^{PD} = \begin{bmatrix} P & T \\ S & R \end{bmatrix},$$

where $T > R > P > S$. We see that for a very wide variety of parameters, PD is a non-zero-sum game : a reward for a player does not enforce a cost for its "opponent".

Player 1 may lead the following reasoning :

- "If player 2 cooperates, then it is preferable to defect, because the *Temptation* of defection T is greater than the *Reward* for mutual cooperation R."
- "If player 2 defects, then it is again better to defect, because the *Penalty* for mutual defection P is greater than the *Sucker*'s punishment S."

This may be represented by drawing two arrows on the table in the following way :

player 2 / player 1	D	C
D	P/P ↑	T/S ↑
C	S/T	R/R

The same reasoning step can be applied to player 2, what leads to the drawing of two symmetric arrows :

player 2 / player 1	D	C
D	N P/P ←	T/S ↑
C	S/T ←	R/R

So, if both player follow the rational reasoning developed above, they both conclude that defection is the best alternative whatever the opponent's choice. It results in the situation (D, D), convergence point of the arrows. This fact is formalised by saying that (D, D) is a Nash equilibrium of the game:

Definition 1. (i, j) is a Nash equilibrium of game ν iff. :

$$\nu_{ij} \geq \nu_{i'j} \quad \forall i' \neq i, \quad \text{and} \quad \nu_{ji} \geq \nu_{j'i} \quad \forall j' \neq j.$$

That is why this situation is marked with a N on the table.

So the defection is the natural choice of two rational, but not speculator at all, agents. However this choice seems somehow non-optimal because both players could obtain a greater gain by cooperating $(R > P)$. We will say that (C, C) strictly dominates (D, D) :

Definition 2. (i, j) strictly dominates (i', j') in the game ν iff. :

$$\nu_{ij} > \nu_{i'j'}, \quad \text{and} \quad \nu_{ji} > \nu_{j'i'}$$

All the dilemma lies in the fact that the Nash equilibrium of the game is strictly dominated by a diametrically opposed situation [3]. For this reason, PD is called a strong cooperation problem.

In PD, the players are supposed to meet only once [4]. In the Iterated Prisoner's Dilemma (IPD), the two same players are supposed to meet and play PD several times successively. They are allowed to have their decision at the nth meeting depend on all the past interactions with the same opponent, i.e. the results of the $n - 1$ previous meetings. The aim of each one is to maximise the summation over all iterations of his rewards [5].

The constraint $T + S < 2R$ is imposed on the utility values. Its role is to guarantee that two players that always cooperate will do better than two players that have agreed to play (C, D) and (D, C) alternatively.

Theoretically, IPD is a two-player symmetric game exactly as PD was, we note uIPD its utility matrix. In IPD, a decision is the choice of a strategy, i.e. of a mapping from the set of all possible sequences of results – the "past histories" – into the set D, C. As we show below, there is a huge combinatorial explosion of the decision space of IPD.

We call "length of the interactions" or "number of iterations" L, the number of times that two same players play PD together. The number of different possible histories $x(L)$ may be iterativelly calculated with the following formulas :

$$x(1) = 1, \quad \text{and} \quad x(L) = 4^{(L-1)} + x(L - 1) \quad \forall L > 1.$$

The first one comes from the fact that, if $L = 1$, then IPD equals PD and there is only one possible history, noted $\emptyset$: "nothing has been played, the game begins". After the first iteration, the game is already finished.

The second formula uses the fact that the result of each iteration takes its value in the set T, R, P, S. There exists so $4^{(L-1)}$ different possible histories of exactly $L - 1$ iterations. The second term of r.h.s. comes from the fact that a situation where less than $L - 1$ iterations have been played is a possible history in a game of length L.

From the definition of a strategy as a mapping from a set of size $x(L)$ into a 2-elements set, it results that the number of decisions of IPD is : $N_{IPD} = 2^{x(L)}$.

---

[3] in the sense that every player must change his decision to go from the dominated situation to the dominating one.

[4] this is equivalent to suppose that the players randomly change of opponent at each turn and never know who they are playing with (anonymous game), or that the players forget all their past experiences between two iterations, or also that they are myopic and cannot anticipate farther than one time-step.

[5] sometimes, it is the discounted cumulated reward (cf. Axelrod 1984) that is to maximise. In regard to the experimental aspect of our work, the non-discounted reward is easier to handle. Moreover, the robustness of our results seems to indicate that the use of discounted rewards would not affect a lot the behaviour of the model.

So it increases exponentially relatively to the exponential of L. For instance, if $L = 4$, two players play only four times together but $N_{\text{IPD}} \simeq 4\,10^{25}$. This number is not computable and so, even with this very short length of interactions, the combinatorial explosion of the search space is getting too much for us.

With the repetition of non-anonymous interactions, a lot of high-level concepts as benevolence, malevolence, susceptibility and indulgence are introduced into the game. In a lot of fields as varied as sociology, psychology, international politic and biology, IPD has proven to be a very useful and powerful model (cf. Axelrod 1984 ; Baefsky and Berger 1974 ; Bethlehem 1975 ; Riker and Brams 1973 ; Schelling 1981). Lately works on IPD are almost all devoted to the emergence of cooperation in an evolving population of players. Since his leading works and the publication of his book "The Evolution of Cooperation" (1984), R. Axelrod's name is almost always associated with this field of study . In the following section, we give a survey of his results and the most relevant of associated works.

2   Previous Works

The first topic presented in R. Axelrod 's book "The Evolution of Cooperation" is the result of computer tournaments opposing several strategies for IPD proposed by researchers from different spheres. As it is well known, the winner of these tournaments was the strategy "Tit for Tat" (TFT), that may be defined in the following way : cooperate at the first iteration and after, always do whatever your opponent did on the previous iteration. The author attributes this success to following special properties of TFT :

benevolence : never to be the first to defect,
susceptibility : to punish a defection of the opponent by another defection,
indulgence : to forgive the opponent after having punished him.

The next step of Axelrod studies is the simulation of what should have happened if he had continued to organise tournaments. Assuming that :

- no new strategy should have been invented and introduced into the tournaments,
- the representation of a strategy in a tournament (i.e. the percentage of people using it) is proportional to its average score at the previous tournament[6],

he leads computer simulations of tournaments. These experiments show that TFT would have continued to dominate the tournaments, in percentage of representation and in score, as long as the two hypothesis above are still respected.

After having verified the superiority of TFT in the artificial environment constituted by the set of strategies proposed by scientists, Axelrod leads a theoretical study of "evolutionary stability" to explain this result. For this purpose, he defines the notion of collective stability in the following way :

---

[6] in a way very similar to Genetic Algorithms'selection operator.

Definition 3. The decision i for game ν is collectively stable iff. :

$$\nu_{ii} \geq \nu_{ji} \quad \forall j \neq i$$

So, a strategy i is collectively stable if (i, i) is a Nash equilibrium (according to definition 1) of the game. Then we say that i is in Nash equilibrium with itself.

The reasoning laid by Axelrod is the following : if the population is uniquely constituted by a single "indigenous" strategy i, then, to be able to survive, an invading strategy j must do a score strictly greater than the natives. As the invader and most of the indigenous play only against native i, because of their great numerical superiority, j will survive if $\nu_{ji} > \nu_{ii}$. Thus strategy i is evolutionary stable if this inequality never holds.

Using this definition, he shows that TFT is collectively stable in IPD. Then he states that collective stability is evolutionary stability and so, TFT is evolutionary stable. This work is closely related to J. Maynard-Smith's studies (1975), although Axelrod's collective stability is less restrictive than Maynard-Smith's evolutionary stability.

A doubt is cast on the utility of these results results by Boyd and Loberbaum (1987). The critic is that it is sufficient that the invading strategy j realises $\nu_{ji} \geq \nu_{ii}$ to be able to survive in an homogenous population of i. As the authors notice, this distinction is important in IPD because a lot of couple (i, j) verify $\nu_{ji}^{\mathrm{IPD}} = \nu_{ii}^{\mathrm{IPD}}$. It is to be noted that, after having argued that the collective stability of TFT is not sufficient to explain the emergence of cooperation, the authors propose another definition of evolutionary stability, more restrictive than Axelrod 's collective stability. Then they show that no deterministic strategy may be evolutionary stable in IPD according to their definition.

It seems to us that this criticism is very well founded. It is particularly pertinent in the case of evolutionary algorithms as Genetic Algorithms (GA) and Evolution Strategies (ES). See (Bäck and Schwefel 1993 ; Hoffmeister and Bäck 1991) for an overview of evolutionary algorithms, (Holland 1975 ; Goldberg 1989 ; Bäck and Hoffmeister 1991 ; Bäck 1992) for GA, and (Rechenberg 1973 ; Schwefel 1977 ; Schwefel 1981 ; Herdy 1991) for ES.

Imagine an evolutionary algorithm (GA or ES) when, at some time, the population is uniquely constituted with TFT. In appearance, it expresses by a population of agent that always cooperate. Evolutionary algorithms working in a somehow blind mode, random mutations may always happen. For instance, an individual playing the strategy "Always Cooperate" (ALLC) may appear. Externally, this modification is not seenable since everybody continues to cooperate. So, the ALLC mutant realises the same score as TFT indigenous, and for this reason, he has the same probability to survive selection. Thus, it is theoretically possible that little by little the TFT population becomes invaded by ALLC, creating so a favourable ground for the later appearance of "Always Defect" (ALLD) individuals. We do not know what will then happen, it seems that the population must tend to an equilibrium point, or enter a cyclic dynamic. In all cases, it is highly improbable that the population still be composed only of TFT after a few iterations of the algorithm. So TFT is not stable under the action of GA and ES [7].

A definition of stability consistent with evolutionary algorithms is the following :

Definition 4. The decision i for game ν is EA-stable iff. :

$$\nu_{ii} > \nu_{ji} \quad \forall j \neq i$$

So, a strategy is EA-stable if it is in strict Nash equilibrium with itself, according to the definition :

Definition 5. (i, j) is a strict Nash equilibrium of game ν iff. :

$$\nu_{ij} > \nu_{i'j} \quad \forall i' \neq i, \quad \text{and} \quad \nu_{ji} > \nu_{j'i} \quad \forall j' \neq j.$$

It is be noted that our EA-stability is more restrictive than Boyd and Loberbaum's evolutionary stability (but this rough definition is enough to illustrate our purpose). So, we can deduce of their results that *no deterministic strategy is EA-stable in IPD*.

Although Axelrod's theoretical results may be put in the balance, he also proposes experimental results to support the thesis that cooperation based on benevolence, susceptibility and indulgence is a natural phenomenon. In (Axelrod 1987), he presents results of simulations where a population of 20 strategies for IPD is submitted to the action of a GA. Each one satisfies :

Hypothesis M_3. The choice of every decision only depends on the three last results with the same opponent (3 steps of memory),

and is encoded on a linear chromosome of length 70. This experiment concludes with the emergence of cooperation, due to the dominance of TFT and TFT-like individuals.

A lot of other studies have been laid to experiment the evolution of a population of IPD-strategies under the action of evolutionary algorithms. The main differences between them is the way to encode strategies, what directly determines the nature of the search space (Fogel 1993 ; Gacôgne 1994 ; Lindgren and Nordal 1994 ; Mühlenbein 1992). Most often, the authors succeed in having cooperation emerge. It seems to us that, because of its simplicity, Axelrod's result is still the most convincing. However, we think that the search space used in this study is too large beside of the population size to allow a deep understanding of the mechanisms in process.

---

[7] all this reasoning may be summarised in the following way : Axelrod's collective stability is consistent with a model of evolution where there is a cost imposed on mutation. This is not the case of GA and ES, where the calculus of the fitness of an individual does not take into account the fact that it is a mutant or an indigenous.

3 Our Approach

As we have seen in section 2, the decision space of IPD is totally out of the range of actual computers. We are concerned with the following of its consequences : a simulated population of strategy for IPD may only cover an insignificant part of the complete search space.This is a issue very important to keep in mind when we approach IPD and evolution.

Even if they restrict their study to a sub-set of the complete strategy space, previous works on evolution and IPD always use search spaces too large for being totally explored. For instance, Axelrod's sub-set of strategies satisfying hypothesis M_3 is of size $2^{70} \simeq 10^{21}$, and he uses a population of only 20 individuals. Moreover, the size of the search space seems to be one of the main limitations to the efficiency of evolutionary algorithms. The following appealing remark is made about the validity of D. E. Goldberg's Schema Theorem by Bäck and Schwefel (1993) :

> "...finite populations often do not contain all instances of a specific schema. Observed schema fitnesses thus might quite mislead the search process."

In order to be able to lead reliable simulations with population sizes not too low beside of the search space, we have chosen to limit ourselves to the set of the 32 strategies that satisfy :

Hypothesis M_1. the choice of every play only depends on the result of the last iteration played with the same opponent (1 step of memory).

It appears to us that this is the easiest way to arbitrarily restrict the search space to a very low sub-set of the set of all strategies, while preserving the symmetry of the original space. Moreover, most of the atypical strategies as TFT, ALLC and ALLD satisfy M_1. An arbitrary numbering for the strategies satisfying M_1 is proposed in table 1. It will be used all over the end of this paper.

When we limit ourselves to the strategies satisfying M_1, IPD becomes a 32-decisions game. Its 32×32 utility-matrix ν^{IPD} may then be easily calculated by simulating, on any computer, all the possible confrontations of two strategies, for a fixed length of interaction L. We can then verify that, as it is announced in section 2, there is no strategy that is EA-stable, whatever the value of L.

However, there may exist mixed situations (i.e. situations where different strategies cohabit) that constitute equilibrium points of the system. The analytical determination of them is possible by a solving 32-unknowns system, after having defined the dynamic of the system (cf. section 4). However we prefer to lead an experimental determination of them. Our idea is to try different evolution schemes on a population of strategies satisfying M_1.

Our work could seem to be "another" IPD-based simulation of evolution. However, the originality of our work lies in the fact that we can use population sizes greater than the search space, and then lead simulations where the observed behaviour of the algorithm is not too far from its expected behaviour. In the limit, we represent the population by a set $\{q_1, q_2, \ldots q_{32}\} \in [0; 1]^{32}$ such that :

$$\sum_{i=1}^{32} q_i = 1$$

(i.e. belonging to the simplex S_{32} of $I\!R^{32}$) where q_i represents the percentage of individual using strategy i in the population.

With this continuous approach to the population, we associate a deterministic handling of chance : all calculus are made by using the expected values of variables with respect to calculated probabilities, instead of by simulating random drawings, as it is the case in classical GA and ES. So, as a consequence of the weak law of large numbers, the continuous approach to the population simulates the theoretical behaviour of the model with an infinite population. It allows a very interesting look at the evolution of the population that would not have been possible in greater search spaces.

To present our model, we develop in section 4 a discussion about the question : "what is an evolutionary dynamic ?".

4 Artificial Evolution

The actual model of natural evolution is the following :

1. each individual metabolism is determined by its genetic code, the DNA, that is proper to him and present in all of its cells.
2. DNA is the support of heredity, it is transmitted from parents to children, on the condition of some random recombinations and mutations.
3. An individual metabolism determines its external characters, the environment selects the individuals possessing the characters the most favourable for survival and reproduction. This phenomenon of natural selection modifies the statistical distribution of genetic codes from generations to generations.

Evolutionary algorithms are based on the idea that there exist an intrinsic "optimising principle" in this scheme, and they try to reproduce it in purpose of optimisation. But, because they only modelise a part of the whole process, the mechanism processed is quite different.

GA basic principle is to replace the problem of optimising a numerical function f of some discrete set X, by the problem optimising a numerical function g of a certain set C of "codes" of X elements. The *Building Block Hypothesis* (Holland 1975 ; Goldberg 1989) states that we wait from the code to satisfy some special properties with respect to the objective function. In particular, this is needed that individuals obtained by recombination (at the code level) of good individuals, are also good individuals. Somehow, things may abstracted the following way : by defining notions of neighbourhood and distances, the recombination operator induces a topology on the code space C ; the function g is then supposed to have some properties close to continuity with respect to this topology.

The *Strong Causality Principle* , one of the two fundamental principles of ES (Hoffmeister and Bäck 1991 ; Herdy 1991), explicitly stipulates that a small variation of the phenotype induces a small variation of the objective (fitness) function. So, ESs directly use a property of continuity of the objective function. The need of a certain continuity of the fitness function in evolutionary mechanisms also appears in (Feistel and Ebeling 1989), that greatly inspired us for building our model. As these authors do, we modelise the evolutionary process with two operators :

mutation : a mechanism of random transformation of an individual into one of his neighbours, with respect to a certain topology. With infinite population a diffusion operator is used.

selection : a mechanism that determines the individuals able to survive to the change of generation, deterministically or stochastically, and in a way that favours the performance.

We tried several alternatives for the choice of the mutation topology and the selection operator. With these choices, we wanted to build a system not to far from GA and ES. However we do not argue that our scheme is an exact reproduction of all these algorithms. This particularly true in the case of GA, because of the absence of explicit recombination in our model.

Even if it does not always constitute a good model of evolutionary algorithms, we argue that our system is a not so bad model of evolution. In all cases it is very close to Feistel and Ebeling's one. Moreover, the surprising robustness of our observations (cf. section 5), allows to conjecture that the behaviour of the system would not have change if we had used an explicit recombinative operator.

There also exist an other point where our model is closer to Feistel and Ebeling's one than to evolutionary algorithms. In GA and ES, the fitness of an individual is constant and deterministically determined by the objective function f. In our model, the fitness f_i of strategy i depends on all the population, and so, it varies through time. Defining q_i as the percentage of individual using strategy i in the population, then we use :

$$f_i = \sum_{j=1}^{32} q_i \, \nu_{ij}^{\text{IPD}}.$$

The underlying hypothesis is explained in the next paragraph. We want to quote now that the use of a formula of type $f_i = f(q_1, q_2, \ldots q_N)$ to replace in Feistel and Ebelling's model and in ours, the less general formula $f_i =$ a constant $= f(x_i)$ of evolutionary algorithms, could be used to differentiate evolutionary computation as a particular field of a more general artificial evolution domain. The artificial evolution approach allows to really modelise evolving interactions, and so, is much closer to the process of natural selection than evolutionary computation models. Previous works on IPD and evolution like (Axelrod 1987 ; Fogel 1993 ; Lindgren and Nordal 1994) are artificial evolution models in the sense defined here. This also the case of artificial ecology models as (Werner and Dyer 1991).

By choosing

$$f_i = \sum_{j=1}^{32} q_i \, \nu_{ij}^{\text{IPD}} \, .$$

we implicitly made the hypothesis of no spatial distribution of the population. With a finite population, it formulates as :

Hypothesis NS. At each generation, every individual meets every other one once, its fitness is its average score over these meetings (no spatialization).

With infinite population, it formulates as :

Hypothesis NS'. At each generation, every individual meets a big number of its colleagues and its opponents are randomly drawn according to the probabilities (q_i).

5 Experimental Protocol

We give below an exhaustive list of all the parameters of the model. Readers desiring more information are invited to contact us.

IPD parameters : i.e. the utilities P, T, S and R, and the length of interactions L. They are used when the simulation begins, to calculate ν^{IPD} by simulating all possible confrontations.

Population size S : that may be infinite, as explained in section 3. It is interesting to see how large must the population be, to have the observed evolution of the system correspond with the expected evolution. We may so verify the accuracy of the quote from Bäck and Schwefel (1993) in section 3

Topology of mutation : we tried two alternatives :
- trivial topology : each strategy is the neighbour of every strategy ;
- structured topology : a neighbour of a given strategy i is obtained by flipping one decision in the rule representation as used in table 1. Thus every strategy has exactly 5 neighbours.

Selection operator : proportional or elitist :
- proportional selection is a GA selection operator. The probability of selection of strategy i is equal to :

$$\frac{q_i \, f_i}{\sum_{j=1}^{32} q_j \, f_j} \, ,$$

where f_i is the fitness of strategy i and q_i is the percentage of individuals using strategy i in the population.
- elitist selection is the selection operator used in $\lambda + \mu$-ES (Bäck and Schwefel 1993 ; Hoffmeister and Bäck 1991). Mutation creates new individuals and increases the population size. The selection brings it back to its initial value by suppressing the worst individuals first.

Temperature θ : the parameter that determines the magnitude of genetic mixing at each generation. With proportional selection, it is a probability of mutation, and with the elitist selection, it is to the ratio μ/λ.

Initial population : uniformly distributed or composed of a single strategy that may be chosen among the 32 possible ones.

With some special sets of value attached to different parameters, our system correspond to existing evolutionary algorithms :

- with elitist selection and trivial topology, our system is a $\lambda + \mu$-ES with no recombination and where $\lambda = S$ and $\mu = S\theta$.
- with proportional selection and structured topology, it is a GA with no crossover and a special mutation operator that works at the scale of individuals. The probability for an individual to mute is $p_m = \theta$.

The real-time needed to run a simulation is very short. The limit behaviour of the system (equilibrium or cycle) is always obtained in less than one minute on a good PC. So, we could try a very large variety of parameters and totally explore the model. As a consequence, an exhaustive description of the totality of our observations is impossible in a small paper like this one. That is why we have chosen to present our results in a descriptive and non-rigorous form. Some representative graphics are added to illustrate our purpose, they are gathered at the end of the paper.

6 Experimental results

This section is divided into two parts. The first describes the behaviour of the model with infinite population. Of course, this behaviour may be obtained with finite but large populations. We develop this point in the second part, that constitutes a description of the influence of all the parameters listed in the previous section.

To describe our observations with infinite population, we also put apart the case where elitist reproduction and structured topology are used together. As a matter of fact, *excepted for this special case, the population always converges to an equilibrium.* Moreover, we verify that the composition of the initial population has no influence on the state of the system at equilibrium. That seems to indicate that *this is the only stable equilibrium of the model.*

So, with proportional selection or/and trivial topology, there exists a unique equilibrium of the model dynamics. The "quality" of the equilibrium is well measured by the average score of individuals over the population when it is reached. Figure 1 represents the evolution of this data as a function of the temperature θ, with proportional and elitist selection, and trivial topology of mutation. This is to be noted that, although, they have been drawn with some special values for the parameters, the two curves have a high level of generality and similar results are obtained with other sets of parameters.

In fig.1 we verify that *the average score at equilibrium tends to 100% of cooperation when θ tends to zero* (no mutation at all). With proportional selection, this value is reached in zero, but nowhere else. In the case of elitist selection, there is a discontinuity in zero, where the score at equilibrium is the one of a uniformly distributed population. This is due to the fact that the $\lambda + \mu$-ES model implemented with elitist selection, does not affect the population in any way when $\mu = \theta S = 0$. Thus the population at equilibrium is the same as the initial population, i.e. uniformly distributed.

We also see in fig.1 that the equilibrium quality rapidly decreases when the temperature raises. In the case of proportional selection, we tend to a random game when the temperature tends to 1. This is not surprising since the GA model behaves almost as a random algorithm when $p_m = 1$. With elitist selection, there is a fall from 100% of cooperation to 100% of defection around $\mu = 2\lambda$. This is a typical instance of the brutal behaviour of the ES model due to its extremist way to carry on selection with the max operator.

We are now interested in the composition of the population at equilibrium. Figure 2 gives two very representative examples with low temperature and so, high level of cooperation at equilibrium. In these barcharts, it appears that strategy 18 is always dominating at equilibrium. Other experiments show that, *each time that cooperation emerges, 18 eventually dominates*. Referring to table 1, we may explicit the behaviour of 18 in the following way : always cooperate until the first defection of the opponent, and then, always defect until the end of the game. Thus, it is an ALLC that turns to ALLD at the first defection of the opponent. So it possesses as TFT the properties of benevolence and susceptibility (cf. section 2), but unlike TFT, it has no indulgence at all. For this reason, and following J. Maynard-Smith's inspiration, we call it "the Retaliator" (RET).

To understand the role of RET in the emergence of cooperation, we must look at the evolution of the population from $t = 0$ to the equilibrium. To present our results, we focus on four strategies that play a primordial role in the phenomenon. They are :

- Always Defect (ALLD) : numbers 1, 2, 3 and 4 ;
- Always Cooperate (ALLC) : numbers 20, 24, 28 and 32 ;
- Tit for Tat (TFT) : number 22 ;
- the Retaliator (RET) : number 18.

Figures 3 and 4 represents the evolution of the proportion of these strategies, and of the average score over the population, starting from a uniformly distributed distribution. In both case, and also in all the other experiments that we laid, the scenario leading to cooperation is the same.

At time zero, when the population is uniformly distributed, the malevolent strategies as ALLD realise the best score. They are so the first to grow, at the expense of dumb benevolents as ALLC that quickly disappears. In the uniformly distributed initial population, the susceptible benevolents as RET and TFT do better than ALLC because they avoid being exploited by malevolents, but worse than ALLD because they do not exploit non-susceptible benevolents. So, there quickly happens a situation very different from the initial one, where defection dominates, the population being mainly composed with ALLD, and a little of RET and TFT.

When opposed to itself ALLD realises the poor score of permanent mutual defection, i.e. LP. When opposed to RET and TFT, it gains a very little more because it exploits the opponent at the first iteration, its score is then $T + (L-1)P$. During this confrontation, the benevolent RET or TFT realises the lesser, but close, score $S+(L-1)P$. It is when two benevolent are opposed that a significant difference appears. Two benevolent realises the score LR, that is the best over the population. We verify that this advantage is enough to have benevolent recover the delay that they gain in front of malevolents. So, this is now the susceptible benevolents that grow, and especially the unforgiving RET.

Because they have destroyed their spring of reward constituted by the ALLC population, the malevolents now decrease at great speed. RET continues to grow and achieves a very wide majority of the population, establishing cooperation. A population almost entirely composed with RET is a very stable equilibrium of the system. Other benevolents strategies may sometimes cohabit. Because of the behaviour of RET, everything seems to them as if they were in an ALLC population. From the point of view of malevolents, the population is mainly composed with individual behaving almost as ALLD. So, they realise the poor score of mutual defection and may not survive. RET is still lightly the best, because he avoids the total rout sustained by ALLC when they are opposed to the rare malevolents.

By this mechanism, the combination of benevolence and susceptibility (cf. section 2) of RET guarantees the stability of cooperation. RET and TFT are the only two strategies that are at the same time benevolent and succeptible [8]. Finer studies are needed to understand why TFT does not play a similar role, but it is a fact that simulations always show an advantage in RET intransigence [9]. Some experiments were laid by preventing the creation of RET individuals, reducing so the search space to the other 31 strategies. We saw then that TFT takes the place of RET, although it takes more time to have cooperation emerge. Moreover cooperation does not emerges, if we suppress RET and TFT from the search space. This confirm the primordial importance of benevolent and susceptible strategies in this phenomenon.

All the results presented above are obtained with proportional selection or trivial topology of mutation. In the special case where elitist selection is associated with structured topology, the scenario is almost the same, excepted that instead of an equilibrium, it is a cyclic oscillation between TFT, RET and sometimes other benevolents, that is the attractor of the dynamic.

For the sake of completeness, we conclude this section with the description of the influence of all the parameters (listed in section 5). It is striking to see that most of them have only a weak influence on the behaviour of the system.

---

[8] benevolence traduces by the schemata $(\emptyset \mapsto C)$ and $(R \mapsto C)$, what leaves 8 benevolent strategies (even numbers between 18 and 32). Susceptibility implies $(S \mapsto D)$ and $(P \mapsto D)$ and so, only two strategies may be benevolent and susceptible. TFT's indulgence is the schema $(T \mapsto C)$ replaced by $(T \mapsto D)$ in RET.

[9] a possible explanation is the better score realised by RET in front of strategy 16 that always cooperate excepted at the first iteration.

IPD parameters : We tried two sets of value for (T, R, P, S) that are $(3, 2, 1, 0)$ and the more classical $(5, 3, 1, 0)$, and a very wide variety of values for L (including odd, even and prime numbers), without a significant modification of the observations. However it is possible to change the behaviour of the system with extreme values for these parameters. The combination

$$S \ll P, \quad P \simeq R, \quad R \ll T$$

and a short length L (e.g. $(T, R, P, S) = (41, 21, 20, 0)$ and $L = 11$), is unfair with susceptible benevolents and stop the scenario at its first stage, when malevolence dominates (cf.previous discussion about the rise of RET and the fall of ALLC).

Population size S : Until now, we have only described the theoretical behaviour of the model with infinite population. As we explained in section 3, it is obtained by representing the population under the form of a set of real numbers $\{q_i\} \in S_{32}$. Of course, the observed behaviour of the system with finite population tends to it when the population size grows. The size needed to have convergence varies with the model used, GA or ES. With elitist selection, a population of 8 individuals, i.e. 25% of the search space, is enough to have the system evolve accordingly to its theoretical behaviour. That is a quite good result when compared to the proportional selection performances. As a matter of fact, a conformable behaviour of the GA model is not obtained until a population size by 96 individuals, i.e. 300% of the search space. We may understand that the proportional selection, because it is somehow more subtle than the rough Max operator of elitist selection, needs bigger populations, but the observed difference was unexpected and strongly confirms the quote from Bäck and Schwefel (1993) in section 3. However, this very low observed performance of the GA model may be explained by the absence of any recombinative operator as crossover, and thus of a real schema processing that is supposed to be the heart of GA. As we argue in section 7, we believe that the use of crossover may reduce the population size needed, but that it would not change the qualitative behaviour of the system.

Topology of mutation : The results presented in the graphics were all obtained with trivial topology of mutation, i.e. when every strategy may be created from every strategy by mutation. As we explained, the use of structured topology with elitist selection modify the behaviour of the model in the fact that the attractor is not an equilibrium anymore, but it becomes a cycle. With proportional selection, the use of structured topology instead of trivial topology has almost no influence, it just modifies a little the distribution at equilibrium.

Selection operator : The difference between the two alternatives for selection appears clearly on the graphics. The GA'proportional selection is a smooth operator that induces continuity in the behaviour of the model. On the contrary, the ES'elitist selection, because it uses the max operator, has a brutal behaviour leading to discontinuity and non derivability of observable statistics. It is also to be noted that elitist selection realises exact optimisation for a large variety of parameters, but when it does not succeed, it falls an the opposite extreme and realises very low performances.

Temperature θ : As explained in the previous discussion, the temperature is a major parameter that determines all the behaviour of the system.

Initial population : Excepted for very special case, we did not see any influence of the initial distribution of strategies in the population. The attractor of the dynamic, equilibrium or cycle, seems always to be the only existing one.

7 Conclusion

We are now interested in the range of our results. We would like to say that :

> If the temperature of the genetic melting is not too high and the population size not too low, cooperation naturally emerges under the action of an evolutionary dynamic. Benevolent but susceptible individuals play a major role in this phenomenon and they ensure the stability of cooperation ;

and so, modify slightly Axelrod's thesis by giving no advantage to indulgence.

The first objection that may be raised is that our system does not represent the generality of evolutionary dynamics, in particular it does not use any recombinative operator as GA'crossover. It seems to us that this critic is well founded with regard to the conclusion that we derived about the algorithm behaviour in general. In particular the observed requirement for a large population size in the GA model (i.e. with proportional selection), may be explained by the absence of crossover and so, of a real schema processing that is supposed to be the heart of a GA. However, we claim that, with regard to the problem of the emergence of cooperation, the use of other genetic operators would not have change the behaviour of the system in a significant way. To support this conjecture, we argue that our system exhibits a surprising robustness, its behaviour not being altered by a change of the selection operator or of the mutation operator. So, we may expect the observed scenario of the emergence of cooperation to be quite general, and valid for every evolutionary dynamics. We are now improving our software by introducing a crossover. We are also putting in place a real implementation of hypothesis NS' (see the end of section 4), that now applies also with finite populations and so replaces hypothesis NS.

The other restriction to the generality of our results is that all our conclusions were derived by restricting ourselves to the small set of strategies that satisfy hypothesis M_1. Although IPD is surely a reliable model for a lot of field of study, its restriction to 32 strategies is only a small game of dubious interest. However, as we explained in section 3, our motivations were to check the validity of the Axelrod 's results with large populations, and it is not more arbitrary to restrict the search space with hypothesis M_1 than with hypothesis M_3. We think that the possibility to cover all the search space, and thus be able to see when the algorithm converge and when it does not, constitute the main interest of our work, that reinforce Axelrod's one.

It is also striking to see how the observed behaviour of the model, i.e. :

1. growth of malevolents and fall of dumb benevolents,
2. raise of susceptible benevolents and fall of malevolents,
3. establishing of cooperation,

is close to the behaviour of the "human" set of strategies of Axelrod's tournaments (Axelrod 1984). It is a pity that no result is provided in (Axelrod 1987), we may wander about the scenario of the raise of cooperation in Axelrod's GA. It seems that the phenomenon that we observe has a certain level of generality, at least when the symmetry of the search space is preserved, as it is the case with M_1 and M_3.

Because of this strong property, IPD may surely be called an "evolution-easy" game. It is tempting to conclude that evolution is a very powerful dynamic that may solve strong cooperation problems. But, if we keep in mind the definition of strong cooperation problems sketched in section 1, i.e. games where Nash equilibriums are strictly dominated by diametrically opposed situations, we are not sure that IPD is a strong cooperation problem. On the contrary, a study that we laid shows us that the situation is not so simple in our 32-alternatives version of IPD. Thus we may wander about the behaviour of evolving populations in strong cooperation problems bigger than the 2-decisions PD. In all cases, it is nice to see that by iterating it, PD becomes an evolution-easy game.

Our theoretical study of IPD and the results of simulations with strong cooperation problems will be the subject of a latter publication. We believe that it is highly improbable that evolution will lead to optimality in a 50-decisions strong cooperation game. If it turns out to be true, no general conclusion about the power of evolutionary dynamics could be derived. Then our future works would be to apply our approach of evolution with large populations, to other not too big paradigmatic games, and find again the nice point of view on evolution that we had with IPD.

References

Aumann R.J. and Hart S. : *Handbook of Game Theory*, North Holland, 1992.

Axelrod R. : Effective choice in the prisoner's dilemma game, *J. Conflict Resolution*, vol. **24**, pp. 3-25, 1980.

Axelrod R. : *The Evolution of Cooperation*, Basic Books, New York, 1984.

Axelrod R. : The evolution of strategies in the iterated prisoner's dilemma, in *Genetic Algorithms and Simulated Annealing*, R.Davis ed., Pitman, London, 1987.

Bäck T. : Self-adaptation in genetic algorithms, *Towards a practice of autonomous systems : Proceedings of the 1st European Conference on Artificial Life*, F. Varela and P. Bourgine eds., MIT press, Cambridge, pp. 263-271, 1992.

Bäck T. and Hoffmeister F. : Extended selection mechanisms in genetic algorithms, *Proceedings of the Fourth International Conference on Genetic Algorithms and their Applications*, R.K. Belew and L.B. Booker eds., Morgan Kaufmann, San Diego, pp. 92-99, 1991.

Bäck T. and Schwefel H.P. : An overview of evolutionary algorithms for parameter optimization, *Evolutionary Computation*, vol. 1, pp.1-23, 1993.

Baefsky P. and Berge S.E. : Self-sacrifice, cooperation and aggression in women and varying sex-role orientation, *Personality and Social Psychology Bulletin*, vol. 1, pp. 296-298, 1974.

Bethlehem D.W. : The effect of westernization on cooperative behaviour in Central Africa, *Int. J. Psychology*, vol. **10**, pp. 219-224, 1975.

Boyd R. and Loberbaum J.P.: No pure strategy is evolutionarily stable in the repeated prisoner's dilemma game, *Nature*, vol. **327**, pp. 58-59, 1987.

Feistel R. and Ebeling W. : *Evolution of Complex Systems : self-organisation*, entropy and development, Kluwer Academic Publishers, Dordrecht, 1989.

Fogel D.B. : Evolving behaviours in the iterated prisoner's dilemma, *Evolutionary Computation*, vol. 1, pp.77-97, 1993.

Gacôgne L. : *Apprentissage génétique d'une extension flou du dilemme itéré du prisonnier*, Rapport n. 94/15, LAFORIA, Université Paris 5, Paris, 1994.

Goldberg D.E. : *Genetic Algorithms in Search, Optimization and Machine Learning*, Addison Wesley, Reading, 1989.

Herdy M. : Application of Evolutionsstrategie to discrete optimization problems, in *Parallel Problem Solving from Nature*, 1st workshop, H.P. Schwefel and R. Männer eds., Springer, pp. 188-192, 1991.

Hoffmeister F. and Bäck T. : Genetic Algorithms and Evolution Strategies : similarities and differences, in *Parallel Problem Solving from Nature*, 1st workshop, H.P. Schwefel and R. Männer eds., Springer, pp. 455-469, 1991.

Holland J. : *Adaptation in natural and artificial systems*, Ann Arbor : the University of Michigan press, Cambridge, 1975.

Lindgren K. and Nordal M.G. : Cooperation and community structure in artificial ecosystem, *Artificial Life*, vol. 1, pp. 15-37, 1994.

Luce R. and Raiffa H. : *Games and Decisions*, John Wiley and sons, New York, 1957.

Maynard-Smith J. : *Evolution and the Theory of Game*, Cambridge University Press, Cambridge, 1975.

Mühlenbein H. : Darwin's continent cycle theory and its simulation by the Prisoner's Dilemma, in *Towards a practice of autonomous systems : Proceedings of the 1st European Conference on Artificial Life*, P. Bourgine and F. Varela eds., MIT press, Bradford Books, pp. 236-244, 1992.

Riker W. and Brams J.S. : The paradox of vote trading, *American Political Science Review*, vol. **67**, pp. 1235-1247, 1973.

Schelling C.T. : *The Strategy of Conflicts*, Harvard University Press, Harvard, 1960.

Rechenberg I. : *Evolutionsstrategie : Optimierung technischer Systeme nach Prinzipien der biologischen Evolution*, Frommann-Holzboog, Stuttgart, 1973.

Schwefel H.P. : *Numerische Optimierung von Computer-Modellen mittels der Evolutionsstrategie*, Interdisciplinary Systems Research, Birkhäuser, Basel, 1977.

Schwefel H.P. : *Numerical optimization of computer models*, John Wiley and sons, New York, 1981.

Werner G. and Dyer M. : Evolution of communication in artificial organisms, in *Artificial Life 2*, Addison Wesley, pp. 659-687, 1991.

Table 1. The 32 strategies for IPD that satify M_1.

1 : **ALLD**	2 : **ALLD**	3 : **ALLD**	4 : **ALLD**
$\emptyset \mapsto D$	$\emptyset \mapsto D$	$\emptyset \mapsto D$	$\emptyset \mapsto D$
$P \mapsto D$	$P \mapsto D$	$P \mapsto D$	$P \mapsto D$
$T \mapsto D$	$T \mapsto D$	$T \mapsto D$	$T \mapsto D$
$S \mapsto D$	$S \mapsto D$	$S \mapsto C$	$S \mapsto C$
$R \mapsto D$	$R \mapsto C$	$R \mapsto D$	$R \mapsto C$
5 :	**6 :**	**7 :**	**8 :**
$\emptyset \mapsto D$	$\emptyset \mapsto D$	$\emptyset \mapsto D$	$\emptyset \mapsto D$
$P \mapsto D$	$P \mapsto D$	$P \mapsto D$	$P \mapsto D$
$T \mapsto C$	$T \mapsto C$	$T \mapsto C$	$T \mapsto C$
$S \mapsto D$	$S \mapsto D$	$S \mapsto C$	$S \mapsto C$
$R \mapsto D$	$R \mapsto C$	$R \mapsto D$	$R \mapsto C$
9 :	**10 :**	**11 :**	**12 :**
$\emptyset \mapsto D$	$\emptyset \mapsto D$	$\emptyset \mapsto D$	$\emptyset \mapsto D$
$P \mapsto C$	$P \mapsto C$	$P \mapsto C$	$P \mapsto C$
$T \mapsto D$	$T \mapsto D$	$T \mapsto D$	$T \mapsto D$
$S \mapsto D$	$S \mapsto D$	$S \mapsto C$	$S \mapsto C$
$R \mapsto D$	$R \mapsto C$	$R \mapsto D$	$R \mapsto C$
13 :	**14 :**	**15 :**	**16 :**
$\emptyset \mapsto D$	$\emptyset \mapsto D$	$\emptyset \mapsto D$	$\emptyset \mapsto D$
$P \mapsto C$	$P \mapsto C$	$P \mapsto C$	$P \mapsto C$
$T \mapsto C$	$T \mapsto C$	$T \mapsto C$	$T \mapsto C$
$S \mapsto D$	$S \mapsto D$	$S \mapsto C$	$S \mapsto C$
$R \mapsto D$	$R \mapsto C$	$R \mapsto D$	$R \mapsto C$
17 :	**18 : RET**	**19 :**	**20 : ALLC**
$\emptyset \mapsto C$	$\emptyset \mapsto C$	$\emptyset \mapsto C$	$\emptyset \mapsto C$
$P \mapsto D$	$P \mapsto D$	$P \mapsto D$	$P \mapsto D$
$T \mapsto D$	$T \mapsto D$	$T \mapsto D$	$T \mapsto D$
$S \mapsto D$	$S \mapsto D$	$S \mapsto C$	$S \mapsto C$
$R \mapsto D$	$R \mapsto C$	$R \mapsto D$	$R \mapsto C$
21 :	**22 : TFT**	**23 :**	**24 : ALLC**
$\emptyset \mapsto C$	$\emptyset \mapsto C$	$\emptyset \mapsto C$	$\emptyset \mapsto C$
$P \mapsto D$	$P \mapsto D$	$P \mapsto D$	$P \mapsto D$
$T \mapsto C$	$T \mapsto C$	$T \mapsto C$	$T \mapsto C$
$S \mapsto D$	$S \mapsto D$	$S \mapsto C$	$S \mapsto C$
$R \mapsto D$	$R \mapsto C$	$R \mapsto D$	$R \mapsto C$
25 :	**26 :**	**27 :**	**28 : ALLC**
$\emptyset \mapsto C$	$\emptyset \mapsto C$	$\emptyset \mapsto C$	$\emptyset \mapsto C$
$P \mapsto C$	$P \mapsto C$	$P \mapsto C$	$P \mapsto C$
$T \mapsto D$	$T \mapsto D$	$T \mapsto D$	$T \mapsto D$
$S \mapsto D$	$S \mapsto D$	$S \mapsto C$	$S \mapsto C$
$R \mapsto D$	$R \mapsto C$	$R \mapsto D$	$R \mapsto C$
29 :	**30 :**	**31 :**	**32 : ALLC**
$\emptyset \mapsto C$	$\emptyset \mapsto C$	$\emptyset \mapsto C$	$\emptyset \mapsto C$
$P \mapsto C$	$P \mapsto C$	$P \mapsto C$	$P \mapsto C$
$T \mapsto C$	$T \mapsto C$	$T \mapsto C$	$T \mapsto C$
$S \mapsto D$	$S \mapsto D$	$S \mapsto C$	$S \mapsto C$
$R \mapsto D$	$R \mapsto C$	$R \mapsto D$	$R \mapsto C$

Fig. 1. Quality of the equilibrium point as a function of the temperature θ,
note that, in the case of elitist selection, the fall from 100% of cooperation to 100%
of defection is at $\theta \simeq 2.06$, and the stage at 100% of defection continues over $\theta = 100$
(experience laid with $(T, R, P, S) = (3, 2, 1, 0), L = 30$, infinite population and trivial
topology).

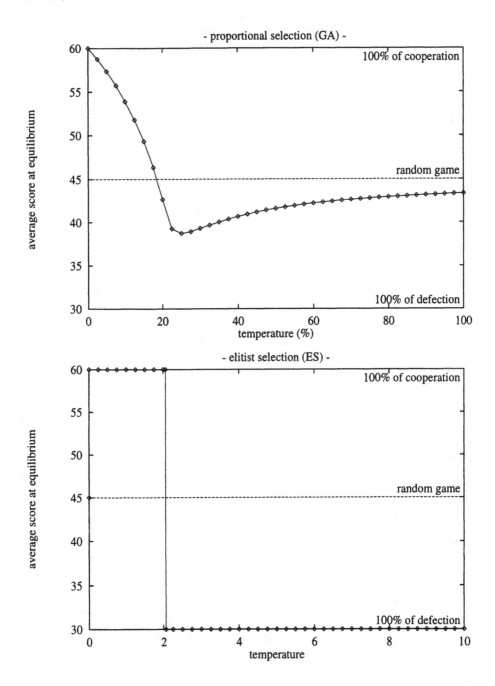

Fig. 2. Population at equilibrium, with proportional selection $\theta = 10\%$ and the average score at equilibrium is 53.89, with elitist selection $\theta = 1$ and the average score at equilibrium is 60 (100% of cooperation) (experience laid with $(T, R, P, S) = (3, 2, 1, 0)$, $L = 30$, infinite population and trivial topology).

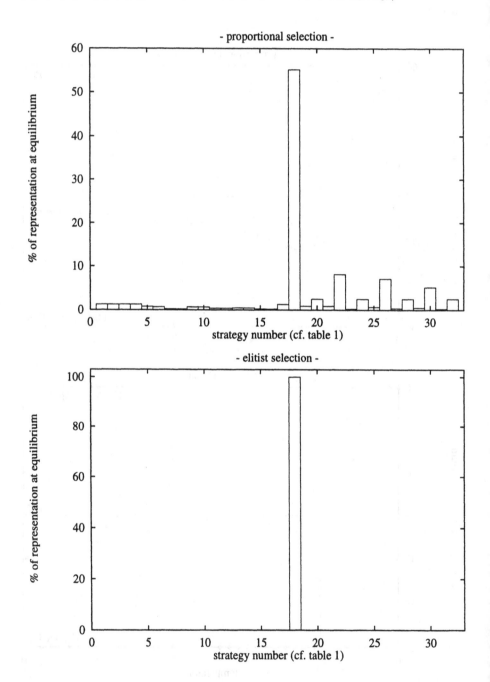

Fig. 3. Evolution of the population through time, starting from a uniformly distributed population and using proportional selection, the average score at equilibrium is 58.76 (experience laid with $(T, R, P, S) = (3, 2, 1, 0), L = 30$, infinite population, $\theta = 2.5\%$ and trivial topology).

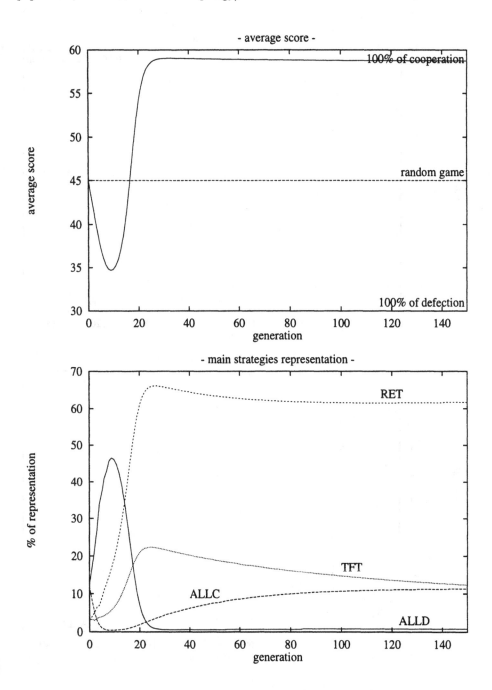

Fig. 4. Evolution of the population through time, starting from a uniformly distributed population and using elitist selection, the average score at equilibrium is 60 (100% of cooperation) (experience laid with $(T, R, P, S) = (3, 2, 1, 0)$, $L = 30$, infinite population, $\theta = 0.75\%$ and trivial topology).

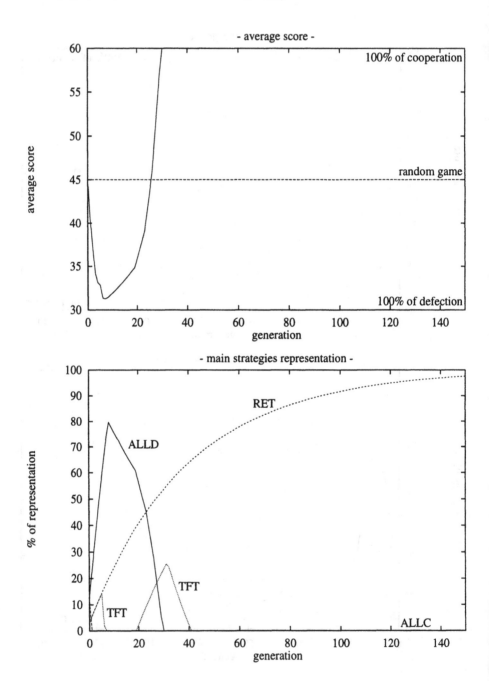

Immune System and Fault-Tolerant Computing

S. Xanthakis[1], S. Karapoulios[1], R. Pajot[1] and A. Rozz[2]

1 : LOGICOM,
Innopolis - Hall B - BP 418
31314 LABEGE CEDEX, FRANCE

2 : VERILOG
150, rue N. Vauquelin
31106 TOULOUSE, FRANCE

Abstract

Immune system (IS) is capable of evolving, learning, recognising and eliminating foreign molecules which invade organisms. Fault tolerant approaches consist in detecting erroneous states of an algorithm and executing recovery procedures. They are generally implemented by adding formal properties to be satisfied by state variables during execution. However, defining fault-tolerant procedures can be error-prone and sometimes equivalent to a formal proof. Moreover, it is always very difficult to make a sound assumption about the typology and frequency of real. We propose an analogy between IS and fault tolerant computing. After a brief presentation of immune algorithms and especially lymphocyte simulation (B-cells, suppressor T-cells, etc.) using genetic operators, we present an immune model for detecting and recovering erroneous states during execution. Program states are examined by procedures playing the role of B-cells. Those properties have previously been learnt, automatically, during testing phase using mutation analysis techniques. When a program state has to be recovered, T-cells procedures are activated in order to recover erroneous states. In this way, each software algorithm may develop automatically its own immune system capable of evolving and stimulating specific responses to software failures. We finish by a brief discussion of the main possibilities of an immune approach to fault-tolerant computing.

Keywords : Fault-Tolerant computing, Immune System, Genetic Algorithms, Mutation Testing, Error-recovery.

1 Introduction

The immune system (IS) has always astounded by its capability to recognise and eliminate molecules or microbes that invade an organism. Among the characteristics of the, there is the function of differentiation of self (molecules that belong to the organism itself) from not-self (external aggressors such as the various antigens or microbes, as well as cells belonging to the organism that have become malignant). Once it has been recognised, the not-self must be eliminated. These two detection and elimination functions, involving complex cellular communication mechanisms, now constitute a very active line of research in immunology and have not yet been completely elucidated. Among the cells actively taking part in IS functioning there is a group of cells called lymphocytes which, according to their function, are classified in two categories: B-cells and T-cells which both come from the same strain cell which has undergone differentiation at the level of the bone marrow. The role played by these two types of cell is not the same:

- The T-cells play an important part in the fight against viruses. They are capable of recognising cells that have been infected by viruses, attaching themselves to them and killing them. An infected cell can, nevertheless, only be recognised by a T-cell if it has previously been marked as belonging to the not-self by a B-cell. The destructive action of T-cells can only be effective therefore in cooperation with the other cells of the IS. The secret of the IS's success lies in this principle of division of tasks and cooperation.

- The first main function of B-cells is therefore to recognise foreign molecules (called *antigens*). This process must be selective , i.e. it must not recognise normal cells that are part of the organism that the IS is supposed to be defending. In order to acquire this recognition capability lymphocytes undergo, throughout embryo gestation, a learning phase. The second function of B-cells is to neutralise antigens. To achieve this, each B-cell has a certain number of receptors on its surface called *antibodies*, capable of recognising, according to a chemical affinity criterion, the proteins present on the surface of the antigens. (Insofar as each lymphocyte has an identical number of antibodies, no particular distinction will be made between the notion of antibody and that of B-cell).

The first consequence of an antigen being recognised by an antibody, is an alteration to the chemical functions of the antigen and, therefore, an inhibition of the latter's infectious nature. But a second, and equally important, consequence is the stimulation (i.e. the proliferation) of the lymphocyte that carried that antibody, which implies an increase in the number of free antibodies secreted in the organism. We call this phenomenon immune response which is followed by the rapid decline of the antigens that initially activated it. By considering this mechanism, the following questions can be asked, the answers to which will be very useful when it comes to determining a software fault tolerance model based on immunology.

1) How does the IS manage to have antibodies capable of recognising all the antigens — and this represents a considerable number — that may invade the organism during the organism life?

2) How is lymphocytes proliferation controlled by the IS while keeping an immune memory that will allow the organism to respond more efficiently to a second infection of the same type?

One of the first answers to be proposed to question 1) considered the IS as a sort of mold on which each antigen, as it passed, left a print which was memorised for ever by the organism. This hypothesis was, unfortunately, found to be false. It has been demonstrated that the IS is capable of reconstructing any antibody by reshuffling the host DNA that codes the antibody genes. Thanks to a set of genetic operators (such as mutation or crossover of sequences of genes), lymphocytes are created randomly, and proliferate according to their ability to recognise antigens. This process could be compared with a natural selection process. But the cells created are not immortal. At all times, a great part of the IS's lymphocytes has to be replaced. The antibodies that have not been sufficiently activated, disappear through a sort of "immune forgetting" mechanism. The IS must nevertheless be capable of memorising (regenerating) the antibodies that have been found to be efficient while avoiding excessive proliferation. This brings us to the second question which concerns the self-regulating aspect of the IS.

A possible solution to problem 2) would be to suppose that the organism keeps the various antigens encountered, in a limited number, and that they periodically stimulate the B-cells. However, several experiments have indicated that this hypothesis is not biologically viable. Among the various self-regulation models, Jerne's [JERNE74] idiotypic network model would seem to impose itself not only due to the elegance of its explanation but also due to the various items of experimental evidence corroborating it.

Jerne's hypothesis is mainly based on the fact that, schematically speaking, an antibody consists of two structures: the paratope and the epitope. Paratopes are responsible for stimulating the lymphocyte activated when an antigen is recognised. Epitopes, on the contrary, give a passive role on the antibody allowing it, in turn, to be recognised and therefore to be inhibited, by another antibody. In other words, the epitope of an antibody A can be recognised by the paratope of another antibody B whose epitope can be recognised by the paratope of another antibody C, and so on. The set of antibodies forms an internal network (the idiotypic network) whose vertices represent the antibodies and the arcs express the recognition relationship. Within this network there are cycles ensuring the dynamic stability of the system.

To illustrate this phenomenon, let us suppose that an antibody is represented by the triplet: (epitope, name, *paratope*). Let us also suppose that the paratope and the epitope are symbolised by integers and that the affinity relationship between them is expressed by the fact that the sum of their codes is equal to 10. With these conventions we can represent, quite schematically, the formation of the following chain:

$$(3, A, 9) \rightarrow (1, B, 8) \rightarrow (2, C, 6) \rightarrow (4, D, 7)$$

In this chain, antibody A is stimulated by its paratope **9**, and inhibits antibody B whose epitope 1 gives a sum equal to 10. But inhibition of antibody B is counterbalanced by another stimulation due to the fact that it, in turn, inhibits antibody C, etc. It can also be noted that at the end of the chain, antibody D in turn inhibits antibody A, thus forming a dynamically stable chain.

Let us now suppose that an antigen X, coded 4, invades the organism. This antigen will be recognised by antibody C, whose paratope is **6**. This stimulation of C will in turn cause stimulation of B, and then of A. In other words, the previous chain will be disturbed and will only return to its stable state when the initial antigen has been eliminated. Once the immune response has been stabilised, it can be understood that antibody C (which recognised the antigen) will not disappear since it is an integral part of the chain's equilibrium. By taking this reasoning a step further, antibody C will not disappear since it is perpetually stimulated by the presence of antibody D. Antibody D (whose epitope is also equal to 4, like antigen X) therefore plays the part of *internal image* of antigen X. In order to react efficiently to an infection caused by antigen X, the IS does not have to physically keep the antigen in question, as was initially supposed. It just has to keep an internal image of this intruder. This internal image is materialised in the form of an antibody that keeps all the pertinent information relative to antigen X (thus allowing a rapid response in the case of re-infection), while being perfectly inoffensive.

We can conclude that the intrusion of an antigen, let's say Y, in the organism will not only result in a proliferation of the corresponding antibodies, but will also have a secondary effect: the creation of an internal image of Y that the IS will keep in memory of that infection. It is surprising to note that the self-regulating properties of the network allow us to suppose that there exist virtual internal images in the sense that they do not correspond to any antigen that has already been encountered. Let us take, for example, the antibody (1, B, **8**). It may be that its presence is only justified by regulatory needs. Which means that it is possible that the IS may keep an internal image of antigens whose code is 1 (such as epitope B) even though it has never encountered such antigens.

It is obvious that the real structure of the idiotypic network is more complex. Several chains can be interlaced, which allows a more sophisticated cooperation between the antibodies in the presence of a group of different antigens (which often occurs in the case of microbial infection).

Various works have been accomplished with a view to either using the IS self-organising paradigm for controlling processes [MAXION90], [BERSINI92] or to verifying, by simulation, the validity of certain immunological hypotheses. In [FARMER85], the dynamics of a set of antibodies (made up of an epitope and of a paratope coded in binary form) have been simulated. The recognition of an epitope by a paratope of another antibody has been achieved by means of a distance function between two binary representations. The process of generating new antibodies has been accomplished by a genetic algorithm (whose characteristics are presented later) which made it possible, by natural selection, to generate antibodies that were increasingly adapted to responding to a set of antigens: in this case we are talking of *immune algorithms.*

The authors have also shown that, even in the absence of antigens, the creation of antibody self-regulating chains occurs spontaneously. In a relatively simple model, it is therefore possible to reveal the recognition, memorisation and immune forgetting functions that are found in the real IS. Their dynamic model has subsequently been refined [FARMER90] where the authors underlined that, thanks to the antibody network created, the immune algorithms would be capable of simulating sophisticated Boolean functions or of accomplishing learning tasks that it would be very difficult to contrive through a deterministic approach.

The purpose of the model that we describe later is to apply the immunological paradigm to a strict perspective of software fault tolerance. Our goal is not to use an algorithmic scheme to simulate a biological phenomenon, but the opposite: to apply a biological scheme to a software system with a view to increasing the latter's reliability. This adaptation will not be accomplished with a strict preoccupation for faithfully reproducing the biological reality, but with the practical preoccupation for applicability.

2 Program Fault Tolerance Techniques

The general framework of Program Fault Tolerance (PFT) that governs our work is a very classic one [AGRAWAL85], [MILI90]. It consists in linking the program to algorithmic processes that ensure two main functions:

1. *Recognition function*: Capability to distinguish between correct states and erroneous states.
2. *Recovery function*: Recover (whenever possible) erroneous states.

The fault tolerance approach is therefore different (but complementary) to the classic test approach. In a test approach (path testing, mutation testing, etc.) the goal is to ensure that certain classes of fault (that are considered to be representative) do not exist in the software tested. In the software fault tolerance approach it is considered that, after all, faults are unavoidable and must be tolerated.

Generally speaking, a program aims to implement a specified function (for example, calculate the sum of two integers x and y and assign it to y). It can be considered as a set of primitives (the algorithm's instructions) that transforms an initial state (the values of x and y) into a final state (the new values of x and y, once the algorithm has terminated).Throughout this process of successive transformations, the algorithm passes through a series of intermediate steps that must verify certain logic conditions, failing which the final state might not be correct.

In order to represent the state in which a program is placed when it has executed a precise instruction (marked by a *label*), the values of the parameters used at that precise moment are tracked. Since an instruction can be executed several times (in the case of an iterative instruction, for instance), an integer, say n, representing the number of times that specific label was visited, is added to the state information (the pair: (label, n) is often called a *milestone*). For instance for a program that handles two parameters a and b, the state S = <13, 7, 45, 89> means, for example that when label No. 13 has been visited for the 7th time, variables a and b respectively took values 45 and 89. Schematically speaking, a state is considered to be strictly correct (*s-correct*) at a precise milestone, and with respect to an initial state, if it meets all the specified and designed logical constraints allowing the successive transformation of the given initial state. On the other hand, a state is considered to be loosely correct (*l-correct*) at a precise milestone and with respect to a given initial state, if later execution of the algorithm allows us to reach the specified final state for that initial state. An s-correct state is necessarily l-correct, but the opposite is not true.

The following example clearly illustrates these two notions. Let us suppose that a program has been specified and designed to calculate the sum of two positive integers, represented by the variables a and b, and to assign that sum to the variable b. The corresponding algorithm could be of the following form:

```
read(a, b):
while (a > 0) do
    begin
    a := a - 1
    b := b + 1
    {Label}
    end;
```

If the algorithm is executed for the initial values $a = 5$ and $b = 2$, and the values of the parameters are observed at the indicated label, we get the following states (for the sake of simplicity we have left out the label code; the first value indicates the execution number of times the indicated label has been visited):

<1, 4, 3> (i.e., at the first execution we have: a = 4 and b = 3)
<2, 3, 4>
<3, 2, 5>
<4, 1, 6>
<5, 0, 7>

All these states are s-correct since they are the result of the required sequencing of the program. More formally, s-correctness of a state <n, a, b> with respect to the initial state a_0 and b_0, is represented here by a logic correctness condition (comma indicates logical *and*):

$$\{a \geq 0, b > 0, a = a_0 - n, b = b_0 + n\}$$

The following condition must be verified by the l-correct states:

$$\{a \geq 0, a + b = a_0 + b_0\}$$

In other words, if the previous state <4, 1, 6> (which is s-correct) was modified to become <4, 4, 3> (which is no longer s-correct but remains l-correct) later execution of the algorithm would all the same allow us to obtain the specified final state. But the state <4, -1, 8> is not l-correct and, of course it is not s-correct either. A state that is not l-correct may be loosely recoverable which means that despite the deterioration to the state, we have sufficient information (such as the initial state for example) to allow a recovery routine to return the program to a state (possibly previous or next) that is l-correct.

PFT approach can be considered as a process for determining a certain number of redundant items of information (s-correctness, l-correctness conditions, recovery procedures, etc.) which allows the software to tolerate faults that may occur during operational use. In this respect, two classes of fault can be distinguished:

1) *Anticipated faults.* These are classic faults that all test approaches attempt to eliminate and which are due to specification errors or programming errors.
2) *Unanticipated faults.* These are faults due to causes "external" to the application and that happen to the host system: hardware malfunction, compilation errors, poor synchronisation between processes, etc.

PFT is an approach that can be found highly useful for critical software: aeronautics, hospitals, banks, etc. Nevertheless it does have its disadvantages. For example, as far as anticipated faults are concerned, it is obvious that in order to determine the various correctness conditions, the program code will have to serve as the basis. However, if the code includes programming faults, it is not possible to make it tolerate those faults since the examination of the deteriorated states will, by definition, be defective. The only solution consists in using properties that are external to the algorithm (i.e., specifications). But, in this case the specified properties will have to be sufficiently precise to make it possible to determine realistically and practically the degree of deterioration of the variables. Verification of the conditions could therefore be long and often resemble a second execution of the algorithm. As far as unanticipated faults are concerned, on the other hand, it is assumed that the programmer has not made any programming errors and it is the program that has to be defended against external deteriorations. Even in this case, the calculation of the conditions can turn out to be a difficult and error-prone task for complex programs without sometimes excluding a notion of circularity insofar as, to make the algorithm fault tolerant, it is necessary to execute another algorithm. In addition, in order to allow activation of a set of adequate procedures, it is sometimes necessary to put forward certain hypotheses concerning the nature of the faults that could affect the software. It is clearly impossible (otherwise the problem of reliability would be solved) to know beforehand the complete typology of faults that will occur in the software's operational cycle. These remarks lead us to the following conclusions:

1) Determining the correctness conditions is a difficult task to automate (it is in fact similar to a sort of symbolic execution) which can sometimes by marred by human errors.
2) PFT can only be applied efficiently if it is combined with a good test process.
3) It is often difficult to make good predictions as far as the typology of faults, that are going to affect the software, are concerned. Evolution of the self-detection and recovery procedures (for example, during a maintenance phase) is sometimes a difficult task. The designer does not have much information concerning how efficiently and frequently the tolerance mechanisms he has implemented, are used.

The immunological approach to PFT can provide the answer to these criticisms by allowing:

1) Automation of the correctness conditions identification process. This automation will be accomplished by means of a process of gradually learning the properties verified by a program's variables after its test phase has been completed. This can be compared to the IS's maturing process during the embryo gestation phase.
2) An evolutive mechanism allowing the information that has been learnt (which will, in definitive, play the part of lymphocytes) to evolve and increasingly adapt itself to the software's operational environment. According to the same principle it is then possible to predict the existence of a certain set of recovery primitives (which are also evolutive and could be compared to T-cells) that will be activated when a deteriorated state (endangering the survival of the system if it continues) occurs.

Now let us try to study in greater detail the implementation of an immunological approach to software fault tolerance. Just as in the IS the basic elements will be the B-cells and the T-cells. A set of B-cells (that will be represented by an antibody) and T-cells will be associated with each precise label of an algorithm. This set of cells will constitute the IS of the program in question. We will look at the immunological mechanism developed at a precise label of the program and, more precisely, at the antibodies representation mechanism that we used in our experiment. The role of the T-cells and their dependence on the antibodies will be studied at the end of this article.

3 Representation of the antibodies

An antibody is formed by an epitope and a paratope. The epitope will consist of a simplified representation of a subset of different properties that have been verified (which will, inductively, become the "properties that *must* be verified") by the variables at the label in question. Symmetrically, the paratope symbolises the properties that are not verified which, by induction, will give us the "properties that *must not* be verified". Let us now try to see in what form the correctness conditions will be represented.

It is obviously impossible to hope for a precise learning of any l-correctness or s-correctness condition whatsoever. Such a process would have an enormous calculatory complexity and would boil down to reconstructing an equivalent program (a task which is formally non decidable). We must therefore focus our attention on weaker properties (such as relations between two variables) in the same way as an antibody only remembers the chemical characteristics that allow it to neutralise the antigen, and not the latter's general configuration. These conditions, determined inductively, are called the *tolerance conditions*. Each antibody will have two tolerance conditions (one for the epitope and one for the paratope) associated with it. The tolerance conditions are formed by *conjunctive* assembly of *primitive relations* that can belong to three different categories:

1. An inequality or equality relationship between two variables that will be written in a classic way, for example: `a < b`. When the variables concern arrays, the relationship is established statistically. For example, to express a condition of the form `(a[i] < b[j])`, it is considered that this relationship is true if k compatible values of i and j verify this inequation. The value of k is a constant determined in advance and depends on the degree of precision required. The notion of compatibility between indices is introduced to provide for the possibility that the indices are also linked by other primitive relations. A primitive relationship can also use at most one "anonymous" variable "?", which can represent any one of the state's variables.

2. A variable's inclusion in an interval. For example, to express the fact that the value of variable a remains between 4 and 10 we will write: `bound(a, 4, 10)`. When the variable is an array this inclusion is also determined statistically. The interval's bounds will be integers.

3. The variation direction and value, of a variable with respect to previous milestone. For example to express the fact that a variable is incremented by 2 each time that the program visits a label, we will write: `incr(a, 2)`. the value will be negative in the case of decrementation. The notations + (or -) will be used when the incrementation (or decrementation) value is not constant.

These primitive relations concern:

- The different variables of the same state. In this case the primitives 1. and 2. will be used.
- The variables of two successive milestones. In this case primitive 3. will be used.
- The variables of two neighbouring labels (in the program structure sense).

In this case we will only compare a variable with its value at the previous label. For example, to indicate that variable a has remained unchanged after execution of an instructions block, we will write: a = a' (where a' indicates the value it had at the previous label). In this case primitive 1. will be used.

For example, for the previous program that calculated the sum of two integers, the tolerance condition (concerning the state <n, a, b> = <2, 3, 4>) is verified:

{bound(a, 0, 4), (a < a'), incr(b, 1), (a > n)}

This condition of course includes primitive relations that are only true for the precise state (such as: bound(a, 0, 4)), but it also includes relations that are true for all the states observed at the label concerned (such as (a < a') for example indicating that the value of a is always smaller than the initial value a_0, supposing that the "previous" label was the beginning of the program). The previous condition can constitute the epitope (since it represents verified properties) of an antibody generated during the learning phase. (In fact, but this is only an implementation detail, this epitope is a binary codification of a tolerance condition).

At this point it could be argued that the primitives chosen are not sufficiently sophisticated and that the loss of information is too important to handle complex correctness conditions (a = 7*c - cos(d) for example). In other words, an l-correct state will always be tolerated, but the fact that a state is tolerated does not imply its l-correctness; erroneous states might not therefore be adequately detected.

The hypothesis that we are putting forward in the immunological model is that a great number of faults affecting the software will often violate at least one tolerance condition that has been learnt.

Furthermore, we will see later that an evolutionary algorithm is capable of recombining these relations to stabilise at a group of tolerance conditions that efficiently recognise most of the l-correct states. The price to be paid for this loss of information is therefore counterbalanced by the automation of the tolerance conditions definition process. Before continuing to a brief presentation of the development of a software tolerance mechanism based on the immunological model, let us try to summarise in a table the analogies between the IS and program fault tolerant framework, that have been adopted.

Immune system	Program Fault tolerance
Recognition of self	Recognition of l-correct states
Recognize not-self or malignant cells	Recognition of erroneous states
Antibodies	Pair of tolerance conditions = Antibody
Epitope	L-correct states tolerance condition
Paratope	Erroneous states tolerance condition
Antigen	Erroneous state
Gene used to create an antibody	Variables forming primitive relations
Antibody Proliferation	Strength of an antibody is augmented
Learning during gestation	Learning of l-correct states after testing
Life of the organism	Software operation

4 Development and Operation of a Software Immune System

Development of the IS associated with a program takes place in two stages (fig. 2): a learning phase (PHASE I: steps 1-3 in the figure 1) and an operational phase (PHASE II: steps 4 to 7).

4.1 Learning phase

The learning phase takes place after the testing phase. It is therefore assumed that there is a set of test cases that provides correct results (otherwise the test phase would not yet be completed).

The tested software is instrumented at specific labels that the tester selects by filtering, if necessary, the set of pertinent variables on which the development of antibodies is going to be grafted. The test cases are executed and the various states of the program (which by construction are l-correct) are stored by means of the instrumentation. We obtain a set of l-correct states each of which is associated with a precise label in the program. Then a set of counter-examples (states that are very probably not l-correct) is generated. For this purpose some, or a combination, of the following states will be used:

1) Randomly generated states,
2) L-correct states on which certain modifications have been made,
3) States that provided an erroneous result during testing,
4) States generated by mutants when a mutation testing technique [DEMILLO80] has been adopted.

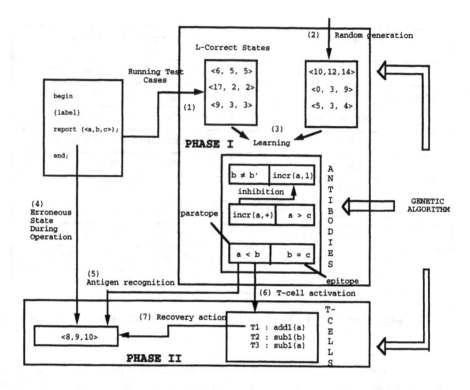

Fig. 1. The development of a software IS.

Once we have these two sets, the learning process can begin. To achieve this a population of antibodies (paratope + epitope) is generated randomly. Each epitope or paratope contains a limited number of primitive relationships thus expressing a tolerance condition. In the model that we have experimented, an antibody's length is fixed but this is not an optimum solution and other, more flexible, models [MICHALEWICZ92] could improve the antibody's expressivity.

Our aim is to filter and make the initial population of antibodies evolve in such a way that the epitope of each antibody recognises the greatest number of 1-correct states whereas its paratope recognises the greatest number of "erroneous" states. We consider that an epitope (or a paratope) recognises a state when its corresponding tolerance condition is verified by the state.

In spite of the fact that there are several deterministic symbolic learning algorithms [MICHALSKI83] based on examples and counter-examples, we have opted for a Genetic Based Machine Learning approach based on an evolutionist algorithm model known as Genetic Algorithms (GAs) [MICHALEWICZ92], [GOLDBERG 89]. However it must be pointed out that, at this stage we do not have the means for obtaining comparative measures capable of definitively justifying our choice. We will try, nevertheless, to explain the advantages of a genetic approach.

GAs are increasingly recognised as being a powerful technique for solving intrinsically complex problems. They have been found to be particularly efficient and robust when combinatorially wide search spaces appear. Their principle consists in making an initial population (randomly generated) evolve according to a natural selection model. Their properties and characteristics can be summarised by the following points:

- GAs work on a coding of the parameters and not on the parameters themselves. Each combination of solutions forms a chromosome (which, in our case, is the antibody) made of a series of genes.
- The search for the optimum using GAs is accomplished starting from a set of points and not from an isolated point. The evolution of the search is not deterministic.
- The behaviour of GAs is *implicitly parallel*. This property, which has been mathematically demonstrated by [HOLLAND73] stipulates that the solution schemata space implicitly covered is a polynomial in the population dimension [BERTONI93]. This property must, nevertheless be moderated by underlying the fact that this property of implicit parallelism is dependent on the coding used. The standard genetic algorithm model can be summarised schematically in four stages:

  1) Random generation of the antibodies making up the first population. The following expressions are two examples of antibodies:
  ```
  Antibody1: (a > f), (b ≠ c) :: (d > 2), incr(e, +).
  Antibody2: (w = b), (b ≠ ?) :: bound(e, 7, 7), (e > x)
  ```
  The sign '::' indicates the separation between the paratope (left side) and the epitope (right side).

  2) Calculation of the adaptation function for each antibody (corresponding to its "concentration" in the organism). In our case the adaptation function is:
  $E_e - C_e + C_p - E_p$, where:
  E_e : The examples (1-correct states) recognised by its epitope
  C_e ; The counter-examples recognised by its epitope
  C_p : The counter-examples recognised by its paratope
  E_p : The examples recognised by its paratope

  3) Application of three genetic operators to create a new population
  - 3.1) *Selection* of the best according to a probabilistic model
  - 3.2) *Crossover* of selected individuals and inclusion of their offsprings in a new population. Crossover of 2 chromosomes is achieved by randomly selecting a crossover site and mutual exchange of genes located at the right. For example, crossover of the two previous antibodies could give birth to the following antibody A3:
    ```
    (w = f), (b ≠ c) :: (d > 2), incr(e, +)
    ```
  - 3.3) Application from time to time of a *mutation* to the genes that are exchanged (e.g.: (b ≠ c) could become (b ≠ ?))

  4) Return to point 2)

In the learning model used, we drew our inspiration from a parasitism model [HILLIS90] thus allowing the counter-examples to evolve in parallel (and still according to a genetic model) trying to deceive antibodies. The more often a counter-example is recognised by epitopes and the less often it is recognised by paratopes, the more it is reproduced and crossed. The population of antibodies is thus constantly obliged to improve its performances. The learning convergence is thus significantly improved which confirms the previous work of [HILLIS90]. This evolution mechanism cannot be implemented on l-correct states since crossing them does not always provide an l-correct state.

But the advantages of an evolutionist approach based on antibodies formed by epitope-paratope pairs does not stop there. In our model it is in fact possible to consider that an antibody's paratope recognises the epitope of another antibody if a recognition function (which depends on the number of primitive relations that they have in common) exceeds a certain threshold. This mechanism reveals the following properties:

- Inhibition (and control) of the antibodies that have over-general epitopes (e.g. : ($w = ?$) or for instance, ($b \neq c$), etc.) and that could proliferate because they recognise a great number of l-correct states.
- Gradual differentiation of the various antibodies.
- Formation of a network (similar to Jerne's idiotypic network) that is capable of memorising and forming internal images of the various configurations of correct or erroneous states, while allowing adequate cooperation between the various recognition functions.

It must also be pointed out that the interest of such a learning phase is not limited to a strict perspective of fault tolerance. Antibodies could be an inestimable source of information for the programmer or the designer studying what types of assertion are verified during execution: why one variable is always greater than another one, why this array is never modified, etc.

4.2 Operational phase

The operational phase corresponds to software operation in its final environment. We are presently in the process of experimenting mechanisms that this phase will implement and we can already give a general outline of its principle.

The initially instrumented software is linked to the IS that it developed during the first phase. During execution the various states are observed by the various antibodies associated with a label. If a state is recognised by at least one of the epitopes in the set of antibodies, it is considered to be tolerated and execution continues. But, if the state is recognised by at least one of the paratopes and is not recognised by any of the epitopes, an immune response is activated. We are not putting forward any hypothesis here as to the origin of the fault which could just as well be an anticipated fault (due to a programming error not detected by testing), as an unanticipated fault. In our preliminary experiments we have found that the detection of anticipated faults was far more efficient when the learning phase was based on states resulting from mutation testing. There is, no doubt, nothing surprising about this phenomenon given that mutation testing provides samples of representative faults.

The immune response can trigger a classic processing procedure (system shutdown, user warning, execution of a recovery or exception procedure, etc.). But the immune model also makes it possible to envisage the automatic activation of a recovery process (that would avoid execution coming to a sudden halt) by returning the erroneous state to the "closest" tolerated state and so allowing execution to continue momentarily until a less critical state.

The part of this recovery procedure is played by a T-cell model. A T-cell is formed by a set of primitive operators (increment a, exchange a with b, etc.) which will be applied to the various primitive relations ((a < b), bound(a, 0, ?), etc.) corresponding to the paratope that recognised the erroneous state.

An example of a T-cell could be: exchange(17, 18), add1(12), sub1(15) which means that the 17th variable of the state will exchange its value with the 18th variable, that the value of the 12th will be incremented by 1, and that the 15th will be decremented by 1. It must not be forgotten that the fact that an antibody recognises a state, by means of its paratope, means that that state must not verify the corresponding tolerance condition. That antibody's paratope is therefore a means of marking that state (in the same way as a cell belonging to not-self is marked by the B-cells) in order to inform the corresponding T-cells that they can intervene. Activation of a T-cell will therefore depend on its capacity to modify the erroneous state. For example, if an erroneous state <n, a, b> = <9, 50, 56> is recognised by a paratope whose expression is of the form (b < a) (i.e., by coding the variables according to their *order* in the state: (3 < 2)) it will be possible to activate a T-cell containing the operators sub1(3), add1(2). In a cooperation process, other T-cells (or possibly the same one) can subsequently be activated. This process of correcting by successive little steps continues as long as the erroneous state is not tolerated.

The set of T-cells evolves according to a classic genetic algorithm scheme (crossover of T-cells, mutation, etc.). A T-cell whose activation has made it possible to re-establish tolerance of a state, is recompensed insofar as it corresponds to a good combination of elementary operations. But the recompense will be all the greater if the resulting state is also recognised by the following label's epitopes. This in fact means that execution has been able to continue in an acceptable manner. For example, in an experiment that we have performed, a fault consisting of a *non-incrementation* of the loop index was simulated during execution of an immunised program. After a certain time, the IS reacted by making a T-cell that incremented that index, allowing normal termination of that iteration. The T-cell that was made was memorised, which means that the IS would react much faster if a similar fault occurred later. This classic fault injection process can therefore be considered to be a "vaccination" process for a software which would become immunised against faults that are considered to be the most frequent. The IS developed would not, however, be frozen and new T-cells could emerge. Furthermore, this is a very useful piece of information for anyone wanting to understand the reasons for an anomaly.

5 Conclusions and perspectives

In this article we have described an immunological model for software fault tolerance. In our opinion such an approach opens up interesting perspectives in this critical area of information technology. Nevertheless, given its innovative nature, several aspects — both theoretical and technical — still have to be determined in order to define its limits and its applicability to more complex programs.

As far as the learning phase based on a set of test cases is concerned, the first results are promising. We would like to insist on this first possible use of the immunological model. Its interest does not simply lie in the automation of the construction process for a tolerant software, but above all in the perspectives of test case reusability. In a classic life cycle, the only possible use that can be envisaged for the test cases is to store them with a view to a non-regression analysis process. In an immunological approach the test cases are, as it were, recycled with a view to increasing a software's fault tolerance. If we consider that production of test cases is by far the most costly phase in a software life cycle, the economic interest of such an approach can easily be understood, especially if it is combined with (but this is not a necessity) a mutation testing technique.

Lastly we looked at the problem of recovery of an erroneous state when in use. We saw that it is possible to create a set of T-cells which, by communicating with the various antibodies, will attempt to modify the defective state so that it will be tolerated. An evolutionary process (initialised by the "initial vaccination") thus allows the software to develop an immune system suited to its operating environment, ensuring a less brutal halt in the case of a problem and even, why not — but this no doubt cannot be envisaged in the present state of our knowledge — allowing execution to continue without the need for any human intervention.

6 Bibliography

[AGRAWAL85] V. Agrawal V. : "Proc, Coll. on Fault Tolerance" Montreal, 1985.

[BERSINI92] H. Bersini : "Immune Network & Adaptive Control" Paris, Décembre 1992.

[BERTONI92] H. Bertoni & M. Dorigo: "Implicit Parallelism in Genetic Algorithms" Artificial Intelligence, 61, 2, 307-314.

[DEMILLO80] R. A. DeMillo & al.:"Mutation Analysis as a Tool for Software Quality Assurance", Proc. COMPSAC 80, (October 1980)

[FARMER85] J. Doyne Farmer, N.H. Packard, A. S. Perelson : "The Immune System, Adaptation, and Machine Learning" in Evolution, Games and Learning, Proc. of 5th Annual Inter. conference, Los Alamos, (May 1985).

[FARMER90] J. Doyne Farmer : "A Rosetta Stone for Connectionism" in Emergent Computation, Ed. S. Forrest, MIT, North-Holland (1990).

[GOLDBERG 89] D.E. Goldberg : "Genetic Algorithms in Search, Optimization, and Machine Learning" Addison Wesley Publ. Company, (1989).

[HILLIS90] W. D. Hillis: "Co-evolving parasites improve simulated evolution as an optimization procedure" in Emergent Computation, S. Forrest, MIT, North-Holland (1990).

[HOLLAND73] J.H. Holland : "Schemata and intrinsically parallel adaptation", Proce. of the NSF Workshop of Learning System Theory and its applications (pp 43-46), Univ. of Florida, (1973).

[JERNE74] N. K. Jerne : "Towards a network theory of the Immune System", Ann. Immunol. (Inst. Pasteur) 125 C (1974).

[MAXION90] R. A. Maxion :"Toward diagnosis as an emergent behavior in a network ecosystem" in Emergent Computation, Ed. S. Forrest, MIT, North-Holland (1990).

[MICHALEWICZ92] Z. Michalewicz : "Genetic Algorithms + Data Structures = Evolution Programs", Springer Verlag, (1992).

[MICHALSKI83] "Machine Learning - An Artificial Intelligence Approach", ed.Michalski R. S. & al, Tioga Publishing Company, Palo Alto, CA, 1983.

[MILI90] A. Mili : "Program Fault Tolerance. A structured programming approach", Prentice Hall International (1990).

Neural Networks

Modular Genetic Neural Networks for 6-Legged Locomotion

Frédéric Gruau

Psychology Department, Stanford University, Palo Alto CA 94305.
email: gruau@psych.stanford.edu

Abstract. This paper illustrates an artificial developmental system that is a computationally efficient technique for the automatic generation of complex Artificial Neural Networks (ANN). Artificial developmental system can develop a graph grammar into a modular ANN made of a combination of more simple subnetworks. Genetic programming is used to evolve coded grammars that generates ANNs for controlling a six-legged robot locomotion. A mechanism for the automatic definition of sub-neural networks is incorporated.

1 Introduction

Why use a developmental process? Two complementary processes contribute to the synthesis and optimization of an animal nervous system: Natural evolution takes place over million of years and it contributes a genetic code. After a complex *developmental process*, this genetic code will be translated into a nervous system. In addition, learning takes place during life time. It tunes the nervous system to the particular environment encountered by the animal.

Nature uses a biological developmental process to transform a genetic code into a nervous system. During the developmental process, cells divide using the genetic information. This allows one to encode incredibly complex systems with a compact code. For example a human brain contains about 10^{11} neurons, each one with an average of 10^5 connections. If the graph of connections was encoded using a list of destinations for each neuron, it would require $10^{11} * 10^5 * \ln_2 10^{11} = 1.7 * 10^{17}$ bits. The information contained in the human chromosome is of the order of $2 * 10^9$. The two numbers differs more than 8 orders of magnitude. How can the developmental process achieve such a compression? We conjecture that the mechanism is similar to the one used in modern programming language. Writing a compact computer program in such languages is done by using a hierarchy of procedures. Each procedure is defined a single time and can be called many times. A procedure corresponds to a sub neural-net, which is encoded a single time on a specific part of the genetic code. During development, that specific part can be read by many different cells, which will develop many copies of the same sub neural-net. We will refer to this property as *modularity*.

Our conjecture implies a prediction: ANNs encoded in a modular way must have a lot of regularities. We should be able to identify neural structures that are repeated many times. The part of the brain responsible for low level vision contains such general structures [Hubel and Wiesel 1979]. Another example of regularity can be found in the nervous system responsible for the locomotion of the six-legged American cockroach. The architecture discovered by Pearson is described in [Beer 1990]. It is made of 6 similar subnetworks, coupled by inhibitory connections. Each subnetwork controls one leg.

In this paper, we show how a simple computer model of biological developmental process allow to generate regular ANNs, for controlling animats.

Artificial developmental systems An Artificial Neural Network (ANN) is a graph of simple computing elements called units, which are an abstract model of the biological neuron in the nervous system. The architecture specifies the graph of interconnections, and each connection is weighted. The computer equivalent of learning is a method for tuning the ANN's weights using gradient descent. Weights are slightly modified over many epochs. Most often, learning is used to optimize the ANN's weights for a fixed architecture. Less frequently, evolutionary algorithms have been used to optimize the ANN's architecture [Cliff, Harvey and Husband 1993]. In this new emerging field, there has been recently some attempts to encode ANNs using inspiration from biological development. The idea is to indirectly represent an ANN, and to use an evolutionary algorithm to evolve high level representations for computational problem solving or animat simulations. Instead of directly describing a graph data structure (like a list of connections from unit to unit), an *artificial developmental system* describes how to build the ANN by applying rules of cell division. Starting with a single cell, an artificial developmental system develops a graph of cells using repeated applications of the development rules. When the development is finished, the graph of cells can be interpreted as an ANN. The system of rules can be modelled as a formal grammar, and the different approaches can be classified depending on the particular kind of formal grammars involved.

[Mjolness, Sharp and Alpert 1988] and [Kitano 1990] proposed the first examples of evolving formal grammars in this context. They use matrix grammars which has some drawbacks explained in [Gruau 1992]. Four different models recently proposed by [Vario 1993], [Parisi and Nolfi 1994] [Belew 1993] and [Dellaert and Beer 94] can be classified as geometric grammars. The object that undergoes division is a point in a discrete 2-dimensional space (or 3D). In general, the goal of those approach is modelling biology. Like Kitano and Mjolness, the problems solved with geometrical grammar can be solved easily by hand. [Boer and Kuiper 1992] use a parallel context sensitive string grammar known as a L-system, that operate on bracketed expressions. Our belief is that L-systems [Prusinkiewicz and Lindenmayer 1992] were proposed to model the growth of plant and trees and they are not well suited to represent an interconnection graph. To our knowledge, Boer and Kuiper have not yet solved a non trivial problem. Sims [Sims 94] proposes a model where a body structure is developed, along with the nervous system. The resulting body is "virtual", it does not correspond to a physical robot. Sims' system seems very effective for the particular task of generating locomotion behavior for virtual robots.

We proposed graph grammars as an efficient and general way to encode graphs [Gruau 1992]. Our method, called cellular encoding, models cell division, where a cell is just the node of a graph. We have implemented a neural compiler called JaNNeT [Gruau, Ratajszczak and Wiber 1994] that compiles a Pascal program into the cellular code of an ANN that simulates the Pascal program. JaNNeT demonstrates the expressive power of cellular encoding. We have proved several other theoretical properties of cellular encoding [Gruau 1994] which are completeness, compactness, closure, modularity, scalability. If one wants to use artificial developmental system for problem solving, it is important to prove theoretical properties of the underlying encoding rather than just do computer simulations.

With cellular encoding, artificial developmental system can be used to generate modular ANNs that exploit the regularity of the problem. In [Gruau 1992] and [Gruau and Whitley 1993], using cellular encoding, the evolutionary algorithm was able to generate recursive graph grammars (or cellular codes) that develop families of arbitrarily large ANNs for computing the Boolean functions parity, symmetry, and decoder of arbitrary large number of inputs.

In this paper, the method is used to solve a more realistic problem: the genetic synthesis of an ANN for locomotion of a six-legged animat. This problem has a regularity that can be exploited by artificial developmental system. [Beer 1990] proposed a model of ANN made of six copies of the same subnetwork, each of which controls one leg. The connections between the subnetworks are also regular. Using these symmetries, Beer and Gallagher were able to collapse the genetic code into a 200 bit string of 50 parameters and to perform the genetic synthesis of the ANN. We solve the same problem, but we do not help the evolutionary algorithm by using our knowledge about the symmetries. Instead, artificial developmental system makes it possible to automatically find symmetries.

Genome splicing We will use a technique called *Genome splicing* to increase the efficiency of the evolutionary algorithm. The structures that are evolved are sliced into a vector with a fixed number of components. During crossover between two vectors, the components are recombined pairwise. Each component may refer to other components many times. In the computation of the evaluation function, the same component may be used many times. The genetic search used in this paper involves a particular kind of genome splicing, where each component of the vector encodes a sub-neural networks. The work of John Koza uses another particular kind of genome splicing [Genetic Programming II 1994]. Here, the component of the vector corresponds to a LISP function that can be called many times. I acknowledge that the idea and the technology of genome slicing come from Koza. Genetic Programming has been demonstrated by Koza [Koza 1992] as a way of evolving computer programs with a GA. In the original GP paradigm the individuals in the population are LISP S-expressions which can be depicted graphically as rooted, point-labeled trees with ordered branches. GP, more generally, includes approaches were the language used can be other than LISP, including cellular encoding. Genome splicing is called by Koza, "automatic definition of function", because each tree encodes an Automatically Defined LISP Function (ADF), and the solution is a hierarchy of functions that call each other many times. We call that "Automatic Definition of Sub Neural Networks" (ADSN), because with cellular encoding each tree encodes a sub neural-network.

2 A non-trivial animat problem

We use the model of artificial six-legged insect described in [Beer 1990]. Each leg controller has 3 motor neurons: The state of the foot is controled by the FS unit. The return stroke is controlled by the RS unit. The power stroke is controled by the PS unit. In order to help coordination of the legs, each leg controller has one sensor unit that records the position of the leg.

Each foot has an internal state. A foot can be either up or down. If the activity of the FS neuron is positive, the foot is pulled off the ground, else it is put on the ground. It takes one time step to move the foot. During this time step, the leg cannot exert a force. This takes into account the inertia of the leg and induces a selective pressure towards using the full range of the leg angle. Without this inertia, there would be a trivial solution to the problem, namely to alternate power stroke and return stroke at each time step. The force exerted by the PS and the RS unit are subtracted. If the foot is up, the resulting force is used to update the leg position relative to the body. If the foot is on the ground, the resulting force pushes the animal backward or forward depending on its sign. Due to friction, a dragging leg exerts a constant force pushing backward, proportional to the speed of the robot. There is also a global friction force pushing backward, also proportional to the speed of the robot. At each time step, we sum the forces exerted by the legs which are down, and the friction forces. This sum is used to update the speed of the robot. If the center of mass of the robot lies outside the polygon formed by the feet which are down, the robot falls down and its speed drops to zero. Otherwise the speed is used to update the position of the robot, and the joint angle of the legs which are down.

We made one modification to Beer's model: we used two discrete sensor neurons [Cruse, Müller-Wilm and Dean 1992]. They are called Anterior Extreme Position (AEP) and Posterior Extreme Position (PEP). AEP 's activity is maximum if the leg is at its anterior extreme position and 0 otherwise. PEP 's activity is maximum if the leg is at its posterior extreme position and 0 otherwise.

We considered a simplified model where each foot is automatically controlled. If the RS unit is activated, and the foot is down, the foot is put up. Else, if the foot is up, the leg is pulled back with speed proportional to RS's activity. If the PS unit is activated, and the foot is up the foot is put down. Else if the foot is down, the leg exerts a force on the body proportional to PS's activity. By convention, if both RS and PS are activated, RS "wins" and PS is ignored. We think this simplified problem is a good benchmark for ADS, because it is not easy to solve directly by hand, it is more realistic than Boolean functions, and it has an internal regularity.

3 The continuous, noisy, neural model

To solve the problem of locomotion of a six-legged robot the ANN must be able to store an internal state. Hence, it needs recurrent links. The activities of all the neurons are updated at the same time using a continuous time update of the neurons activities, as advocated in [Beer and Gallagher 1993].

$$\tau_i * (a_i(t + \delta t) - a_i(t))/\delta t = s_\alpha(netinput_i) - a_i(t) \tag{1}$$

The net input is the weighted sum of the neighbour's activities minus the threshold of the neuron, τ_i is a time constant whose value is always 3 in our hand coded ANN. The activities are integers that ranges from -2048 to $+2048$. to more precisely represent the sigmoid. We updated the neuron's activities three times before updating the body state. The sigmoid of the neuron is called s. With this model of neuron, for some particular symmetric leg position, our hand-coded solution would get stuck in a wrong attractor. The animat used four legs instead of six. We added random noise in the model. Each time a unit computes its activity, it adds a random number uniformly distributed between -10 and $+10$. Noise perturbed the system out of the wrong local attractor. The global attractor is more stable and is not affected by noise. During the rhythmic activity, when the legs switch from RS to PS or vice versa, the ANN's state must come very near to the boarder between the wrong attractor and the right attractor. A little noise is then enough to allow the system to move from the wrong attractor into the right attractor. When the network is in the right attractor is never returns to the wrong attractor.

4 Cellular encoding revisited

Cellular encoding is a method for encoding ANNs.. In this paper we present an updated and improved version of cellular encoding compared to the one in [Gruau 1994b]. Cellular encoding uses an abstract notion of cells. A cell has an input site and an output site. It is linked to other cells, with directed and ordered links that fan into the cell at the input site and fan out from the cell at the output site. A cell also possesses a list of internal registers that represent a local memory and store labels. The data structure of an ANN is a directed labeled graph. The cell concept is simplified to provide only what is needed to describe a directed and labeled graph. The cellular code is based on local graph transformations or graph rewriting rules that act upon cells. Examples of possible graph transformations used in cellular encoding are represented in Figure 1. Picture (a) represents an initial graph of cells. It is composed of one central cell connected to 6 input neighbours and 6 output neighbours. The remaining pictures, (b) to (l), show the effect of different graph transformations acting upon the central cell. The graph transformations can be classified into cell divisions, local topology transformations, and modifications of weights.

- Picture (b) to (h) represent the effect of different cell divisions. A cell division replaces one cell called the mother cell by two cells called child cells. One can imagine many possible cell divisions depending on the way the links of the mother cell are inherited by the child cells. The links are ordered and it is possible to refer to a link by using its number. In picture (a) the number of the links are represented. A sublist of consecutive links is specified by the number of the first link and the number of the last link in the sublist. A particular cell division is implemented by copying one or many sublists of links, from the mother cell to each of the child cell. A cell division must also specifies whether the two child cells will be linked or not. For practical purposes, we give a one-letter name to the graph transformation, and the set of letters will be the set of alleles used with the genetic programming algorithm. The particular letters we use do not have a particular meaning. Division "S" represented in picture (b) is the *sequential* division. In the sequential division, the first child cell inherits the input links, the second child cell inherits the output links and the first child cell is connected to the second child cell. Division "P" represented in picture (c) is the *parallel* division. Both child cells inherit both the input and output links from the mother cell. Hence, each link is duplicated. The child cells are not connected. Divisions "S" and "P" are canonical divisions, because they are the most simple: all the other divisions do not handle all the links in a uniform way independently from their position. Division "T" is like "S", except that the input link number one and the output link number one are duplicated. Picture (e) and (f) represent two possible effects of the same cell division "A". Division "G" and "H" are symmetric to one another with respect to the inputs and outputs. They allow a cell to be inserted on the output or the input link number one.
- Picture (i) and (j) represent graph transformations that locally transform the topology. The first one called "R" adds a recurrent link to the cell, the second one called "C" has an argument 3 — it removes link number 3.
- The remaining two pictures describe graph transformations that have an argument, and modify the weights. A weight -1 is represented by a dashed line. The first is "D", the value of the argument is 3: it sets the input link number 3 to -1. The second is "K" with argument 2: it sets all the output links starting at 2 to -1.

We use four other program symbols that are not illustrated: The program symbol "I" sets the input weight n to the value $+1$, where n is the argument. The program symbol "Y" sets the time constant of the neuron to n. The program symbols "U" and "L" set the sigmoids to s_{256} and s_{2048} respectively, and the threshold to n. (s_α) is a family of piece-wise linear functions that takes values $-\alpha/2$ at infinite negative value, $+\alpha$ at infinite positive value, and crosses the origin. We choose the particular set of graph transformation presented here because this set allows the user to provide a hand-coded solution.

Development consists of successive graph transformations on a graph of cells that causes it to grow into an ANN. In order to combine many graph transform-ations into a cellular code we used an ordered list of labeled trees instead of a set of grammatical rules. An example of cellular code is represented in Figure 2. The trees are labeled by names of graph transformations and are called grammar trees. Each cell carries a duplicate copy of the cellular code (i.e., the set of grammar trees) and has an internal register called reading head that points to a particular position of the grammar tree. At each step of the development, each cell executes the graph transformation pointed to by its reading head. ANN units are cells that have terminated their development and lost their reading-head.

The instructions are called *program symbols*. Program symbols indicating cell division label nodes of arity two; when a cell divides, the first child cell goes to read the left sub tree, and the second child cell goes to read the right sub tree. Program symbols indicating other graph transformations label nodes of arity one, and once the transformation is executed, the cell will simply move its reading head on the unique sub tree. Cells can also execute instructions for piloting the reading head and instructions to finish the development and produce an ANN unit. Since at each step of the development all the reading heads move one level down in the tree; they come to reach the leaves. When a reading head reads the leaf of a tree, two events may happen.

- It can encounters a program symbol such as "U" or "L" in figure 2 which finishes the development. The cell can be considered as an ANN unit with distinct particular features.
- It can read a program symbol "n" which has an argument d. If the number of the tree that is currently read is x, the cell moves its reading head on the root of the grammar tree $x + d$. For example, the program symbol "n 1" moves the reading head on the root of the next grammar tree. The program symbol "n 0" backtracks the reading head to the root of the grammar tree that is currently read, the program symbol "n 2" jumps on the next-to-the-next grammar tree. This is a reference mechanism using relative addresses.

During a step of the development, the cells execute their program symbols one after the other. The order in which cells execute program symbols is determined as follows: once a cell has executed its program symbol, it enters a First In First Out (FIFO) queue. The next cell to execute is the head of the FIFO queue. If the cell divides, the child which reads the left subtree enters the FIFO queue first. This order of execution tries to model what would happen if cells were active in parallel. It ensures that a cell cannot be active twice while another cell has not been active at all. In some cases, the final configuration of the network depends on the order in which cells execute their corresponding instructions. The waiting program symbol denoted "W" has no effect: it makes the cell wait for its next rewriting step. It is necessary for those cases where the development process must be controlled by generating appropriate delays.

4.1 An example

As an example, consider the problem of finding an ANN for the 6-legged locomotion problem. The input units are sensory inputs that test the position

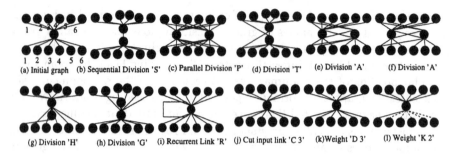

Fig. 1. Example of local graph transformations used in the hand coded solutions.

of the legs. The output units command the legs and the feet. In Figures 3 we show the development of the following hand designed cellular code: tree 1 defines the whole ANN, it is R(T(n1)(n1)). Tree 2 encodes a row of three legs, it is A((W(n1)))(A(n1)(n1)). Tree 3 encodes an individual leg controller. It is: K2(G(P(L) (D1(L))) (H(R(U)) (P(L) (P(D1(L)) (D1(L))))))).

Each cell is represented by a circle. Inside the circle, we write the name (one letter) of the program symbol currently read by the reading head.

The development of an ANN starts with a single cell called the *ancestor cell* connected to an input pointer cell and an output pointer cell. A pointer cell is represented as a square. Consider the picture at the top left of Figure 3. Initially, the reading head of the ancestor cell is located on the root of tree 1, and reads the program symbol "R". Its registers are initialized with default values. For example, its threshold is set to 0. As this cell repeatedly divides, it gives birth to all the other cells that will eventually become a finished neuron and make up the ANN. The input and output pointer cells to which the ancestor is linked do not execute any program symbol. Rather, at the end of the developmental process, the upper input pointer cell is connected to the set of input units, while the lower output pointer cell is connected to the set of output units. These input and output units are created during the development, they are not added independently at the end. After development is complete, the pointer cells can be deleted and are replaced by duplicate input and output units, as shown in the picture in the lower right corner of Figure 3.

The hand-coded ANN is made of six oscillators coupled by inhibitory connections. Each oscillator controls one leg. It is inspired by the Pearson model described in [Beer 1990]. Thanks to the modularity, each oscillator is encoded a single time in the tree number 3, Figure 2. The tree number 2 encodes a line of 3 leg controller (half a body). Then, the tree number one puts together two half bodies to produce a complete controller.

5 The parallel Genetic Algorithm

We used the mixed parallel Genetic Algorithm described in [Gruau 93]. This model combines the stepping-stone model and the isolating-by-distance model.

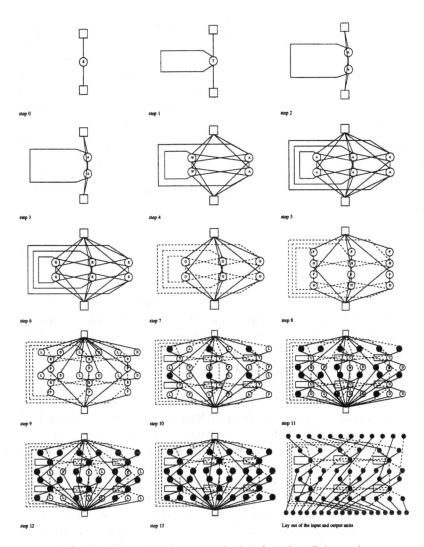

Fig. 2. Steps of development of a hand-made cellular code

Thus the advantages of both models are combined. Local mating allows to achieve a high degree of inbreeding between mates [Collins and Jefferson 1991]. Isolated islands help to maintain genetic diversity [Muehlenbein 1991]. The migration rate is parametrized using a single parameter. Our GA is steady state. Each time a new individual is to be created, it can be built either through the recombination of two other individuals, or by mean of an exchange. The exchange is chosen with a probability called the exchange rate which is fixed to 0.01. The MIMD parallel machine used is an IPSC860, with 32 processors. One node of this machine is 40 MIPS which is roughly 1.5 SPARC-2. The communications between the processor are asynchronous, and are overlapped by the computations. So the communication cost is null. This theoretical prediction is confirmed by experiments. The subpopulation on each processor is 64 individuals, so the total population is 2048.

The set of alleles used was $\{S, P, T, A, G, H, R, C, D, I, U, L, W, n\}$. These alleles correspond to program symbols of the cellular encodings explained in Section 4. Using these alleles, an initial random population is created, in which all the grammar trees have an equal and fixed number of nodes. As explained in section 4, some of these program symbols have an argument. For example the argument of 'C' is the number of the link to be cut. Program symbols $\{C, D, I, U, L\}$ have arguments. The arguments are set to random values between -9 and +9. The GA is applied until a genetic code is found that develops an ANN meeting the termination criterion explained in section 6, or the allocated time of two hours has elapse. Each time an offspring is created, all the alleles are mutated with a small probability 0.005.

6 Approaching generalization by distributing 32 fitness cases

The fitness of a given ANN for the problem of the six legged locomotion robot is the average of the distance covered by the robot on a certain sample of initial leg positions. Our aim is not merely to produce an ANN for six legged locomotion, we also want our ANN to be general. Whatever the initial condition of the six legs, we want our robot to perform the tripod gait after a transient of reasonable length. So in all of our experiments, we carefully test the ANNs produced by the Genetic Algorithm (GA) for 64 possible initial positions of the leg, where each of the legs is initially either on Anterior Extreme Position (AEP) or Posterior Extreme Position (PEP). This tests whether the ANN can generalize.

One of the most delicate tasks was to devise a fitness quick to evaluate, and that measures robustness. We carefully selected a fixed set of six difficult initial positions of the leg plus one that is easy. An ANN may be evaluated up to these 7 positions. However, in order to proceed to the next one, it must have been able to walk the threshold distance on the preceding one. Time is saved because very often the evaluation will stop at the first initial random position. As individuals become better, more time is spent evaluating them. The termination criteria was to generate an ANN having the same performance of the hand coded solution on the seven positions. The threshold distance was chosen as the distance walked by the hand coded solution, multiplied by 0.9. If some individuals get more fitness cases than others because they exceed the threshold, they get the sum over all the fitness cases they tried and not the average. We divided the population into 32 sub-populations, and gave a different evaluation function to each sub-population. Individual mate mainly within a sub-population and very occasionally across two neighbouring sub-populations. This is easy to implement, because we use a parallel GA that runs on a 32 processor parallel machine, mating across subpopulation means exchanging individuals. A given evaluation function was built as follows: choose a first position among the seven as a function of the number of the processor; if the distance walked exceeds a threshold distance, choose another particular initial position also as a function of the number of the processor, and so on, until all the set has been tried. Each sub population chooses initial positions in a particular sequence. depending on the processor number. For a given sub population on a given processor, the sequence stays fixed during all the runs. During the first generation, each sub population concentrates on one of the seven initial positions. Hence, there are seven different fitness functions. When the ANNs walked passed the threshold distance, the sub-population begins to learn ANNs that can walk starting from two different initial positions. At this stage there are $7*6/2 = 21$ fitness functions. The maximum number of fitness functions is 35. The processors are on a 2D grid which dimensions are 4X8. No two adjacent processors have the same first initial leg position among the 7 possible. Hence, adjacent processors will evolve different genetic material. When mating across adjacent processors occurs, new genetic material can be created.

Using these different fitness functions, the solution found by the GA passed the generalization test with 100% success, and the average evaluation time of one individual was not significantly different from the average evaluation time we obtained using a single initial position.

7 Stochastic Hill-climbing on the weights

We apply a stochastic hill-climber on the weights of each ANN produced by the GA, in order to speed up the genetic search. Deterministic hill-climbing is not possible because it would involve too much computer time, due to the large number of weight change to try. We randomly choose a weight w, and modify it to a random value in $\{0, 1, -1\}$. We then modify the cellular code of that ANN so that the modified cellular code produces the same ANN where w is modified. We call this back-coding. Many authors [Gruau and Whitley 1993] [Ackley and Littman 1992] have considered such techniques which originates from ideas back to Lamark and Baldwin. We compute the performance of the ANN developed with the modified code. We compare that performance with the performance before hill climbing. If the performance increases, we accept the weight change, and put the modified code back into the population (Lamarkian strategy). If not, we still accept the weight change with a probability $e^{-0.1}$. We use a single epoch because hill-climbing is expensive. In this context it nearly doubles the time of fitness evaluation.

Back coding has an interesting side effect when combined with Automatic Definition of Subnetwork. The genome is spliced in two trees. The tree number two encodes a subnetwork. The tree number one encodes the general structure of the network and specify how many occurrence of the subnetwork to include and how to connect them. Suppose that the back coding modifies the tree number two, wherever this sub network will be instantiated in the final ANN, the weight modification will be reproduced.

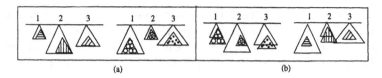

(a)   (b)

Fig. 3. The genome is spliced in three trees. (a) before the crossover of two genomes (b) after the crossover.

8 Simulation with the simplified model

8.1 Comparison between runs with and without ADSN

In this section we report simulation and results on the simpler problem of 6-legged locomotion, where the the feet need not be controlled. The experiments are focussed on showing the interest of genome splicing (see the introduction). We did two kinds of experiments for a comparative study. In the first kind, the genome consists of a single grammar-tree and the crossover is done by exchanging sub trees. In the second kind, the genome represented in figure 3 is a vector of three grammar trees. Cross over is done by exchanging sub trees between pairwise component trees. Three subtrees are exchanged for one cross over. Each of the subtree can be only a leaf, or it can be the whole grammar tree. We use the program symbol "n", that has an argument. The argument of "n" is a relative address for moving the reading head from one tree to another tree of strictly higher number. It can be 1 or 2 for tree 1, it is always 1 for the tree 2, it does not exist in tree 3. The argument of "n" cannot be 0, hence we did not allowed for recursion. The experiment with ADSN and without ADSN had exactly the same parameters, except for one difference: in the run without ADSN, the initial population consisted of individuals having 60 nodes, and the upper bound on the size of the trees was 600 nodes, and in the run with ADSN, the initial population consisted of individuals having three trees of 20 nodes each, with an upper bound of 100 nodes. In both case, the genetic material (number of nodes) is the same in the initial population. We gave the possibility to grow grammar trees with two times more nodes to the run without ADSN. Since for those runs, genetic material cannot be reused, more genetic material is needed.

	number of evaluations	time
without ADSN	18805	12752
with ADSN	5152	1968
relative gain	3,65	6.5
ADSN, hill-climbing disabled	10724	2589

Table 1. Comparison between runs with and without ADSN for the simplified problem

We ran two trials with ADSN, and two trials without ADSN. The program symbol "Y" which sets the time constant was not used. The time constant τ_i defined in equation 1 was chosen to be 3 for all the neurons. We used the stochastic hill-climbing method described in section 7. In all the runs, the robot first learns to generate oscillatory movements of the legs, and thereafter learns to coordinate the movements of the leg to avoid falling. With ADSN the GA found a solution in both of the two trials. Without ADSN, it found a solution only in one run. In all cases, the solution found by the GP was general: the ANN produces tripod gait over all the 64 possible initial conditions of the legs, and the average distance made by the robot is the same as the hand coded ANN. Table 1 summarizes the results of the runs. The speed up produced by ADSN is 6.5 and the run with ADSN uses 3.65 times fewer individuals. These numbers show that on average, the evaluation of one ANN takes twice as much time with ADSN. This is because ADSN tends to generate networks with many units, because subnetworks are repeated many times. On average, networks generated without ADSN are twice as big. The genetic code found without ADSN is on average more than four times bigger than the one found with ADSN. We reran the experiment with ADSN and the stochastic hill-climbing disabled. The result are indicated in the fourth row of table 1 and show that learning produces a speed-up of 30%. Two trials are not sufficient to provide statistical information. However, there appears to be a real difference between runs without ADSN and runs with ADSN.

8.2 Analysis of the ANNs found by the GP

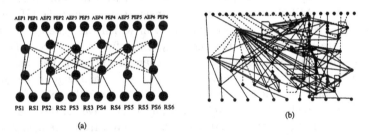

Fig. 4. ANNs found by the GP for the simplified six legged locomotion problem: (a) with ADSN (b) without ADSN

In figure 4, we compare ANNs found with the GP, with and without ADSN for the simplified locomotion problem solved in the previous subsection. Whereas our hand-coded solution encompasses 6 subnetworks, the ANN generated by the GP with ADSN has a subnetwork which is repeated only 3 times. Each subnetwork controls two legs. It contains a neuron with a recurrent connection that we call LR because it also acts like a latch register. For each LR neuron, there is an inhibitory connections to another neuron N, and an inhibitory connection from N to the LR neuron of an adjacent controller. Since $-1 * -1 = 1$, the resulting coupling between the two LR neurons is +1. Hence, the three LR neurons in each controller are connected between pairs, by excitatory connections. As a result, the three controllers are in phase. On the other hand, the two legs controlled by one controller are anti-phase. The ANN found by the GP has 12 units, It is much less than the hand coded solution shown in the lower right corner of figure 2. Our solution has 30 units for the simplified problem, and 36 units for the complete problem. The ANN shown if figure 4 (b) has been evolved without ADSN, and therefore, it has no structure. It is almost impossible to understand how it functions.

```
left pos. _      _      _    _    _    _    _    _    _    _    _
left mid. _             __   __   __   __   __   __   __   __   __
left ant. _      _    _    __   _    __   __   __   __   __   __
right pos. __     __   _    _    _    _    _    _    _    _    _
right mid. _     _    __   _    _    _    __   __   __   __   __
right ant. __     __   _    _    _    __   _    __   __   __   __
```

Fig. 5. Illustration of foot steps for the 6 legged locomotion without foot motor neurons. Foot steps are represented in the following way: whenever a leg is up, we plot a dot; otherwise nothing is plotted.

8.3   Analysis of a GP run

A run of the GP produces 32 files recording the genetic code of the best-so-far individual found by the GP, for each processor. We can afterwards run a program that displays the phenotype (ANN) of the best solutions at different stages of the search. Figure 6 reports the sequence for a GP run with the simplified model of locomotion with ADSN. It provides useful insights as to how the GP proceeds to builds the ANNs. There are some periods in which the general structure of the architecture remains the same and only a careful scrutiny reveals the few connections that have been changed. These changes are made by mutation and crossover towards the leaves of the trees. The impact of a change in the genotype is big if the change is made early in the development, near the root of the tree; it is small if it is made late in the development, near the leaves of the tree. At other periods, a deep change in the structure can be observed from one best individual to the next. This is due either to crossover near the root of the tree, or to the receipt of a good genetic material from neighbouring processors. Sometimes we can see two species of ANNs reappearing alternatively (for example at generation 19 and 21). With ADSN, the impact of crossover or mutation also depends on the tree to which it is applied. The crossover on the first tree changes the global structure of the ANN, how many copies of the sub network are included and how they are combined. During the run with ADSN, at generation 17 , we see a marked transition toward a general structure which is composed of two oscillators. These structure will remain and be improved eight times before another structure takes over at generation 49. The new structure is made of three oscillators. After two improvements the three oscillators structure leads to a solution that satisfies the termination criteria. The two improvement have operated a simple pruning of misleading weights.

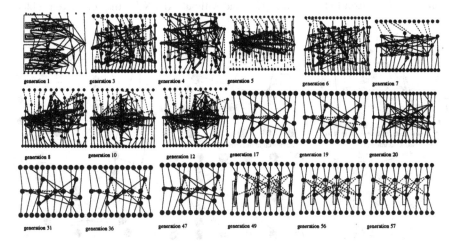

Fig. 6. Evolution of the best individual, with automatic definition of subnetworks, simplified model of locomotion

9 The complete model used by Beer and Gallagher

9.1 Comparison between runs with and without ADSN

In the second set of experiments, we used the model described in [Beer and Gallagher 1993]. Each foot is now controlled by a foot motor unit. The GP was the same as for the simplified model except that now the time constant of the neuron is genetically determined, by using the program symbol 'Y' to set the time constant. We run five trials with and without ADSN using the same GP setting as in the experiments with the simplified model. We report here all the trials we have done. We could not do trials to tune parameters, because it was too time consuming. The time limit is two hours and the population size per processor is 64. The GP was successful two times out of five with ADSN, but could not find a solution without ADSN. The results are reported in table 2. The two solution found by the GP are represented in figure 7 (a) and (b). Unlike the hand-coded solution, they generalized over the 64 initial leg positions.

9.2 Analysis of the ANNs found by the GP

Figure 7 analyses the two ANNs (a) and (b), found by the GP. They are made of 6 copies of the same subnetwork. By freezing the development of the cell that jumps to ANNs, we can generate the general architectures (c) and (b). It shows which

	number of evaluations	time
without ADSN	-	-
with ADSN	16234	6480

Table 2. Comparison between runs with and without ADSN for the complete problem

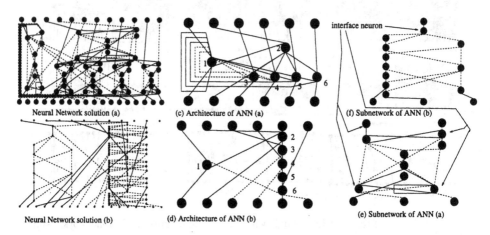

Fig. 7. (a) and (b): ANNs found by the GP for the complete problem. (c) and (d): General architecture of respectively ANN (a) and (b). The number indicates the leg controlled by the subnetwork which will be included at that particular position. (e) and (f): Subnetwork of ANN (a) and (b). The interface neuron are those which will make connections to other subnetworks.

subnetwork is connected to which other subnetwork. By developing the ADSN separately, we can develop the subnetworks. The computations of ANN (b) are controlled by a ring of 6 interface neurons. Each node of the ring is coupled to the next node with a weight -1[1]. The activities flow along this ring, and change signs at each node. Each subnetwork is feed-forward, there are no recurrent connections. Each node in the ring is at the root of this feed-forward subnetwork. It receives input from the AEP and PEP sensors, and directly controls the RS, PS, and FS actuators. The tripod gait simply emerges from the fact that the interface neurons are connected by -1 weights. We had not thought about this solution to the problem. It is a "GP surprise". However, this architecture is not robust, if one node in the ring is suppressed, synchronization between the subnetworks is broken.

The mechanics of ANN (a) is intermediate between the one of the ANN found for the simpler problem figure 4 a, and the one of ANN (b). Each subnetwork has a latch register, however, the interconnections between subnetworks also flows in a ring. This ring has three steps: the first step is subnetwork 1, the second step

---

[1] The weights are not the final weights on figure 7 (c) and (d). The final weights are determined by the genetic code of the subnetworks, and not by the general architecture.

is subnetwork 2, and the third step is subnetwork 3, 4, 5 and 6 in parallel. The activities flows through all the sub-networks, and not only through the interface neurons. The tripod gait as usual is induced by correctly placed −1 weights that ensures anti-phase locking between pairs of adjacent subnetwork's latch register. However, not all pair of subnetwork for adjacent legs are connected. This architecture is more robust, but less concise than the preceding one. Three of the subnetworks in {3, 4, 5, 6} may be removed without breaking the ring and the synchronization.

10 Conclusion and a future direction

Regularity occurs whenever the same pattern is repeated, possibly at different scales. It is possible to make a lengthy list of examples of regularity: atoms, molecules, polymers, cells, body limbs, any fractals, the homogeneity of space and time, the cut and paste principles of a word processor, the words of a language, the signs of any code... We believe that regularity is a basic characteristic of information, and no communication or coupling, no any interesting phenomenon can happen without it. Natural evolution has found a way to exploit and produce regularity. The body architecture of the creatures built through natural evolution have some internal regularities that are coupled with the external regularity of this world. For example, the body symmetry maps to the isotropic nature of the space. Similarly, the nervous system of animals may be highly regular. An example is the part of the brain responsible for low level vision. Here, different pixels of an image are treated in the same way. The regularity of this network also maps to the homogeneity of space. Nature generates regular animal bodies or nervous system by using a developmental process. A structure genetically encoded a single time can be developed many times. This paper shows that this principles can be turned into an efficient computer technique called *artificial developmental process*. This technique can generate automatically Artificial Neural Networks (ANN) for non-trivial animat control. These ANNs have an internal structure that exploits the regularities inherent in the problem.

Our artificial developmental system scheme uses explicit genome splicing. The genome is spliced into a list of trees which can refer to each other in a hierarchical way, and each subtree encodes a subnetwork, so that the total ANN is a hierarchy of many copies of these smaller subnetworks. We did two kinds of experiments. In the first kind, the structure of the chromosome is a single tree. In the second kind, the structure is a list of three trees. We solved two problems, the first problem is a simplified model of six-legged locomotion, the second model is a complete model used in [Beer and Gallagher 1993], where a motor neuron controls the foot. With the second experimental setting an ANN solution to the simplified six-legged locomotion problem is found 6.5 faster than with the first setting. Furthermore, the size of the genotype and the number of hidden units of the best phenotypes (ANN) are both more than four times smaller in the second setting. Concerning the complete problem, the first setting could not even find solutions out of 5 trials, whereas the first setting did 2 times. Despite the small number of trials, we believe the difference between the two settings is sufficiently high to show the advantage of artificial developmental system which makes it possible to automatically solve the same problem Beer and Gallagher solved, but without using any information about the symmetries of the problem. This problem is another problem in the list of the non-trivial problems we have solved with cellular encoding. The problems considered up to now always have a certain amount of regularities. Cellular encoding is efficient whenever the problem to solve is regular. Regularity means that the problem is composed of a hierarchy of subproblems that can be concisely expressed in algorithmic form.

11 Acknowledgment

This work has been supported by the French Institut National de Recherche en Informatique et Automatique We thanks Oak Ridge National Laboratory for providing access to their 128 nodes Ipsc860. We are very indebted to Darell Whitley, Eric Siegel, Melanie Mitchell, Rajarshi Das for their precious comments and corrections.

References

1. D. Ackley and M. Littman. The interaction between learning and evolution. In *Artificial life II*, 1991.
2. Randall Beer. *Intelligence as adaptive behavior*. Academic Press, 1990.
3. Randall Beer and John Gallagher. Evolving dynamical neural networks for adaptive behavior. *Adaptive Behavior*, 1:92–122, 1992.
4. R. Belew. Interposing an ontogenic model between genetic algorithms and neural networks. In *Neural Information Processing System 5th*, 1993.
5. E.J.W. Boers and H. Kuiper. *Biological Metaphor and the design of modular artificial neural networks*. Master Thesis, Leiden University, the Netherlands, 1992.
6. Rodney Brooks. Intelligence without representation. *Artificial Intelligence*, 47:139–159, 1991.
7. Dave Cliff, Inmahn Harvey, and Phil Hubands. Explorations in evolutionary robotics. *Adaptive Behavior*, 2:73–110, 1993.
8. R. Collins and D. Jefferson. Selection in massively parallel genetic algorithm. In *Proc. of 4th International Conf. on Genetic Algorithms*, 1991.
9. F. Dellaert and D. Beer. Co-evolving body and brain in autonomous agent using a developmental model. Ces94-16, Case Western University, 1994.
10. F. Gruau. Genetic synthesis of boolean neural networks with a cell rewriting developmental process. In *Combination of Genetic Algorithms and Neural Networks*, 1992.
11. F. Gruau. The mixed parallel genetic algorithm. In *Parallel Computing 93*, 1993.
12. F. Gruau. *Neural Network Synthesis using Cellular Encoding and the Genetic Algorithm*. PhD Thesis, Ecole Normale Supérieure de Lyon, 1994. anonymous ftp: lip.ens-lyon.fr (140.77.1.11) directory pub/Rapports/PhD file PhD94-01-E.ps.Z (english) PhD94-01-F.ps.Z (french).
13. F. Gruau, J. Ratajszczak, and G. Wiber. A neural compiler. *Theoretical Computer Science*, 1994. to appear.

14. F. Gruau and D. Whitley. Adding learning to the the cellular developmental process: a comparative study. *Evolutionary Computation V1N3*, 1993.

15. U. Muller-Wilm H. Cruse and J. Dean. Artificial neural nets for controlling a 6-legged walking system. In *Second International Conference on Simulation of Adaptive Behavior*, 1993.

16. D. Hubel and T. Wiesel. Brain mechanisms of vision. *Scientific American*, 1979.

17. H. Kitano. Designing neural network using genetic algorithm with graph generation system. *Complex Systems*, 4:461–476, 1992.

18. John R. Koza. *Genetic programming: A paradigm for genetically breeding computer population of computer programs to solve problems*. MIT press, 1992.

19. John R. Koza. *Genetic programming II: Automatic Discovery of reusable programs*. MIT press, 1994.

20. Eric Mjolness, David Sharp, and Bradley Alpert. Scaling, machine learning and genetic neural nets. La-ur-88-142, Los Alamos National Laboratory, 1988.

21. H. Muehlenbein, M. Schomish, and J.Born. The parallel genetic algorithm as function optimizer. *Parallel Computing*, 17:619–632, 1991.

22. Parisi and Nolfi. Morphogenesis of neural networks. Technical report, University of Rome, 1992.

23. P. Prusinkiewicz and A. Lindenmmayer. *The algorithm beauty of plants*. 1992.

24. Karl Sims. Evolving 3d morphology and behavior by competition. In R. Brook and P. Maes, editors, *4th Intern. Conf. on Artificial Life, MIT Ress*, 1994.

25. J. Vaario. *An emergent modeling method for artificial neural networks*. PhD Thesis, University of Tokyo, 1993.

An Artificial Life Approach
for the Synthesis of Autonomous Agents

Olivier MICHEL

E-mail: om@alto.unice.fr, Web: http : //alto.unice.fr/ ∼ om/

University of Nice–Sophia Antipolis, Laboratoire I3S — CNRS, bât. 4
250, av. A. Einstein 06560 Valbonne, France

Abstract. This paper describes an evolutionary process producing dynamical neural networks used as "brains" for autonomous agents. The main concepts used : genetic algorithms, morphogenesis process, artificial neural networks and artificial metabolism, illustrate our conviction that some fundamental principles of nature may help to design processes from which emerge artificial autonomous agents. The evolutionary process presented here is applied to a simulated autonomous robot. The resulting neural networks are then embedded on a real mobile robot. We emphasize the role of the artificial metabolism and the role of the environment which appear to be the motors of evolution. The first results observed are encouraging and motivate a deeper investigation of this research area.

1 Introduction

Most of artificial life researchers try to obtain synthetic forms of organization inspired from biological principles of life. The interdisciplinarity of this research area induces the integration of biological concepts inside artificial constructions. Although most of the vocabulary and inspirations are borrowed from biology, we do not claim to model biological processes. Our aim is to apply fundamental principles of evolution and self-organization to autonomous agents.

The evolutionary approach applied to autonomous agents is today a rising research area. Various methods might be mentioned which make use of evolutionary processes on different structures. Rodney Brooks [3] proposed a high level language, GEN, which may evolve under genetic control, and which can be compiled into BL (Behavior Language), a lower level language, dedicated to a real mobile robot. He outlines the danger of the lack of realism of simulators but also acknowledges that such tools seem to be necessary for evolutionary processes.

Marco Colombetti and Marco Dorigo [4] make use of classifier systems for the control of mobile robots. Different kinds of classifier systems are used for different kinds of behaviors, the overall system being also supervised by a classifier system. Experiments are presented on a simulator and on a real robot.

Many authors addressed the evolution of neural structures dedicated to autonomous agents. The different approaches present various chromosome codings and morphogenesis processes. Hugo de Garis proposed a cellular automaton based morphogenesis involving CAM machines (parallel computers dedicated to cellular automata) [5]. Frédéric Gruau defined a rather complex grammar tree based process [7]. Stefano Nolfi, Orazio Miglino and Domenico Parisi designed a neural network morphogenesis by modeling dendritic growth and allowing the environment to influence the morphogenesis process. Finally, Dario Floreano and Francesco Mondada [6], as well as Inman Harvey, Phil Husband and Dave Cliff [8] have been among the first ones to test evolutionary neural networks on real robots.

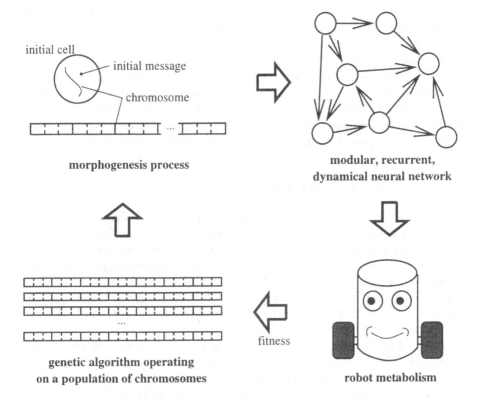

Fig. 1. The Evolutionary Loop

The evolutionary loop we propose (see figure 1) involves successively a genetic algorithm evolving chromosomes, a morphogenesis process allowing to decode a chromosome into a neural network, a dynamical neural network driving a mobile robot and finally an artificial metabolism defining the viability domain of the robots and returning a fitness value to the genetic algorithm. This methodology was applied in order to observe the emergence of mobile robot behaviors. The genetic evolution occured in simulation and the resulting neural networks were then embedded on the real robot.

2 Models

2.1 Genetic Algorithm and Morphogenesis Process

We used a genetic algorithm operating on a population of variable length chromosomes. The genes are production rules used by the morphogenesis process. The genuine structure of the rules, the genetic algorithm and the morphogenesis process are not detailed in this paper and will be presented in a future paper.

At the beginning, the initial population is randomly generated. The particular structure of the chromosomes led us to make use of a special set of genetic operators including gene addition or deletion.

We designed a morphogenesis process allowing to build recurrent neural networks using a chromosome organized in a pseudo-linear structure. It takes inspiration from the biological explanation of protein synthesis regulation [10]. This recurrent process allows an easy generation of modular neural networks (where the same sub-structures exist at different places in the same network). Moreover, due to a strong epistasis, it features some properties of dynamical systems that could permit to generate complexity at the edge between chaos and order [9].

2.2 Artificial Neural Networks

Many reasons led us to use dynamical neural networks. This family of networks includes multi-layer perceptrons as well as recurrent networks. Consequently, it seems to be rather universal. Moreover, the properties of such networks are very interesting for autonomous agents (temporal processing, sequence generation and recognition, models of memory, etc.). It is possible to implement local learning algorithms, cheap in computation time and friendly to parallel computation. Their very complex dynamics make their structural design very hard. Genetic algorithms may be suitable for such a task.

Here, all the neurons have the same transfer function (linear thresholded). When the state of a neuron is close to 1, it will be said to be excited. If the state of a neuron is close to 0, it will be said to be inhibited (or at rest). Let $x_i(t)$ be the state of the neuron i at iteration t and ω_{ij}, the weight of the link between neuron j and neuron i. The state of neuron i updates as described here :

$$x_i(t+1) = \begin{cases} 0 & \text{if } \sum_j \omega_{ij} x_j(t) \leq 0 \\ 1 & \text{if } \sum_j \omega_{ij} x_j(t) \geq 1 \\ \sum_j \omega_{ij} x_j(t) & \text{otherwise} \end{cases}$$

Different kinds of links exist. In the current version, all the links have a given weight equal to a positive or a negative real number. It remains possible to add other kinds of links whose weight could evolve according to a specific learning rule (Hebb rule, Anti-Hebb rule, etc.) A collection of different kinds of links has been carefully designed, allowing to build various neural schemes (see figure 2) that can be seen as building blocks for a larger neural network.

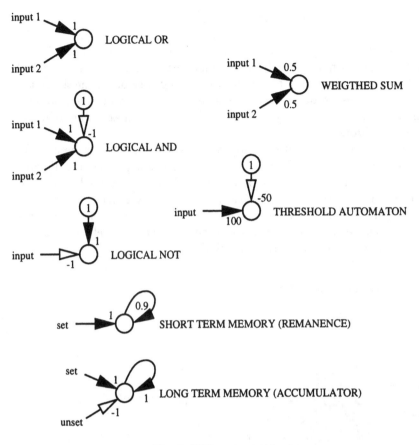

Fig. 2. Different neural schemes

2.3 Metabolism based Fitness Value

The fitness value used by genetic algorithms to perform selection is computed using the concept of artificial metabolism.

At the beginning, each robot receives an initial amount of energy points. The definition of the interactions between the robot and its environment will rule the variation of energy points of the robot. For example, in the case of a mobile robot, one could choose to define the loss of one energy point at regular period of time to penalize slowness or idleness ; the loss of one energy point at each move would penalize useless moves and finally the gain of several energy points would reward satisfactory behaviors. An example of satisfactory behavior could be something like *wall following, moving while avoiding obstacles, target pursuit,* or *object seeking.*

Each robot is set in the environment with its initial amount of energy points during a given time t. At the end, the remaining energy points will be used to assign a fitness value to the robot. This value will be used by the genetic algorithm to select the best individuals.

3 Artificial Life Simulator

This is the name of the program we developed in order to test our models. It features a mobile robot simulator including a simulated environment, a dynamical neural network simulator connected to the sensors and effectors of the simulated robot, a morphogenesis process used to generate neural networks from chromosomes and a genetic algorithm using a metabolism based fitness value to evolve a population of chromosomes.

This simulator was designed to evolve neural networks that can be directly embedded on the real robot. Even if the simulator is not perfect, i.e., if some divergence between the simulated robot and the real one are observed, it remains possible to continue evolution on the real robot for a few generations in order to fine tune parameters of the neural network. Our first series of experiments did not necessitate such an operation, which demonstrates the reliability of the simulator concerning the behaviors we evolved.

3.1 Description of the robot

Fig. 3. Khepera (5 cm diameter) and its simulated counterpart

Dedicated to *Khepera* [11], the simulated mobile robot includes 8 infrared sensors (small rectangles) allowing it to detect the proximity of objects in front of it, behind it, and to the right and the left sides of it. Each sensors return a value ranging between 0 and 1023. 0 means that no object is perceived while 1023 means that an object is very close to the sensor (almost touching the sensor). Intermediate values may give an approximate idea of the distance between the sensor and the object. Sensors are connected to the inputs of the neural network according to the following description : The state of the first input neuron I_0

is always 1 to ensure at least one active input in the neural network (necessary for some neural schemes), the states I_x of the other input neurons are computed according to the values of the sensors S_x, $x \in [1; 8]$.

$$\begin{cases} I_0 = 1 \\ I_x = S_x \div 1023 \text{ with } x \in [1; 8] \end{cases}$$

Each of the right and left motors are driven by two output neurons. Each motor can take a speed value ranging between -10 and $+10$. Each neuron can take a state value ranging between 0 and 1. The speed value of a motor is given by the state value of the first neuron minus the state value of the second one, the resulting value being multiplied by 10. This choice makes that, on one hand, when both neurons are inhibited, no motor action occurs (no excitation induces no move), and on the other hand, when the "move-forward" motor neuron and the "move-backwards" motor neuron are simultaneous active, the actual move results of the competition between these two neurons. The direction of the speed is given by the winning neuron, and the amplitude of the speed corresponds to the amplitude of the success of a neuron against the other. Let Vl (resp. Vr) be the velocity of left motor (resp. right motor) and let O_x, $x \in [0; 3]$ be the states of the output neurons of the network, the velocity of the motors is given by the following equations :

$$\begin{cases} V_g = (O_0 - O_2) \times 10 \\ V_d = (O_1 - O_3) \times 10 \end{cases}$$

3.2 Description of the environment

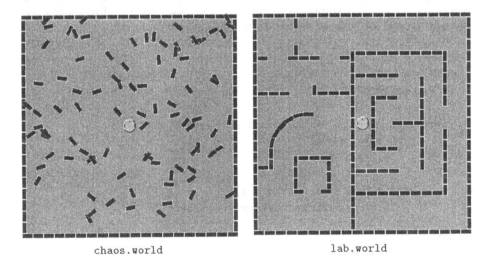

chaos.world   lab.world

Fig. 4. Two examples of simulated environments

Bricks (rectangular objects) are laid by the user in the environment et allow to design a maze a more or less complex (see figure 4). The real dimensions of this simulated environment (comparing to Khepera) are $1m \times 1m$.

3.3   Obstacle avoidance experiment

The fitness function. The fitness function used here is not metabolism based, as described earlier. It optimizes a criterion (a distance). Anyway, the robot has an artificial metabolism which is not used for the calculation of the fitness function, but which allow to optimize the time of evaluation for a robot (a good robot could "live" longer that a bad one). Like in our previous work [1], our future research will make use of a metabolism based fitness function.

A robot staying motionless avoid perfectly obstacles ! A robot that turn round on itself indefinitely will probably avoid obstacles as well. Finally, a robot that go forward, and then backwards and then forward again, etc. will certainly avoid surrounding obstacles. All these solutions have been discovered by our genetic algorithm and have led us to define a fitness function forcing the robot to big moves in order to observe more elaborated behaviors than those observed earlier.

Consequently, we chose to reward (gain of energy points) the robots that get as far as possible from their initial position. The reward is proportional to the difference between the new position of the robot and the last position the more away from the initial position. A punishment (loss of energy points) is inflicted to the robot as soon as it hits an obstacle. It consists in a reduction to zero of its energy points. This punishment will allow evolution to design robots that avoid the obstacles they meet. Moreover, a regular slow diminution of the energy points disadvantages motionless or idle robots. As soon as the energy of a robot reaches 0, the robot is stopped and assigned a fitness value equal to the d_max distance it reached. Here is the principle of the corresponding algorithm :

```
d_max  = 0;
energy = 100;
while(energy > 0)
{
  RunRobot();        /* compute d and HitObstacle */
  energy = energy - 1;
  if (d > d_max)
  {
    energy = energy + (d - d_max);
    if (energy > ceiling) energy = ceiling;
    d_max = d;
  }
  if (HitObstacle) energy = 0;
}
fitness = d_max;
```

d is the distance between the robot and its initial position, d_max is the maximal value of d, *energy* is the amount of energy points of the robot (artificial metabolism), *ceiling* is the maximal amount of energy points the robot can reach, *HitObstacle* is a boolean variable saying if the robot hits an obstacle or not, and finally *fitness* is the fitness value returned.

The environment being closed, the robot could not go away indefinitely (d_max is bounded), even if it doesn't hit any obstacle, it will gradually loose all its energy points, so the evaluation will be over.

The simulation results. We tested the evolutionary process in two different environments depicted on figure 4 : lab.world and chaos.world. In each environment, the evolutionary process was stopped after 1000 generations. Various results are observed in each environment.

In lab.world, which is a maze with rectilinear walls and right angles, evolution generated robots that follow the walls on their right. This behavior drives them significantly far away from their initial position. The figure 5 (a) illustrates the convergence of the fitness function while the figure 6 (a) shows the best neural network resulting from evolution.

In chaos.world, where the obstacles are randomly laid in the environment, the evolution generate robots that move forward and strongly turn to the left as soon as they detect an obstacle, then they start again moving forward in their new direction (see figures 5 (b) and 6 (b)). The fitness jump observed around generation 150 shows the random discovery of a neural connection giving to the robot a better behavior. This feature illustrates the ruggedness of the fitness landscape, making it harder for the genetic algorithms.

In both cases, robots are finally able to avoid obstacles most of the time, they are also able to turn back in a dead end. However, robots generated in lab.world are "lost" if set in chaos.world far from any obstacle. Indeed, if they perceive no obstacle, those robots will turn round, with the "hope" of finding a wall to follow. This strategy, which is very efficient in lab.world because the obstacles are always sufficiently close to the robot, appears to be inefficient in chaos.world because it may happen that the robot is far away from any obstacle, and so the robot cannot find any wall to follow using the "turn round" strategy. The "lost" robot will then turn round indefinitely.

A second test performed in chaos.world led to the emergence of robots whose behavior resembles the behavior of those obtained during the first test, but that remain anyway slightly less efficient (not so fast, fitness value not so good, see figure 5 (c) et 6 (c)). This result is a consequence of the fact that genetic algorithms use random values. So, the evolutionary process seems to be very sensitive to this random aspect of genetic algorithms.

One could notice that the structures of the resulting neural networks is rather simple and could be compared to the simplest Braitenberg's vehicles [2] (see figure 6 (d)). Even if they are perfectly adapted to the given problem, such networks exhibit only reactive behaviors and are unable of learning.

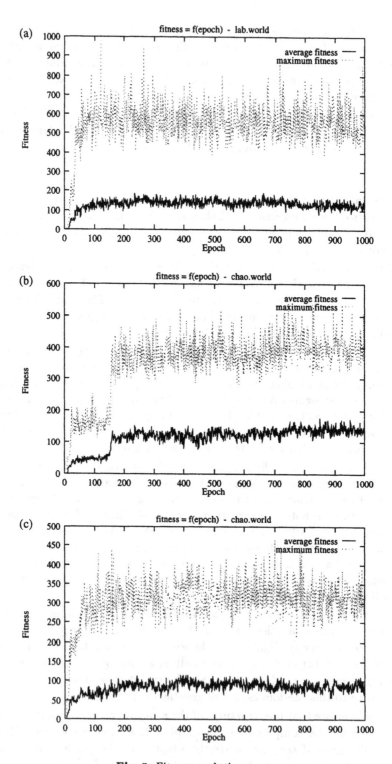

Fig. 5. Fitness evolution

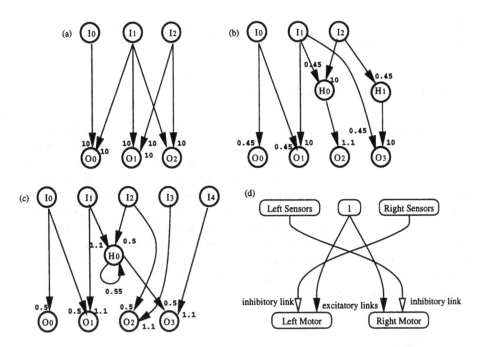

Fig. 6. Resulting neural networks (a, b, c), looking like Braitenberg simplest vehicles (d)

The real robot. The neural networks obtained with the simulation were directly transfered to Khepera. The behaviors observed on the real robot seem to be similar to the ones observed on the simulator : wall following for the first robot and forward move with left turn if an obstacle is seen for the second and third robots. These first behaviors are very simple, thus it is difficult for the moment to make a precise evaluation of the reliability of the simulator. The only observable differences between simulated and real behavior are only quantitative. For example, we observed in the wall following behavior that the real robot slightly rubed the walls while its simulated counterpart didn't touch the wall, although it came very close.

Discussion. In both cases, the evolutionary process discovered solutions that correspond to the given problems. These results outline the important role of the environment which, associated with the definition of fitness function, seems to be one of the motors of the evolutionary process.

Moreover, several runs of the same evolutionary process in the same initial conditions underlined the random aspect of genetic algorithms leading to different solutions. This fact might be explained by the shape of the fitness landscape which is very rugged, so the genetic algorithm get caught in local optima. Adding ecological niches to our evolutionary model could be an answer to that problem. Evolution would occur independently in different ecological niches and migrations would be rare to allow the really best individuals to settle in neighboring niches and to explore from there various directions of the search space.

4 Conclusion and Future Work

The evolutionary process presented in this paper seems able to generate simple solutions to the problems given here. The reactive behaviors observed here are a consequence of a phylogenetic adaptation of the robots. However, the neural network model as well as the morphogenesis process we used, allow to obtain much more complex neural structures and especially neural structure able to learn during the "lifetime" of the autonomous agent (ontogenetic adaptation). The emergence of this kind of adaptive neural network seems to be strongly constrained by the environmental dynamics and the definition of the artificial metabolism of the agent. Our current research focuses on the determination of such constraints that would allow the emergence of more complex neural structures and behaviors.

References

1. Joëlle Biondi and Olivier Michel. Evolution de structures neuronales. application à un robot mobile autonome. In *Actes des journées de Rochebrune*, ENST, 46 rue Barrault - 75634 Paris cedex 13, Mars 1995. ENST 95 S 001.

2. Valentino Braitenberg. *Vehicles: Experiments in Synthetic Psychology*. MIT Press, Cambridge, 1984.

3. R.A. Brooks. Artificial life and real robots. In F. Varela and P. Bourgine, editors, *Towards a Practice of Autonomous Systems, Proceedings of the First International Conference on Artificial Life, Paris*. MIT Press, 1992.

4. Marco Colombetti and Marco Dorigo. Learning to control an autonomous robot by distributed genetic algorithms. In Jean-Arcady Meyer, H. Roitblat, and Wilson Stewart, editors, *From animals to animats 2: Proceedings of the Second International Conference on Simulation of Adaptive Behavior*. MIT Press, Bradford Books, 1992.

5. Hugo De Garis. Cam-brain: The evolutionary engineering of a billion neuron artificial brain by 2001 which grows/evolves at electronic epeeds inside a cellular automata machine (cam). In D. W. Pearson, N. C. Steele, and R. F. Albrecht, editors, *Proceedings of the International Conference on Artificial Neural Networks and Genetic Algorithms (ICANNGA'95)*. Springer-Verlag Wien New York, 1995.

6. Dario Floreano and Francesco Mondada. Automatic creation of an autonomous agent: Genetic evolution of a neural-network driven robot. In Dave Cliff, Phil Husband, Jean-Arcady Meyer, and W. Stewart Wilson, editors, *From animals to animats 3: Proceedings of the Third International Conference on Simulation of Adaptive Behavior*. MIT Press, 1994.

7. Frédéric Gruau and Darrell Whitley. The cellular developmental of neural networks : the interaction of learning and evolution. Technical report, Ecole Normale Supérieure de Lyon, 46, Allée d'Italie, 69364 Lyon Cedex 07, France, January 1993.

8. Inman Harvey, Philip Husbands, and Dave Cliff. Seeing the light: Artificial evolution, real vision. In Dave Cliff, Phil Husband, Jean-Arcady Meyer, and W. Stewart Wilson, editors, *From animals to animats 3: Proceedings of the Third International Conference on Simulation of Adaptive Behavior.* MIT Press, 1994.

9. Stuart A. Kauffman. *The Origins of Order: Self-Organisation and Selection in Evolution.* Oxford University Press, 1993.

10. Olivier Michel and Joëlle Biondi. Morphogenesis of neural networks. *Neural Processing Letters*, 2(1), January 1995.

11. F. Mondada, E. Franzi, and P. Ienne. Mobile robot miniaturisation: A tool for investigation in control algorithms. In *Third International Symposium on Experimental Robotics*, Kyoto, Japan, October 1993.

How Long does it Take to Evolve a Neural Net ?

Marc Schoenauer and Edmund Ronald

CMAP, CNRS-URA 756, Ecole Polytechnique, 91128 Palaiseau, France
{Marc.Schoenauer,Edmund.Ronald}@polytechnique.fr

Abstract. This paper deals with technical issues relevant to artificial neural net (ANN) training by genetic algorithms. Neural nets have applications ranging from perception to control; in the context of control, achieving great precision is more critical than in pattern recognition or classification tasks. In previous work, the authors have found that when employing genetic search to train a net, both precision and training speed can be greatly enhanced by an input renormalization technique. In this paper we investigate the automatic tuning of such renormalization coefficients, as well as the tuning of the slopes of the transfer functions of the individual neurons in the net. Waiting time analysis is presented as an alternative to the classical "mean performance" interpretation of GA experiments. It is felt that it provides a more realistic evaluation of the real-world usefulness of a GA.

1   The Usefulness of Automatic Parameter Tuning

The operator of a heuristic program spends a lot of time predicting what his program will do. Now and then a test run actually validates the programmer's prediction and life, science, and everything is wonderful – yet more often, the program goes off to do its own thing and the programmer is left to scratch his head.

In this context, manual tuning of parameters is one of the least rewarding facets of heuristic programming. For instance, the authors have spent hours in front of computer screens, when investigating neural net training. These hours were occupied by running the same program time and again with just the change of a single real parameter like the slope of the neural transfer function.

Computer programming is of necessity experimental; however every worker in the field of genetic algorithms has been brought to conjecture that the experimenter's action could be automated. In the case of GA research, the manual variation of GA parameters – e.g. search for good mutation or crossover rates – could be replaced by the action of a meta-GA. Unfortunately, the computational expense of running a population of GA's in parallel usually discourages the GA experimenter from pursuing such a course of automatic experimentation, although object-oriented software engineering makes a 2-level GA easily feasible. Indeed, the possibility of running a meta-GA is a design goal of the authors' next although object-oriented software engineering makes a 2-level GA easily feasible. Indeed, the possibility of running a meta-GA is a design goal of the authors' next generation GA software.

In the special case of the authors' research in neural net control, some of the parameters originally subject to tuning could be varied by the same GA employed to train the net by searching the space of weight matrices. In this document the results of this automated research is confronted with the results originally published in [Ronald & Schoenauer 93] .

In summary, by confronting the results of our previous work with this automated tuning by GA, we show that the GA improves on a human operator in tuning some of the net parameters, namely the transfer functions. On the other hand, automatic renormalization, as presented in section 5 does not improve mean precision, and indeed produces barely acceptable results; but it holds the promise of obsoleting the painful data pre-processing steps which hinder real-world and industrial applications of neural nets.

We have included a waiting-time interpretation of our results; we believe that this interpretation methodology, while unusual, is more appropriate in an industrial context than mean performance: GAs are stochastic, and estimations of their performance must perforce be formulated in probabilistic terms. A waiting time estimation provides us with a confidence factor whereby a control problem can be solved with a predetermined amount of computation.

The plan of the paper follows: In section 2 below we recall the neural net formalism and its application to control, and summarize earlier work. Section 3 presents an investigation into the automatic tuning of the neural transfer function slopes. Section 4 introduces the waiting time analysis, and applies it to the results of section 3. Section 5 describes experiments in generalized tuning of both data renormalization coefficients and transfer functions. Section 6 discusses the significance of the results, and highlights some possible applications.

2 Genetic Training of Neural Net Controllers

In order to make this paper self-contained, this section summarizes the methods employed by the authors for neural net control in [Ronald & Schoenauer 93], [Schoenauer & Ronald 94]. This establishes the background for the numerical experiments on the lunar lander simulator, whose results form the body of sections 3 and 4 below. The reader desiring an overview of the field may refer to [Yao 93], a broad survey of evolutionary methods as applied to neural net training and design.

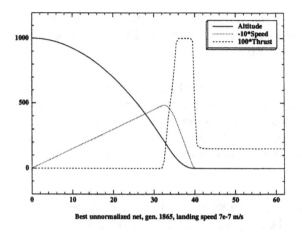

Best unnormalized net, gen. 1865, landing speed 7e-7 m/s

Figure 1: *Best non-normalized control action in study.*

2.1   Lunar lander dynamics

This paper deals with training a neural net to land a simulated rocket-driven lunar-lander module, under a gravity of 1.5 m/s^2. This simulation has given rise to numerous computer games since the advent of interactive computing, and will be familiar to most readers. We have simplified the model already employed by [Suddarth 88].

The lunar module is dropped with no initial velocity from a height of 1000 meters. The fuel tank is assumed to contain 100 units of fuel. Burning one unit of fuel yields one unit of thrust. Maximum thrust is limited to 10 units and the variation of the mass of the lunar module due to fuel consumption is not taken into account. The simulation time-slice was arbitrarily fixed at 0.5 seconds.

All the experiments reported in this document were conducted with the given initial height; the net can be trained to solve the general control problem by the expedient of supplying random starting points. In fact, the authors performed experiments solving the general control problem for lunar lander. But the general problem was deemed too expensive computationally to allow for such exhaustive comparative experimentation as the experiments whose results are reported here.

The input parameters relayed to the neural net once every time-step are the speed, and the altitude. The net then computes the desired fraction of maximal thrust, on a scale from 0 to 1, which is then linearly rescaled between 0 and 10 units.

A very nice lunar landing effected in this study is shown in Figure 1. This achieved a landing speed of $7\ 10^{-7}m/s$, i.e.. $0.7mm/s$, hardly enough to mark the lunar surface! Such excellence would hardly be necessary in practice. This result was obtained by straightforward application of neuro-genetic control, with no data renormalization.

It was attained at the price of a long run, namely almost 2000 generations. The control action is also exemplary in the economy of fuel in the deceleration phase: For the main deceleration burst, thrust is pulsed in what amounts to almost a square wave to its maximal permissible value. The remarkable landing softness is attained by means of a long - and very smooth- hover phase which begins immediately after deceleration.

The best landing we achieved in this study is shown in Figure 2. The landing speed of $7\ 10^{-9}m/s$, i.e.. $0.002mm/s$ is a 2 orders of magnitude improvement over the previous result, and was attained in half as many (1035) generations. This is an achievement of data renormalization: The net was presented with the two previously cited state inputs, namely speed and altitude, and another pair consisting of the same variables pre-multiplied by a factor of 10.

The interesting features of the best "Armstrong", as displayed in Figure 2, are its surprising precision, and the fact that this precision is attained either in spite of, or more probably, because of the displayed sawtooth shape and roughness of the thrust control. Of course, control by rocket to fit a speed tolerance of 20 Angstroms/s seems rather implausible in reality.

2.2   The Networks

A classical 3-layered net architecture was employed, with complete interconnection between layers 1 and 2, and 2 and 3. The neural transfer function (non-linear squashing) was chosen to be the usual logistic function $\mathcal{F}$, a sigmoid defined by

$$\mathcal{F}(x) = \frac{1}{1 + e^{-\alpha x}}$$

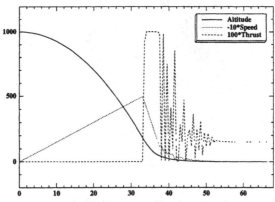

Best 10-normalized net, gen. 1035, landing speed 2e-9 m/s

Figure 2: *Best overall control action in study.*

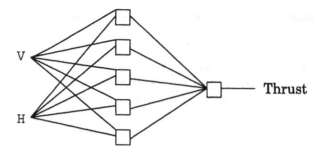

2-input Net

Figure 3: *Totally connected feedforward 2-5-1 architecture.*

In our work the parameter α was fixed, $\alpha = 3.0$. Each individual neuron j computes the traditional [PDP 86] squashed sum-of-inputs

$$o_j = \mathcal{F}(\sum_i w_i^j x_i)$$

Regarding the sizing of the middle layer, we chose to apply the Kolmogorov model [Hecht-Nieslen 90] which for a net with n inputs and 1 output requires at least 2n+1 intermediate neurons.

Only two net architectures occur in this paper. Both types have only one output (controlling the lunar module's thrust). The canonical method for solving the control problem entails 2 inputs (see Figure 3), namely speed and altitude, appropriately normalized between 0 and 1.

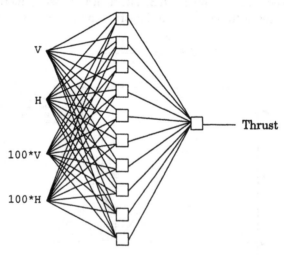

100-renormalised Net

Figure 4: *The actual inputs have been duplicated and renormalized.*

However, the optimization by input renormalization which forms the core of [Ronald & Schoenauer 93] entails adding two inputs to the above cited net, therefore employing 4-input nets (see Figure 4). As in both cases we have adhered to the Kolmogorov paradigm, we are studying 2-5-1 and 4-9-1 nets, which respectively have 21 and 55 weights/biases.

With the topology fixed, training these nets for a given purpose is a search in a space of dimension 21 or 55, to which must be added real numbers representing the parameters α, and those for the renormalization, when these parameters are left to the GA to find. In the neuro-genetic approach, net training is thus a search in a 21 or 55-dimensional space, to which must be added real numbers representing the parameters α, and those for the renormalization, when these parameters are left to the GA to find.

2.3 The Genetic Model

Our genetic algorithm software subjects a small population of nets to a crude parody of natural evolution. This artificial evolutionary process aims to achieve nets which display a large *fitness*. The fitness is a positive real value which denotes how well a net does at its assigned task of landing the lunar module. Thus our fitness will be greater for slower landing speeds. The details of the calculation of the fitness are found below.

For the purpose of applying the genetic algorithm, a net is canonically represented by its weights, i.e. as a vector of real numbers. During the initial stages of this work, two distinct home-brew GA software packages were employed. One package followed the first methods presented by John Holland in that it uses bit-strings to encode floating-point numbers. The second software package, described below, was a hybridized GA which directly exploits the native floating-point representation of the workstations which it was run on. The hybridization towards real numbers is described in [Radcliffe 91] and [Michalewicz 92]. The results obtained with both programs were consistent, and only the experiments with the floating point package are detailed in this document.

The genetic algorithm progresses in discrete time steps called generations. At every generation the fitness values of all the nets in the population are computed. Then a new population is derived from the old by applying the stochastic selection, crossover and mutation operators to the members of the old population.

- Selection is an operator that discards the least fit members of the population, and allows fitter ones to reproduce. We use the roulette wheel selection procedure as described in [Goldberg 89], with fitness scaling and elitism, carrying the best individual over from one generation to the next.
- Crossover mixes the traits of two parents into two offspring. In our case, random choice is made between two crossover operators: Exchange of the

weights between parents, at random positions. Or assigning to some of the weights of each offspring a random linear combination of its parents' weights.
– Mutation randomly changes some weights by adding some gaussian noise.

The experiments described in the next section used the following parameters in the GA: The net connection weights forming the object of the search were confined to the interval $[-10, +10]$ thereby avoiding overflow conditions in the computation of the exponential. Population size was held to 50 over the whole length of each run, fitness scaling was set to a constant selective pressure of 2.0, crossover rate was 0.6, mutation rate 0.2, the gaussian noise had its standard deviation set to half the weight space diameter, i.e.. 10.0, decreasing geometrically in time by a factor 0.999 at each generation.

3 Adjusting Transfer Functions

3.1 The family of logistic transfer functions

A great deal of work in the neural net community has employed logistic sigmoid transfer functions of the form

$$\mathcal{F}_\alpha(x) = \frac{1}{1 + e^{-\alpha x}}$$

where the parameter α, usually set to 1, can be seen to be defining the slope of the function for $x = 0$, i.e..

$$\alpha = \mathcal{F}_\alpha'(0)$$

In the Figure 5 below, we show the behavior of $\mathcal{F}_\alpha$ for values of α, taken from the set $\{0.1, 0.5, 1.0, 2.0, 5\}$, with the slopes steepness increasing with α from a quasi-linear function, through a sigmoid to a very steep staircase.

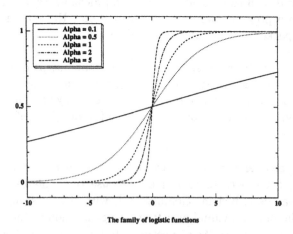

The family of logistic functions

Figure 5: *The sigmoïd function for various values of* α.

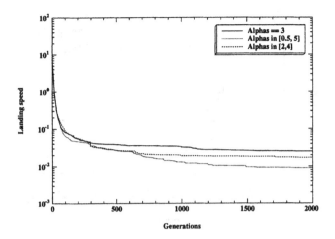

Figure 6: *Mean performance with varying* α.

In an analogy to biological systems, α can be interpreted as a chemical parameter defining the sensitivity of a neuron. However, researchers in back-propagation usually fix this parameter, although Yao, in his survey [Yao 93] cites [Mani 90], as having experimented with a modified form of back-prop which performs gradient descent learning by adjusting the α slopes as well as the synaptic connection weights themselves. However, Yao also cites [Stork &al. 90] as having applied a GA approach to the definition of both neural connectivity and transfer functions, an investigation similar to the one reported here.

3.2 Experimental Results

In our case, we investigated how allowing a GA to individually tune the transfer functions affects the precision of control in the "genetic lander" experiment. Our results in allowing the GA to vary the α slope for each individual neuron in the controller net may be compared with work reported in [Ronald & Schoenauer 93], where the slopes of the transfer functions of all neurons were held constant at $\alpha = 3$.

In Figure 6 we have graphed the performance of the GA search, comparing the mean of a set of 60 [1] runs of 2000 generations length with fixed α, as reported in [Ronald & Schoenauer 93] with two new experiments of identical size. In these new experiments the GA was permitted to adjust the slope of each neuron, within the limits indicated in the figure.

It can be seen that the mean maximum precision of the runs greatly improves when the GA is allowed to search a larger range of α slopes. The computational cost involves adding just one degree of freedom per neuron, and can thus be considered negligible. In summary, we find the technique of varying the slopes fully successful here, and we believe it deserves to be investigated in other contexts.

---

[1] 4 runs on each CPU of a 15 workstation computer farm

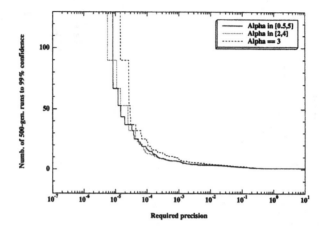

Figure 7: *Waiting time estimation with varying α.*

4 Waiting Time Estimation

In the real world, a user who wishes to exploit a GA for a given task is not interested in *average* performance over a number of runs – he may rather wish for an estimation of how much it will take to achieve a result of a given quality: Such a result is an acceptable solution to the control problem, which can be acted upon.

Now, if we assume a really bizarre genetic algorithm which yields a very bad result e.g. a crash at 10 m/s for half the random seeds, and an excellent result e.g. $10e^{-3}$ m/s for the other half, then the *average result* is a 5 m/s crash! However, just a very few runs of the algorithm will usually yield one of the "good" controllers, capable of piloting a safe flight.

We may wish to formalize the above reasoning by asking the following question like someone who must run the GA in batch processing: If I know that on average one run yields an acceptable result with probability a, then how many runs N must I schedule in order to have probability p or more of finding a good run in the schedules runs?

Elementary probability theory tells us that the confidence number N(p,a) is given by the equation

$$N(p,a) \geq log(p)/log(1-a)$$

As an example of this formula, assuming we get it right one time out of two as above, and wish to be 99.9% sure of getting it right, we will substitute $a = 0.5$ and $p = 0.001$ in the formula, and obtain $N = 10$. So the time required for 10 runs ought to be budgeted for.

The above reasoning allows us to exploit the results of our computer runs with the aim of graphing the confidence factors. In this way, we transformed the data of the runs illustrated in Figure 6 above, yielding the graphs of Figure 7.

The results of comparing the graphs in Figures 6 and 7 – created from the same experimental data – are nothing short of amazing! When we peer at Figure 6, the mean landing speed of the 60 runs of 2000 generations' duration never reaches 10^{-3}. The waiting time graph in Figure 7 however tells us that 10^{-3} is a soft landing which we can reach with a confidence factor of 99% by effecting just 10 runs of 500 generations duration! The waiting time graph can be interpreted as showing that precisions of 10^{-4} are perfectly attainable in practice with any of the investigated learning methods, be it manual or automatic tuning of the slopes.

5   Generalized Tuning

In Figures 9 and 10 below we have reproduced the mean performance and corresponding waiting time graphs obtained from 4 experimental batches of 60 runs each, with and without GA search for input renormalization coefficients, i.e. using the network of Figure 4 or that of Figure 8 below. It can be seen that having no renormalization at all is better than leaving the search of renormalization coefficients $R_1, \ldots, R_4$ entirely to the GA to find, within the interval $[10^{-10}, 10^{10}]$.

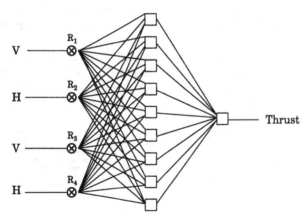

Generalised 4-input Net

Figure 8: *Generalized renormalization network: the renormalization coefficients are adjusted by the GA search.*

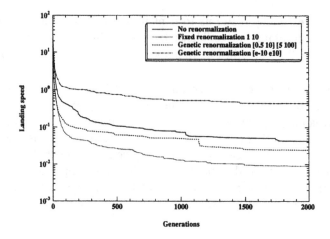

Figure 9: *Mean performance with varying α and renormalization.*

However, these results may be considered less discouraging when the reader is told that the inputs for "full genetic renormalization" were not pre-processed in any way! Thus the neural net, as trained by the GA, was left to deal with input values of e.g. starting altitudes of 1000 m/s and free-fall speeds to 500 m/s, whereas in all other experiments neural input data had been folded (by hard code) into [0, 1]. Hence the "full genetic renormalization" results are still promising - they yield a case of functional albeit imprecise control, as might be exploited by a prototype application. However such a prototype is obtained without any human expertise whatsoever being applied in the problem-domain.

In this way one might imagine this "full genetic renormalization" experiment to represent an industrial context, in which a feasibility study would be done for GA learning, without any prior attempt at data pre-processing. The authors would not believe their lives in danger even if landing is effected at 10 cm/s, a

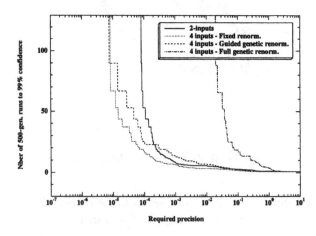

Figure 10: *Waiting time estimation with varying α and renormalization*

speed which the graph shows the GA-trained nets routinely achieve. Of course, a hand–tuned net able to land at 10^{-7} m/s or so, would be more appropriate for piloting the director of a funding agency visiting a scientific lunar base.

6 Discussion

The results of the numerical experiments reported in this paper are mixed. On the one hand our data indicates that the slopes of the neural transfer functions can be profitably trained by the same GA which trains the net. On the other hand the experiments on training renormalization coefficients by GA proved surprisingly disappointing, and will be investigated further. This should be contrasted with past experiments in which tuning these coefficients by hand yielded excellent results.

The waiting time metric which we in introduced in section 4 above seems more suited to the logic of industrial application of GA control than the more conventional mean performance metric. Moreover, the counter-intuitive nature of this metric might inspire some statistical re–interpretation of existing data. Indeed, it may be profitable to unearth old experiments, deemed failures by the mean fitness method, and re-evaluate them. Let us remember the dictum of Edgar Allan Poe, that there be lies, damned lies, and statistics!

7 Acknowledgments

We would like to thank Steve Suddarth and Michèle Sebag for their entertainment and advice.

References

[Angeline, Saunders & Pollack 94] P. J. Angeline, G. M. Saunders and J. B. Pollack, An evolutionary algorithm that construct recurrent neural networks. To appear in *IEEE Transactions on Neural Networks*.

[Goldberg 89] D. E. Goldberg, *Genetic algorithms in search, optimization and machine learning*, Addison Wesley, 1989.

[Harp & Samad 91] S. A. Harp and T. Samad, Genetic synthesis of neural network architecture, in *Handbook of Genetic Algorithms*, L. Davis Ed., Van Nostrand Reinhold, New York, 1991.

[Hecht-Nieslen 90] R. Hecht-Nielsen, *Neurocomputing*, Addison Wesley, 1990.

[Holland 75] J. Holland, *Adaptation in natural and artificial systems*, University of Michigan Press, Ann Harbor, 1975.

[Kitano 90] Empirical studies on the speed of convergence of neural network training using genetic algorithms. In *Proceedings of Eight National Conference on Artificial Intelligence, AAAI-90*, Vol. 2, pp 789-795, Boston, MA, 29 July - 3 Aug 1990. MIT Press, Cambridge, MA.

[Mani 90] G. Mani, Learning by Gradient Descent in Function Space, in *Proceedings of IEEE Conference on System, Man, and Cybernetics*, pp. 242-247, Los Angeles 1990.

[Michalewicz 92] Z. Michalewicz, Genetic Algorithms + Data Structures = Evolution Programs, Springer Verlag 1992.

[Nguyen & Widrow 90] D. Nguyen, B. Widrow, The truck Backer Upper: An example of self learning in neural networks, in *Neural networks for Control*, W. T. Miller III, R. S. Sutton, P. J. Werbos eds, The MIT Press, Cambridge MA, 1990.

[Radcliffe 91] N. J. Radcliffe, Equivalence Class Analysis of Genetic Algorithms, in *Complex Systems* **5**, pp 183-205, 1991.

[Ronald & Schoenauer 93] E. Ronald, M. Schoenauer Genetic lander: An experiment in accurate neuro-genetic control, in *PPSN 94*, to appear.

[PDP 86] D. E. Rumelhart, J. L. McClelland, *Parallel Distributed Processing - Exploration in the micro structure of cognition*, MIT Press, Cambridge MA, 1986.

[Schaffer, Caruana & Eshelman 1990] J. D. Schaffer, R. A. Caruana and L. J. Eshelman, Using genetic search to exploit the emergent behavior of neural networks, *Physica D* **42** (1990), pp244-248.

[Schoenauer & Ronald 93] M. Schoenauer, E.Ronald S.Damour, Evolving Networks for Control, in *Neuronimes 93* , EC2, Paris 1993.

[Schoenauer & Ronald 94] M. Schoenauer, E.Ronald Neuro Genetic Truck Backer-Upper Controller, in *IEEE World Conference on Computational Intelligence* , Orlando 1994.

[Stork &al. 90] D. G. Stork, S. Walker, M. Burns & B. Jackson, Preadaptation in Neural Circuits, in *Proceedings of International Joint Conference on Neural Networks* **8**, pp I-202-1-205, Erlbaum, Hillsdale 1990.

[Suddarth 88] Steve C. Suddarth, The symbolic-neural method for creating models and control behaviors from examples. Ph.D. dissertation. University of Washington. 1988.

[Yao 93] Xin Yao, A Review of Evolutionary Artificial Neural Networks, in *International Journal of Intelligent Systems* **8**, pp 539-567, 1993.

Image-Processing

Mixed IFS: Resolution of the Inverse Problem Using Genetic Programming

Guillaume CRETIN, Evelyne LUTTON, Jacques LEVY-VEHEL,
Philippe GLEVAREC, Cédric ROLL

INRIA - Rocquencourt
B.P. 105, 78153 LE CHESNAY Cedex, France
Tel : 33 1 39 63 55 23 - Fax : 33 1 39 63 53 30

Abstract. We address here the resolution of the so-called inverse problem for IFS. This problem has already been widely considered, and some studies have been performed for affine IFS, using deterministic or stochastic methods (simulated annealing or Genetic Algorithm) [9, 12, 6]. When dealing with non affine IFS, the usual techniques do not perform well, except if some a priori hypotheses on the structure of IFS (number and type functions) are made. A Genetic Programming method is investigated to solve the "general" inverse problem, which permits to perform at the same time a numeric and a symbolic optimization. The use of "mixed IFS", as we call them, may enlarge the scope of some applications, as for example image compression, because they allow to code a wider range of shapes.

Keywords : Fractals, Genetic Programming, Inverse problem for IFS

1 Introduction

IFS (Iterated Functions System) theory is an important topic in fractals. A major challenge of both theoretical and practical interest is to solve the so called inverse problem.

From the practical viewpoint this problem may be formulated as an optimization problem. In this framework, a lot of work has been done and some solutions exist, based on deterministic or stochastic optimization methods. Most of them make some a priori restrictive hypotheses : affine IFS, with a fixed number of functions [3, 1, 2, 7, 5, 14, 9]. Solutions based on Genetic Algorithmes (GA) or Evolutionary Algorithms have recently appeared for affine IFS [6, 12, 13, 11].

As will be seen in section 3, non-affine IFS provide an interesting variety of shapes, whose practical interest might be large. However, in that case, the inverse problem cannot be addressed using the classical techniques. We propose to make use of Genetic Programming in that framework. As far as we know, this is the first attempt to use Genetic Programming to solve that problem.

2 IFS theory

An IFS (Iterated Function System) $\mathcal{U} = \{F, (w_n)_{n=1,..,N}\}$ is a collection of N functions defined on a complete metric space (F, d). Let W be the Hutchinson operator, defined on the space of subsets of F:

$$\forall K \subset F, \; W(K) = \bigcup_{n \in \{1..N\}} w_n(K)$$

Then, if the w_n functions are contractive (this IFS is then called an *hyperbolic* IFS), there exists a unique set A such that: $W(A) = A$. A is called the **attractor** of the IFS.

Recall: A mapping $w : F \to F$, from a metric space (F, d) into itself, is called **contractive** if there exists a positive real number $s < 1$ such that: $d(w(x), w(y)) \le s.d(x, y) \quad \forall x, y \in F$

The uniqueness of a hyperbolic attractor is a result of the Contractive Mapping Fixed Point Theorem for W, which is contractive according to the Hausdorff distance:

$$d_H(A, B) = \max[\max_{x \in A}(\min_{y \in B} d(x, y)), \max_{y \in B}(\min_{x \in A} d(x, y))]$$

From a computational viewpoint, an attractor can be generated according to two techniques:

- **Stochastic method (toss coin):** Let x_0 be the fixed point of one of the w_i functions. We build the points sequence (x_n) as follows: $x_{n+1} = w_{i_n}(x_n)$ where i_n is randomly chosen in $\{1..N\}$. Then $\bigcup_n x_n$ is an approximation of the real attractor of $\mho$.
- **Deterministic method:** From any kernel S_0, we build the sets sequence $\{S_n\}$ such that:
$S_{n+1} = W(S_n) = \bigcup_{n \in \{1..N\}} w_n(S_n)$. When n tends to ∞, S_n is an approximation of the real attractor of $\mho$.

The inverse problem for 2D IFS can be stated as follows: *for a given 2D shape (a binary image), find a set of contractive maps whose attractor resembles more this shape.*

3 Mixed IFS

In the case of affine IFS, each contractive map w_i of $\mho$ is represented as:

$$w_i(x, y) = \begin{bmatrix} a_i & b_i \\ c_i & d_i \end{bmatrix} \cdot \begin{bmatrix} x \\ y \end{bmatrix} + \begin{bmatrix} e_i \\ f_i \end{bmatrix}$$

The inverse problem corresponds to the optimization of the values $(a_i, b_i, c_i, d_i, e_i, f_i)$ in order to get the attractor which resembles more the target.

In order to model the general problem when the w_i are not anymore restricted to be affine functions, we have defined the notion of **Mixed IFS**. The first point we have to address is the one of finding an adequate representation of these mixed IFS: the more natural one is to represent them as trees.

$$w_1(x,y) = \begin{pmatrix} \sqrt{|\sin(\cos 0.90856 - \log(1+|x|))|} \\ \sin y \end{pmatrix}$$

$$w_2(x,y) = \begin{pmatrix} \cos(\cos(\sqrt{|x|})) \\ \cos(\log(1+|y|)) \end{pmatrix}$$

$$w_3(x,y) = \begin{pmatrix} \log(1+|\cos(\log(1+|y+x|))|) \\ \sqrt{|\sin 0.084698|} \end{pmatrix}$$

$$w_4(x,y) = \begin{pmatrix} \log(1+|\sin(\sqrt{|0.565372|})|) \\ \sqrt{|0.81366 - ((\log(1+|0.814259|)) * \cos y)|} \end{pmatrix}$$

$$w_5(x,y) = \begin{pmatrix} \log(1+|\sqrt{|0.747399 + \cos y|}|) \\ \sin \frac{0.73624}{0.0001+|0.264553 * y + 0.581647 + x|} \end{pmatrix}$$

Fig. 1. A Mixed IFS, left, and its attractor, right.

Fig. 2. Other examples of attractors generated with mixed IFS.

The attractors of figures 1 and 2 are random mixed IFS: the w_i functions have been recursively built with help of random shots in a set of basic functions, a set of terminals (x and y), and a set of constants. In our examples, the basic functions set is: $\{ +, -, \times, div(x,y) = \frac{x}{0.0001+|y|}, \cos, \sin, root(x) = \sqrt{|x|}, loga(x) = \log(1+|x|) \}$, and the constants belong to $[0,1]$.

We thus represent each w_i as a tree (see figure 3). The trees of the w_i are then gathered to build the main tree which represents the IFS $\mho$ (figure 4). This is a very simple structure which allows to code IFS with different numbers and different types of functions. The evaluation of such a structure is that of a simple mathematical expression. However, note that the evaluation is recursive, and thus may be time consuming.

As we have seen, generating a mixed IFS is done via simple recursive random shots. The set of possible IFS depend on the choice of the basic functions set and constants set. The main problem for mixed IFS is to verify that the w_i are contractive, in order to select *hyperbolic* IFS. On the contrary to affine IFS, this verification is not straightforward. We thus propose to use some heuristics that reject strongly non-contractive functions. The simplest way to do that (see section 4.3 for a finer criterion) is to verify the contractivity on some sample points, for example vertices of a grid placed on the domain (typically 25 points).

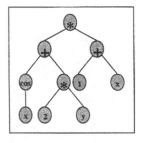

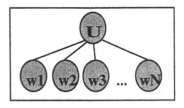

Fig. 3. The function $((cos(x) + 2 * y) * (1 + x))$.

Fig. 4. Representation of a mixed IFS.

4 Genetic Programming to address the inverse problem

4.1 Introduction

Since John Koza [8] first suggested to extend the GA model to the space of computer programs, in order to create programs able to solve problems for which they haven't been explicitly programmed, a lot of very different applications have arised: robotics, control, symbolic regression for example.

Compared to Genetic Algorithms approaches (GA), the individuals in a GP population are not any more bit strings of fixed length, but are programs that, when they are executed, give a a possible solution to the problem. Typically, these programs are coded as trees (this is why most of the applications are developed in LISP).

The populations programs are built from elements of a set of functions and of a set of terminals which are typically symbols selected as being appropriate to the kind of problems to be solved. The "crossover" operation is performed by exchanging sub-trees between the programs and generally the "mutation" operation is not used in GP. When it is used, mutation consists in modifying (with a weak probability) a symbol of the tree.

The evolution of a program inside a GP algorithm is done simultaneously on its size, its structure and its content: the search space is the set of all recursively possible structures (sometimes according to some restriction rules), built from the functions, terminal and constant sets.

When applying GP (or GA) to the resolution of a given problem, one generally have to deal with several points, namely:

- coding of the individuals,
- evaluation function of the individuals (fitness),
- definition of the genetic operators,
- choice of the parameters.

Concerning the first point, as we have already seen, the individuals of the population (i.e the Mixed IFS), are coded as trees. It allows to code a variable number of functions (dynamically), and it is an appropriate data structure for the mutation and the crossover.

In the following, we will address the other points, and insist on the original ones for our application: the use of two different types of mutation and the integration of the contractivity constraints in the fitness.

4.2 The fitness function

From a general viewpoint, the fitness function is a major procedure in GP or GA applications, because fitness is evaluated a large number of times at each generation. Moreover, in most complex problems, as the one we deal with, the fitness evaluation step is time consuming. For these reasons, the fitness evaluation procedure must be very carefully implemented: it can severely influence the computational time and results accuracy.

In our application, we have to characterize the quality of an IFS, i.e. to evaluate how far is its attractor from the target image.

Fitness based on Collage theorem vs fitness based on toss-coin algorithm: Among people dealing with inverse problem for IFS with GA, it is largely admitted that the fitness function based on the so-called collage theorem yields better results than a fitness based on a direct evaluation of the attractor via the toss-coin algorithm. Indeed, the collage theorem is very attractive and can be less time consuming than the toss-coin evaluation algorithm.

Collage theorem: Let A be the attractor of the hyperbolic IFS $\mho = \{w_1, ..., w_n\}$:

$$\forall K \subset F \qquad d_H(K, W(K)) < \varepsilon \qquad \Rightarrow d_H(K, A) < \frac{\varepsilon}{1 - \lambda}$$

λ being the smallest number such that: $\forall n, \forall (x, y) \in F^2, \ d(w_n(x), w_n(y)) < \lambda.d(x, y)$

This theorem means that the problem of finding an IFS $\mho$, whose attractor is arbitrary close to a given shape I, is equivalent to the minimization of the distance: $d_H(I, \bigcup_{i=1}^{n} w_i(I))$.

If $d_H(I, \bigcup_{i=1}^{n} w_i(I))$ is to be used as the fitness function in a GA (or a GP algorithm), then:

- The fitness depends on the contractivity of the maps; if one of the maps is weakly contractive, then the term $\frac{1}{1-\lambda}$ is very large, and the bound becomes meaningless.
- The Hausdorff distance itself is CPU-time consuming, and may also appear as counter-intuitive in some cases: on the figure 5 are represented two couples of shapes [(a), (b)] and [(a'), (b')] who have the same Hausdorff distance, while (a) and (b) are perceived as similar and (a') and (b') not.

These drawbacks led us to use the toss-coin fitness, which provides more precise results, and to base the images differences measures on simple pixels comparisons.

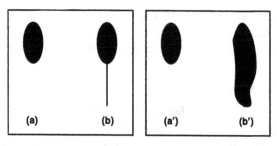

Fig. 5. Hausdorff distance may be counter-intuitive.

Practical fitness computation: In order to improve the algorithm efficiency, we have modified the fitness computation in two ways:

- The toss-coin algorithm is time consuming, it generally needs a lot of iterations to create the attractor. But we noticed that the population quickly converges to a rough approximation of the target. We thus made the iteration number linearly increase during the generations, in order to provide a quickly computed approximation at the beginning of the GP, and then progressively provide details along the computation.

- Instead of using the Hausdorf distance as a measure between 2D shapes, and in order to guide more precisely the research of the optimum, we use distance images. This allows to consider "smoother" functions to be optimized, as in [10]. A distance image is the transformation of a black & white image into a grey-level one, where the level affected to each image point is a function of its distance to the original shape. It can easily be computed by a simple algorithm (see [4]), based on the use of two masks (see figure 6): the resulting images are parameterized by $d1$ and $d2$ which represent the two elementary distances on vertical/horizontal and diagonal directions. This parameterization allows to use distances which are more or less "abrupt".

The computation of the fitness of the current IFS is thus based on a measure of the difference between its attractor and the distance image of the target. The simple byte-to-byte difference (i.e. a counting of coinciding pixels) is thus completed by a mean intensity of pixels on the trace of the w_i'attractor. This yields to the algorithm more "local" information about the resemblance between the attractor and the target.

We improved this technique by varying the distance image parameters ($d1$ and $d2$) along the generations: we begin with a very fuzzy distance image, then every x generations we modify it so that in the end it becomes the real B&W attractor. Tolerance to small errors, and computation times have thus been improved.

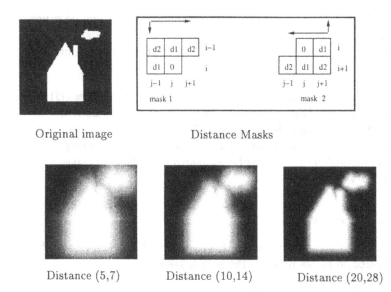

Original image	Distance Masks

Distance (5,7)	Distance (10,14)	Distance (20,28)

Fig. 6. Distance images.

4.3 Contractivity constraints

Before each individual evaluation, we have to verify that it is an hyperbolic IFS (thus yielding an unique attractor). As we have seen before, this verification is difficult on mixed IFS, mainly because of the non linearity of the mappings. We have proposed in section 3 to simply verify the contractivity conditions on some sample points of the domain, and reject the individuals which does not verify it. This is a way to discard a lot of non-contractive IFS from the current population. But we have to notice that it may not discard some pathological mappings, even if we use a lot of sampling points.

We propose to address this problem in a different way, which will allow at the same time to use an a priori information from the target image, and to reduce the computation time. Our approach is based on the fixed point theorem.

For a hyperbolic IFS $\mho$ whose attractor is A, each mapping w_i is contractive, and thus admits a unique fixed point X_i. We must then have:

$$\forall i, X_i \in A$$

The verification of the existence of the X_i's and their estimation can easily be performed: we built two sequences of points $x_{n+1}^i = w_i(x_{n+1}^i)$ starting from two points of the domain (for example $(0,0)$ and $(1,1)$). Within a few iterations we can estimate the fixed point or decide that the function is not contractive. The use of two sequences allows to speed up the fixed points estimation. We then verify if the X_i's belongs to the target shape.

Practically, we compute a constraints function: $C(\mho)$ which is the mean value of the distance (measured on the distance image of the target) of the X_i's to the target. If $C(\mho)$ has too low a value, the toss-coin algorithm can be pruned. The fitness is thus directly given by $C(\mho)$.

The fitness computation integrates the contractivity constraints in the following way:

1. *If there exists a w_i which is not contractive, then fitness$(\mho) = -1$ and the individual is directly discarded from the population.*
2. *If $C(\mho) < C_0$ then fitness$(\mho) = C(\mho)$*
3. *If $C(\mho) \geq C_0$ then the attractor A of $\mho$ is computed using the toss-coin algorithm, and fitness$(\mho)$ measures the difference between A and the target.*

4.4   Genetic operators

Crossover : we use the classical GP crossover which performs exchanges of randomly selected nodes between the parent-trees.

Mutation : we decided to use mutation in our algorithm, which is a common operator in GA, but a quite rare one in classical GP.

Indeed, mutation in a GA is a small change in the genetic code of the chromosome. For example, in the case of binary codes, mutation is a bit flip of one of the genes. In the case of GP, mutation has to slightly perturb a tree structure. For that purpose, we have to differentiate the nodes and the leaves of the tree:

- The nodes belong to the basic functions set, which is finite. A node mutation could be to replace one node by another basic function randomly chosen in the basic functions set. Since such a perturbation may have too drastic effects, we have decided not to use it.
- The leaves are chosen in a terminals set (x or y) or in a constants set, which is a continuous interval ($[0, 1]$). We also have to separate the mutation of constants to the mutation of variables, because they are of different nature. Of course we could also imagine a mutation process which would transform a constant into a variable and reversely, but it seems to be too violent, except in the case of variables, as we will see.
  - *Constants :* mutation is the only mean to make constants evolve, and it is very important in our case, because we need to perform a numerical optimization of the constants. We perturb the constants with a parameterized probability. A constant is replaced by a new value obtained from a uniform random shot inside a disk of fixed radius (another parameter of our algorithm) around it (see figure 7).
  - *Variables :* an "internal" mutation, i.e. a change from x to y and reversely is again possible, but we preferred a mutation which changes a variable into a randomly chosen constant (see figure 8). We have made this choice on an empirical basis : we noticed that, in some cases, constants tend to disappear from the current population. Once they have disappeared, they cannot reappear in the offsprings populations. We thus propose to use

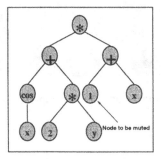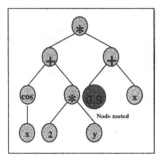

Fig. 7. Mutation of constants.

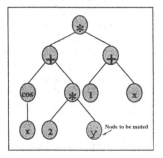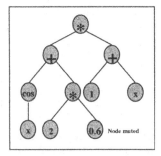

Fig. 8. Mutation of variables.

a constants creation process, via mutation of variables, to maintain a minimal proportion of constants in the population.

The constants vanishing effect we have experimentally noticed may be explained as follows: the numerical optimization of the constants is a more difficult task than the symbolic optimization of the other nodes. The selection operator thus tends to eliminate too rapidly IFS having bad constants. This difference is due to the fact that the search spaces of the nodes and variables are finite while the search space of the constants is theoretically infinite. Other techniques (that we have not tested) to avoid this fact may be to reduce the size of the constants search space by allowing only a finite set of constants (via sampling for example), or to separate the symbolical and the numerical optimization (i.e. having a subprocess which optimizes the constants before each IFS evaluation).

5 Results

We have tested our algorithm on shapes that were actual attractors of IFS. The choice of basic functions for the GP is the one presented in section 3. Initial populations are randomly chosen. The only constraints set on the structures of

Fig. 9. Example #1. From left to right: original image and best images after 10, 100, 300 and 1500 generations.

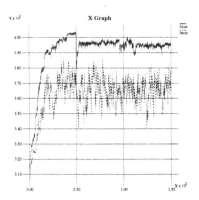

Image size	64 pixels
Population size	30
Max number of generations	1500
Crossover probability	0.7
Mutation probability for constants	0.2
Mutation probability for variables	0
Range of the constants	[0,1]
Perturbation radius for the constants	0.1
Max and min number of contractive maps	3 to 6

Fig. 10. Example #1: fitness evolution. The maximum fitness of the current population is the continuous curve, the mean fitness is the dotted one.

Fig. 11. Example #1: parameters setting.

the evolved IFS's trees is a maximal and a minimal number of functions. No depth restrictions are imposed. However, we experimentally verified that their structures do not excessively expand along the evolution. We present here two good convergence results. For each example, we present: the target attractor, the best image obtained after convergence, the fitness evolution curve and the parameters setting.

The parameters adjustment remains a difficult task, but we empirically noticed the following facts:

- The distance images are very efficient. It is particularly obvious on the fitness evolution curves (figure 10 and even more on figure 13): when updating the distance image, the curve suddenly falls down and then grows up again. For the new distance image, the value of the fitness becomes lower, because it is computed on a distance image with larger d_1 and d_2 parameters. This corresponds in fact to a more precise evaluation of the difference between the current IFS and the target.
- The mutation of the constants is important, it brings diversity and cannot be set to zero.
- The target images which yield good results are rather compact: the convergence to line-shaped targets is more difficult.

Fig. 12. Example #2. From left to right: original image and best images after 50, 260 and 1300 generations.

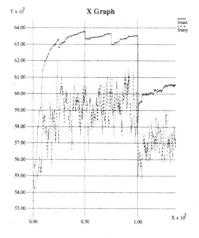

Fig. 13. Example #2: fitness evolution. The maximum fitness of the current population is the continuous curve, the mean fitness is the dotted one.

Image size	256 pixels
Population size	30
Max number of generations	1300
Crossover probability	0.85
Mutation probability for constants	0.25
Mutation probability for variables	0.001
Range of the constants	[0,1]
Perturbation radius for the constants	0.1
Max and min number of contractive maps	4 to 7

Fig. 14. Example #2: parameters setting.

6 Conclusion

We have proposed a method to solve the inverse "general" problem for mixed IFS within a reasonable computation time (a few hours on Sparc 10 and Dec 5000 stations). This computation time is similar to computation times of GA applied to the inverse problem for affine IFS [12], although in the case of mixed IFS the size of the search space is much more larger. This fact may be explained by the use of variable sized structures in the GP algorithm, which seem to perform a more efficient search in a large space.

The method may be improved in several directions:

- test a "smoother" transition between distance images: a re-computation of distances images at every generations would allow to let the parameters d_1 and d_2 vary more smoothly,
- test other mutation strategies, as suggested in section 4.4,
- test an adaptative radius for mutation of constants, in the same way as in evolutionary programming techniques, where mutation variance is dynamically adapted, according to the performance of the individual,
- make the iteration number of the toss coin evaluation algorithm be more adaptative (we can theoretically fix the iterations number and the probabilities of the toss coin algorithm in order to more rapidly approximate the attractor within a fixed error),
- modify the storage structure of the IFS in order to reduce the computation time (mainly by avoiding some useless computations)

Such an approach might be interesting in the field of image compression. IFS compression techniques are generally based on affine IFS. The use of mixed IFS may yield more flexible spatial and grey-level transformations, and thus may allow to improve the compression ratio for the same number of functions.

References

1. M. Barnsley and S. Demko. Iterated function system and the global construction of fractals. *Proceedings of the Royal Society*, A 399:243–245, 1985.
2. M. Barnsley, V. Ervin, D. Hardin, and J. Lancaster. Solution of an inverse problem for fractals and other sets. *Proc. Natl. Acad. Sci. USA*, 83, 1986.
3. M. F. Barnsley. *Fractals Everywhere*. Academic Press, N Y, 1988.
4. G. Borgefors. Distance transformation in arbitrary dimension. *Computer Vision, Graphics, and Image Processing*, (27), 1984.
5. Y. Fisher. Fractal image compression. *Siggraph 92 course notes*, 1992.
6. B. Goertzel. Fractal image compression with the genetic algorithm. *complexity International*, (1), 1994.
7. A.E. Jacquin. Fractal image coding: a review. *Proceedings of the IEEE*, 81(10), october 1993.
8. J. R. Koza. *Genetic Programming*. MIT Press, 1992.
9. Jacques Levy-Vehel. *Analyse et synthese d'objets bi-dimensionnels par des méthodes stochastiques*. PhD thesis, Université de Paris Sud, Decembre 1988.
10. Evelyne Lutton and Patrice Martinez. A genetic algorithm for the detection of 2d geometric primitives in images. In *12-ICPR*, 1994. Jerusalem, Israel, 9-13 October.
11. D. J. Nettleton and R. Garigliano. Evolutionary algorithms and a fractal inverse problem. *Biosystems*, (33):221–231, 1994. Technical note.
12. J. Lévy Véhel and E. Lutton. Optimization of fractal functions using genetic algorithms. In *Fractal 93*, 1993. London.
13. L. Vences and I. Rudomin. Fractal compression of single images and image sequences using genetic algorithms. *The Eurographics Association*, 1994.
14. E.W. Jacobs Y. Fisher and R.D. Boss. Fractal image compression using iterated transforms. *data compression*, 1992.

Interactive Evolution For Simulated Natural Evolution

Jeanine Graf and Wolfgang Banzhaf

Informatik Centrum Dortmund (ICD)
44227 Dortmund, Germany

Abstract. Evolutionary algorithms of selection and variation by recombination and/or mutation have been used to simulate biological evolution. This paper demonstrates how *interactive evolution* can be used to study the evolution of simulated natural evolution. Since interactive evolution allows the user to direct the development of models of natural systems, it can be used to direct the evolution of models of animals and plants. We show that interactivity of artificial evolution can serve as a useful tool in the ontogenesis and phylogenesis of simulated models. This may help paleontologists solve problems in identifying likely missing links and provides a technique to generate constrained conjectures regarding gaps in evolutionary data.

Keywords: Growth, Paleontology, Evolutionary Algorithms, Simulation of Natural Evolution.

1 Motivation

Our world view is influenced by the knowledge that the universe, earth and all living things have evolved through a long history of continual, gradual change shaped by more or less directional natural processes consistent with the laws of physics (E. Mayr 1978).

Despite a rich fossil record chronicling the evolutionary process, there are still gaps in this record. Since paleontologists would like to discover morphologic intermediates in the fossil record, this motivated us to develop a simulation framework to evolve simulated models of animals and plants and to visualize their evolution. Specifically, we have developed a system that presents progressively evolving natural models by interactive processes. The application of interactive evolutionary algorithms to computer graphics helps us to demonstrate continuous evolutionary change.

In addition to problems in paleontology, biological sciences face the more general problem of concluding how the growth and expansion process evolved due to problems in observing change in nature. We thus show that interactive evolution can serve as a useful tool for achieving flexibility and complexity in simulated evolution with a certain amount of scientific feedback and detailed knowledge.

2   Evolution in Nature

Natural selection acts nearly without fail on individuals or on groups of related individuals. It is only during the last fifth of the history of life on earth that multicellular organisms have existed. They seem to have taken shape from unicellular organisms on numerous occasions (J. Valentine, 1978)

Animals and plants are multicellular, made up of millions or billions of individual cells. Even the simplest multicellular organisms include diverse categories and types of cells. At least two million multicellular species exist today, and many others have come and gone over the ages. The important advantages of multicellularity stem from the repitition of cellular machinery it implies. From this property or attribute derives the ability to live longer, to produce more offspring, to be larger, and so to have a greater internal physiological stability and to construct a variety of body types.

Just as change in the environment can involve organisms, so too the activities of organisms can involve the environment to create new conditions. The history of life shows evidence of interaction between environmental change on the one hand and the evolutionary potential of organisms on the other. It is therefore of interest to briefly examine and to keep in mind the major causes of environmental changes of biological relevance. and empty in the next, and no great ecological roles have been long unfilled. different limited adaptions, all in moderate equilibrium to each other (Robert May 1978).

Therefore, we have tried to derive general principles from the evolution of many different kinds of animals and to find out the continuity of their evolution.

3   Artificial Evolution

Natural evolution is a mechanism that can be simulated, a process described by the term *artificial evolution*. In nature many shapes change spectactularly in form. Evolution, however, implies change by continuity. Therefore complex processes of evolution generally can be sufficiently represented by different continuous transformations. D'Arcy W. Thompson (1942), for example, recognized that the concept of geometric transformation could be helpful in identifying morphological change.

Some researchers in biology are trying to find the ancestors and interconnections between species. We can observe that there is a line of continuity in the morphology of some animals. In that case, we try to observe the transition from one animal to another or from one species to another. It is, therefore, of interest to apply this knowledge as background for simulated evolution.

We are not able to determine whether two or more animals, which look morphologically similar, are descendants from the same ancestor. The evidence of convergent evolution is demonstrated by the fact that related groups of animals do respond to similar selective pressures with similar adaptations. But, let us try with the help of interactive evolution to simulate some elements of biological evolution. We shall observe that individuals of related groups resemble each other. Much of evolutionary biology is the working out of an adaptative program. Evolutionary biologists assume that each aspect of an organism's morphology, physiology and behavior has been modeled by natural selection as a solution to a problem posed by the environment (R. Lewontin 1978). From these facts, and from the wealth of accumulated evidence in the comparative embryology and morphology of living representatives of fossil groups, it is possible to build a picture of the rise of major animal groups. It is difficult to know the descendants of these animals for certain, hence we use interactive simulated evolution to examine such problems.

4 Evolutionary Algorithms

Interactive evolution is a technique from the class of evolutionary algorithms (EAs), which are based upon a simple model of organic evolution. Most of these algorithms operate on a population of individuals that represent search points in the space of the decision variables (either directly, or by using coding mappings). Evolution proceeds from generation to generation by exchanging genetic material between individuals (*recombination*), i.e. by trying out new combinations of partial solutions, and by random changes of individuals (*mutation*). New variations are subjected to *selection* based on an evaluation of features of the individuals according to certain (*fitness*) criteria.

The best-known representatives of this class of algorithms are evolution strategies (ESs), developed in Germany by I. Rechenberg (1994) and H.–P. Schwefel (1995) evolutionary programming (EP), developed in the U.S. by L.J. Fogel (1996) and genetic algorithms (GAs), developed in the U.S. by J.H. Holland (1975)

Evolution strategies (ES) were first applied in the field of experimental optimization and used discrete mutations (Rechenberg 1994). When computers became available, algorithms were devised that operated on continuous-valued variables. ES uses random mutation as its main search operator. But in addition, recombination operators are also included, at the discretion of the human operator. Parameters for self-adaption of the step sizes are also provided. Evolution strategies are often described in the form $(\mu+\lambda)$, in which μ parents give rise to λ offspring and the best μ solutions from both parents and offspring become parents of the next generation, or (μ, λ), in which the best μ solutions are chosen solely from the λ offspring, and the old parents are replaced (see Rechenberg 1973; Schwefel 1981; Baeck 1994).

Evolutionary programming (EP) was originally proposed as a method to optimize finite state machines for time series prediction (Fogel et al. 1966); this as a prerequisite for evolving intelligent behavior. More recently, EP has been used to address real-valued optimization problems. A population of candidate solutions to the task at hand is subjected to random variation and selection over successive iterations. For real-valued problems, the variation usually takes the form of zero mean Gaussian perturbations, with the variances of the perturbations either being set as a function of the parent's error score (Fogel 1995), or as a self-adaptive value (Fogel et al. 1992); Recombination is not used in typical implementations of EP. Selection is accomplished probabilistically using a competition procedure that allows less-fit solutions some probability of being included in the subsequent generation, while also guaranteeing that the best solution found to that point will be retained.

Genetic algorithms (GAs) emphasize the genotypic level of evolution, primarily operating on binary encodings (a choice that is founded by the argument of maximizing the number of schemata available; see Holland 1975; Goldberg 1989; Davis 1991). Selection is often made probabilistically in proportion to a solution's relative fitness: the greater the fitness, the greater the probability of generating offspring. Emphasis is placed on using crossover operators to search for new solutions by bringing together pieces of different parents. The effectiveness of many of these crossover operators depends on the degree to which the components of a solution are independent. A low rate of mutation is applied to ensure that all solutions are possible in the limit.

In all of these cases, the selection criteria are traditionally fixed and are held constant from the start of the simulation. Therefore, these criteria must be detailed explicitly beforehand. This constitutes a significant problem in many realistic applications (apart from optimization), because an explicit fitness function may not be available in closed form. Recently, various work-arounds have been tried, one of the most prominent being *co-evolution*. In this method, rather than using one population to search for the best solution, two or more antagonistic populations compete against each other. The realized fitness in this case is in part determined by the relationship of one population to the other, and does not have to be defined explicitly beforehand (Hillis 1990).

This paper makes use of an alternative method to generate fitness by incorporating the scientist into the selection process of artificial evolution. In the following section, we present this method in detail.

5   Interactive Evolution

Dawkins (1986) demonstrated convincingly the potential of Darwinian variation and selection in graphics. He evolved *biomorphs* 2-D graphic objects, from a collection of genetic parameters with the user. Recently, much research has been directed into the application of genetic algorithms to image and graphics problems. Sims (1991) used genetic algorithms for interactive generation of color art; Todd and Latham (1991) have considered similar ideas to reproduce computer sculptures through structural geometric techniques.

The idea offered in this paper is to provide a user with a new technique to evolve biological models in the field of biology. We have developed a system that presents progressively evolving solutions for paleontology and problems of ontogony by means of interactive processes. In *interactive evolution*, the user selects one or more individuals that survive and reproduce (with variation) to constitute a new generation. These techniques can be applied to the production of computer graphics, animation, creating forms, textures and motion . Potential applications of interactive evolution include artificial life design, e.g., development of components of biological nature and engineering construction design. In addition to its application to simulations of natural evolution, interactive evolution can also be used in many real-world application (Graf et al. 1995).

Phenotypes and Genotypes

We shall need to discern between genotypes and phenotypes in interactive evolution; both terms are also basic concepts for biological evolution. The biological genotype is the information that codes the development of an individual. Genotypes most often consist of DNA sequences. In this interactive evolution, genotypes are represented as numerical data and real values. The phenotype is the realised behavior of the individual itself, i.e., the product of an interpretation of the underlying genotypic representation. In our case, the phenotype is the resulting graphical model.

Fitness and Selection

The term *fitness* in interactive evolution is the capability of an individual or model to survive into the next generation, and therefore is tied directly to selection. Usually, in interactive evolution, fitness is not defined explicitly but is instead a relative measure of success following from the selection activity of a human scientist. Here, it is also based on non-quantifiable measures like certain predefined criteria (for example: the historical knowledge about existing recombination between two defined animals) help to sort candidates for survival from the set of all variants, among which a human user ultimately determines the next generation.

Variation

In interactive evolution, one of the main benefits is the automatic generation of variants. Variation is accomplished by defining problem-specific mutation and recombination operators that constantly propose new variants to the presently existing population of simulated graphic models on the screen. An important amount of knowledge has to be invested in order to find appropriate operators for an application domain. The next section provides appropriate operators for manipulation of simulated animal or plant models.

5.1 Application of Interactive Evolution to the Simulation of Natural Evolution

The fossil record illustrates many phenomena of natural change that can be addressed by interactive evolution. These aspects must be explained by any satisfactory model of the evolution of life on earth. Animal groups appear in the fossil record in a certain unmistakable order. For example, primitive fish appear first, amphibians later, then reptiles, then primitive mammals, then (for example) legged whales, then legless whales. In the next section we show an example of this transformation from legged to legless whales (see figure 6). This temporal morphological reciprocality is very spectacular and impressive. It is important to apply this predicate order in interactive evolution. Many chains of groups contain one or more species which appear to link early, primitive genera with much more recent, radically different genera (e.g. reptile- mammal transition, hyenids, horses, elephants), through major morphological changes. Even for the spottiest disparities, there are a few separated intermediates that show how two apparently very different groups could, in fact, be related to each other (e.g. -Archeopteryx-, linking reptiles to birds).

Paleontologists have observed transitions of at least 30 genera in nearly perfect morphological order, with most of the reptilian first and most of the mammalians last, and with only relatively small morphological differences separating each successive genus.

There are still unknown species transitions, and the chain of genera is not complete, but we now have at least a partial lineage, and sure enough, the new whale fossils have legs. (for discussions see Berta, 1994; Gingerich et al. 1990; Thewissen et al. 1994; Discover magazine, Jan. 1995).

6   Evolving Simulated Models

In simulated evolution, one approach is to consider a growing object as a set of points. Each point can be defined using three coordinates, x, y and z. In artificial evolution, the process of altering or of growth can be systematically implemented using the concept of geometric transformations. A transformation is determined by defining new variables, x', y' and z', as functions of x, y and z. A geometric transformation can often be considered as a general way of change, independent of the special object to which it is then applied.

We describe methods for evolving plant and animal models based on distributive data interpolation. The methods use corresponding points in the images to be interpolated. Interactive evolution techniques which can be used for object metamorphosis include solid deformations. In such a case, the 2-D or 3-D model of the first object is transformed in order to assume the shape and properties of the second model, and the resulting animation is recorded. The interactive evolution process helps us to construct bodies with varying architectures, which are potentially real shapes from nature.

Our approach also evolves *two-dimensional* images, specified either directly as bitmaps or as parameterized geometric models, such as those provided by vector graphics. Bitmaps and other forms of direct encoding of images have found an excellent niche in computer graphics, video composition and image rendering (Foley 1992 and Kirik 1992). Any 2-D shape can be represented as a sequence of points or vertices, with each vertex consisting of an ordered pair of numbers (x, y), its coordinates. The array of pixel values of a 2-D image, however, has nothing to do with the structure being represented in the image. This constitutes the challenge of finding appropriate operators for the generation of new variants of an existing image, because structural or functional conservation of the image content is of primary importance.

We solve the problem of realizing evolutionary operations by using *image interpolation* to create variations. Warping and morphing are methods which, by using tiepoints in two images "A" and "B", allow for the creation of intermediate images (Woldberg 1990 and Ruprecht 1994). Basically, these intermediate images are interpolations along an abstract axis from image "A" and image "B". In our interactive evolution, the tiepoints are the genotype and are constraints for the interpolation process, since corresponding tiepoints in "A" have to be transformed into those of "B". The method requires distributing tiepoints over two or more models in such a way as to conserve essential structures in interpolated (intermediate) models. In this way, an arbitrary initial model can evolve via intermediate steps of interpolation into a final model.

We adopt this approach for simulated evolution. By specifying tiepoints, sufficient control can be exerted about structures in 2-D images so as to provide useful variants to a series of images.

Let us look more closely at the operations usually implemented in EAs. We begin by describing the concept of an individual or a model. Roughly, it is any class of objects, which may have a functional form. The formal model is based on an object-oriented data model, which describes a model hierarchy. This hierarchy is defined by the is-part-of relation, which relates each model to its components. Any model $\mathbf{M}$ is usually composed of salient components ϑ and non-salient components $\Re$ of the model; these can be described as follows. The components ϑ of a model are a set of parts, features or bodies. A polygon mesh consists of a structure of vertices, or points. The non-salient component $\Re$ are excluded from the mutation operator, although recombination may be applied to them.

The following, then, presents our model definition.

$\mathbf{M} := (\vartheta, \Re)$, model

$\vartheta := (\vartheta_1, \ldots, \vartheta_c)$, Salient components of the model, where $c \in \mathbf{N}$ denotes the number of components, parts, or features

$\vartheta_i := (X_{1_i}, \ldots, X_{n_i})$ vertices or points represented each component ϑ, with $i \in \{1, \ldots, c\}$ and $n_i \in \mathbf{N}$;

$\Re := (\Re_1, \ldots, \Re_d)$, Non-salient components. $d \in \mathbf{N}$;

$\Re_i := (X_{1_j}, \ldots, X_{m_j})$ vertices or points of the non-salient components $\Re$.

The elements of ϑ and $\Re$ are linked to $\mathbf{M}$ by the is-part-of relation of the hierarchy model.

In computer graphics the model has as components $\vartheta := (\vartheta_1, \ldots, \vartheta_c)$, salient components of the model where denotes features, or polygons of graphical models;

$\vartheta_i := (X_{1_i}, \ldots, X_{n_i})$ vertices or points of each component ϑ, with $i \in \{1, \ldots, c\}$ and $n_i \in \mathbf{N}$;

The variation process may be applied to the entire model representing the whole structure $\mathbf{M}$, or it might be applied to its components ϑ.

Let us look closely at a top-level description of interactive evolution:

Step 1: Start with an initial time T := 0;
        Initialize an initial population P (0)
        of models $\mathbf{M}$.
        Compute the fitness function $F(\mathbf{M}_i)$
        for each object $\mathbf{M}_i$.

Step 2: Apply the selection with the
        interactive and automatic
        system by selecting the best
        parents for offspring production.

Step 3: Select pairs of objects and apply
        recombination and/or mutation.
        Create new generation of models
        by mating current models;

Step 4: Compute the fitness function $F(\mathbf{M'}_i)$
        for each object $\mathbf{M'}_i$.

step 5: Increase the time counter; set T: = T+1
        repeat until you get the best models
        solution.

6.1 Population

One or more individuals of simulated animals or plants constitutes a population. Many groups containing one or more species are able to constitute one or more populations. Structurally, the content of a simulated model is usually composed of components. In our example, a bird is composed of a body, bill, breast, wing, claw, leg, foot, tail, feather, etc. and a fish is composed of a tail, anal fin, dorsal fin, pelvic fin, scales, pectoral fin, gill, snout or mouth, eye, etc. The variation process can be applied to the entire model representing the whole structure, or might just be applied to its components. By using many tiepoints in a component, influence can be exerted to any necessary degree about the details of the evolving structure.

6.2 Selection

The primary advantage of interactive evolution is that the researcher may select any simulated individual from any population, which subsequently reproduces with the variation process.

Before presenting the entire interactive evolution system, we briefly discuss other considerations in the visualization of simulated individuals.

Procedural models of images can be characterized by certain parameters, constituting the genotype of an image, that must be interpreted in the appropriate context. Its interpretation is the genotype-phenotype mapping in which the resulting image is the phenotype. Because the number of structural elements usually varies from image to image, it is necessary to allow for variable-length genotypes. Variation supports evolution to produce more offspring and to construct different body shapes. Variation takes place here at the level of parameters that are subject to normally distributed random mutations as well as to intermediate or discrete recombination operations.

6.3   Recombination

Recombination is implemented as a global operation by which two models exchange information. We use as many tiepoints as necessary to conserve the underlying structure in two models "A" and "B". A recombination would then be quantified in the model space between "A" and "B" by a certain parameter indicating the degree of "intermediateness" of a variant. Figure 2 demonstrates the method of recombination, through the example of variants of fish. Note that recombination always operates within the present generation.

An algorithm for smoothly blending between two models supports the process of recombination between individuals. These models are composed of 2-D or 3-D polygonal shapes. The algorithm uses a physical model in which one of the shapes is considered to be created by wire and the result is observed. The first shape can be streched or bent into the second shape. This process is known as metamorphosis, shape interpolation, or shape blending. It has important application in change and expansion in biological processes.

A further use of recombination lies in its simultaneous application to different animal species to generate novel descendants. The flexibility of this operator then, allows us to explore a large range of morphological forms.

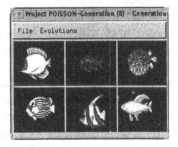

Fig. 1.: This figure shows the first generation of six individuals of fish forms.

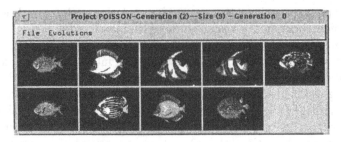

Fig. 2.: This figure shows an example recombination of fish forms
after two generations.

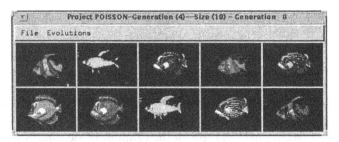

Fig. 3.: The ten fish forms are generated by recombination after
four generations.

6.4   Mutation

Mutation is commonly considered to be a local operation that does not radically
change the resulting phenotype. In order to provide this feature in bitmap image
evolution, we use a very small number of tiepoints. A one-point mutation would
select one tiepoint in a model. The corresponding tiepoint in a second model
would then be used as a source for novelty, by providing information regarding
which direction to evolve the original model. Structure is conserved because
tiepoints in both models correspond to each other. A parameter would then be
used to quantify the degree of substitution in the model.

Note that the second model, from which novelty is gained, is not necessarily in the present generation of the evolutionary process. Instead, a generation
0 of models equipped with a number of tiepoints is used for mutation. Some
domain knowledge must be used in the process of tiepoint selection for generation 0. By selecting one tiepoint in a model of generation n that corresponds
to a tiepoint in a model of generation 0 and constraining the effect to a local
neighborhood, we provide a path for evolving between the two models. The generation 0 model helps to form equivalence classes between structures expressed
as tiepoints. Figure 4 demonstrates the mutation process through using whale
models. A sequence of local variations take place among legged whales as is
displayed in a transition form.

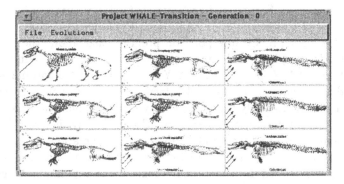

Fig. 4.: This figure shows the transition form of legged whales to leggless whales.

6.5 Reproduction

Selected models reproduce to constitute the next generation. In each new generation, every individual receives its special genotype, which defines it. Figure 6 demonstrates the recombination and reproduction process. These genotypes are inherited from the previous generation. The reproduction algorithm is given by:

Fig. 5.: This figure shows images of two different fox parents, an ice fox and a desert fox

Fig. 6.: This figure shows the result of artificial recombination between images of an ice fox and a desert fox, resulting in different kinds of fox variations including shapes that actually exist in nature, such as the red fox.

7 Jardin2 – A System for Simulated Evolution

Jardin2 is a digital image interpolation program that allows the evolution of simulate models of plants and animals, and runs under the X Window System (Culter et al. 1993, Gaskin 1992 and Young et al. 1992). Jardin2 loads and saves model populations and provides facilities to store points in images, to evolve images, and to apply the evolutionary process. Points are inherited from generation to generation, with generation "0" provided by the user. With a very small population, between two and 20 models per generation, and over a short time, a scientist can select new generations of simulated animal or plant models. This process is repeated until some individuals in the population are generated.

8 Summary and Conclusions

This paper demonstrates how interactive evolution might be applied to natural evolution. The main idea is to combine the concepts from interactive evolutionary algorithms with the concepts of image interpolation from computer graphics. Structure in animal and plant models is substituted by a collection of tiepoints, manually provided by the user or automatically generated. Evolution then proceeds along paths constrained by the set of tiepoints either in all models.

In our version of *interactive evolution*, the user selects individuals which then reproduce to constitute the next generation. This technique can be applied to simulation and animation, for scientific visualisation in the biological domain, and the like in paleontology and ontogeny.

Acknowledgements

Funding from the German Bundesministerium für Forschung und Technologie (BMFT) under project EVOALG is gratefully acknowledged. Thanks to Steve Quartz and Marc Schonauer for their helpful comments and suggestions.

References

1. J. Arvo. *Graphics Gems II*. Academic Press, California, 1991.
2. Th. Bäck. *Evolutionary Algorithms in Theory and Practice*. Oxford University Press, New York, 1996.
3. A. Berta. What is a whale? *Science 263*, pages 180–181, 1994.
4. E. Cutler, D. Gilly, and T. O'Reilly. *The X Window System in a Nutshell*. The Definitive Guides to the X Window System. O'Reilly & Associates, Inc., Sebastopol, CA, 1993.
5. L. Davis. *Handbook of Genetic Algorithms*. Van Nostrand Reinold, New York, 1991.
6. R. Dawkins. *The Blind Watchmaker*. Longman, Harlow, 1986.

7. D. B. Fogel, L. J. Fogel, W. Atmar, and G. B. Fogel. Hierarchic methods of evolutionary programming. In D. B. Fogel and W. Atmar, editors, *Proc. of the First Annual Conference on Evolutionary Programming*, pages 175–182, La Jolla, CA, 1992. Evolutionary Programming Society.

8. D.B. Fogel. *Evolutionary Computation: Toward a New Philosophy of Machine Intelligence*. IEEE Press, New York, 1995.

9. L. J. Fogel, A. J. Owens, and M. J. Walsh. *Artificial Intelligence through Simulated Evolution*. Wiley, New York, 1966.

10. T. A. Foley. *Computer Graphics Principles and Practice*. Addison-Wesley, MA, 1992.

11. T. Gaskin. *DGXWS - PEXlib Programming Manual 3D Programming in X*. O'Reilly & Associates, Sebastopol, CA, 1992.

12. P.D. Gingerich, B.H. Smith, and E.L. Simons. Hind limb of eocene basilosaurus: evidence of feet in whales. *Science 249*, pages 154–156, 1990.

13. A. Glassner. *Graphics Gems I*. Academic Press, California, 1990.

14. D. E. Goldberg. *Genetic Algorithms in Search, Optimization and Machine Learning*. Addison-Wesley, MA, 1989.

15. Jeanine Graf. Interactive evolutionary algorithms in design. In *International Conference on Artificial Neural Networks and Genetic Algorithms*, pages 227–230, Ecole des Mines d'Alès, France, 1995. Proceedings of the ICANNGA, Springer-Verlag, Vienna.

16. Jeanine Graf and Banzhaf Wolfgang. Interactive evolution of images. In *Proceedings of the Fourth Annual Conference on Evolutionary Programming*, pages 53–65, San Diego CA, 1995. The MIT Press, Cambridge, MA.

17. W. D. Hillis. Co-evolving parasites improve simulated evolution as an optimization procedure. In *Physica D 42, 228-234*, 1990.

18. J. H. Holland. *Adaption in Natural and Artificial Systems*. Ann Arbor, The University of Michigan Press, 1975.

19. K. De Jong. *An analysis of the behavior of a class of genetic adaptive systems*. Doctoral dissertation, U. of Mich, 1975.

20. D. Kirik. *Graphics Gems III*. Academic Press, California, 1992.

21. R. Lewontin. Adaptation. *Special Issue on Evolution*, pages 105–114, 1978.

22. E. Mayr. Evolution. *Special Issue on Evolution*, pages 2–13, 1978.

23. I. Rechenberg. *Evolutionsstrategie: Optimierung technischer Systeme nach Prinzipien der biologischen Evolution*. Frommann–Holzboog, Stuttgart, 1994.

24. D. Ruprecht. *Geometrische Deformationen als Werkzeug in der graphischen Datenverarbeitung*. Doctoral dissertation, University of Dortmund, 1994.

25. H.-P. Schwefel. *Numerical Optimization of Computer Models*. Wiley, NY, 1995.

26. K. Sims. Artificial evolution for computer graphics. *Computer Graphics*, (25):319–328, 1991.

27. J.G.M. Thewissen, S.T. Hussain, and M. Arif. Fossil evidence for the origin of aquatic locomotion in archaeocete whales. *Science 263*, pages 210–212, 1993.

28. D'Arcy Thompson. *On Growth and Form* . Cambridge University Press, NY, 1942.

29. S.P. Todd and W. Latham. *Mutator, a Subjective Human Interface for Evolution of Computer Sculptures*. IBM United Kingdom Scientific Center Report, 1991.

30. J. Valentine. The evolution of multicellular plants and animals. *Special Issue on Evolution*, pages 49–66, 1978.

31. A. Watt. *3D Computer Graphics.* Addison-Wesley, Reading, Massachusetts, second edition, 1993.

32. G. Woldberg. *Digital Image Warping.* IEEE Computer Society Press, Los Alamitos, CA, 1990.

33. D.A. Young and J.A. Pew. *The X Window System Programming & Applications with Xt.* Prentice Hall, Englewood Cliffs, NJ, 1992.

Building New Tools for Synthetic Image Animation by Using Evolutionary Techniques

Jean LOUCHET†‡, Michael BOCCARA†, David CROCHEMORE†, Xavier PROVOT‡

† ENSTA, Laboratoire d'Electronique et d'Informatique
32 boulevard Victor   75739 PARIS cedex 15, France
33-1-45 52 60 75

‡ INRIA, projet SYNTIM
Rocquencourt, B.P. 105   78153 LE CHESNAY Cedex, France
33-1-39 63 54 38

e-mail: louchet@bora.inria.fr

Abstract

*Particle-based models and articulated models are increasingly used in synthetic im-
age animation applications. This paper aims at showing examples of how Evolutionary Algorithms can be used as tools to build realistic physical models for image animation.*

*First, a method to detect regions with rigid 2D motion in image sequences, without
solving explicitly the Optical Flow equation, is presented. It is based on the resolution
of an equation involving rotation descriptors and first-order image derivatives. An
evolutionary technique is used to obtain a raw segmentation based on motion; the result of segmentation is then refined by an accumulation technique in order to determine more accurate rotation centres and deduce articulation points.*

*Second, an evolutionary algorithm designed to identify internal parameters of a mass-
spring animation model from kinematic data ("Physics from Motion") is presented
through its application to cloth animation modelling.*

Keywords

Computer vision, motion analysis, image animation, evolutionary algorithms.

1 Introduction

In this paper, we shall examine through two examples, how evolutionary techniques
can contribute to the resolution of problems at the border of the image analysis and
image synthesis domains, where optimisation methods tend to play an increasing role
and, in our opinion, evolutionary techniques bring interesting new possibilities, notably through their abilities to cope with large numbers of unknown variables or noisy
data, to use heterogeneous cost functions, and to find families of solutions rather than
a single optimum.

Both examples shown are inspired by an image synthesis point of view: how to achieve realism of motion in synthetic image animation. Our basic assumption is that motion realism may only be achieved through the explicit use of physical laws to generate animated image sequences. A long-term goal is to help the human animator build physically realistic image sequences without the task of evaluating himself visually the physical consistency of motion. It is therefore important to develop physical model-based animation techniques as the basis of animation tools. To be realistic, such a model should be identified from *real world* image sequences.

In this scope, we devised a family of mechanical models based on points with masses, and interactions (bonds) modelled by energy potentials. We developed ([L94], [L94b]) a general method to identify the set of parameters of such a structure, from given particle kinematic data, based on an evolutionary algorithm technique: being given a set of particles and their positions in function of time, it consists in tuning up a set of mechanical parameters in order to give the object (i.e. its set of geometric primitives) a behaviour which minimises a quadratic norm of the difference of the generated trajectories from the observed ones.

The second part of this paper will show an application of this evolutionary technique to identify internal mechanical parameters of a cloth sample from its kinematics.

The first part of the paper is devoted to another important question still unanswered: how is it possible to build the kinematic data themselves from real image sequences? We describe an algorithm which combines evolutionary methods with more classical approaches to detect solid 2D motion components in image sequences (corresponding to the bottom right arrow).

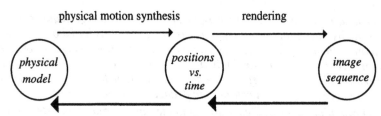

The cloth application described in the second part of the paper, corresponds to the bottom left arrow.

2 Detecting Rigid Motion in Image Sequences: a phenomenological analysis.

2.1 Instantaneous 2-D rotation centres

Let $I(x, y, t)$ an image sequence. I is the intensity (grey level) of pixel (x, y) at time t.

To each point (x, y, t) corresponds a speed vector with coordinates (V_x, V_y).

The classical Optical Flow hypothesis assumes that the image of each physical point moving in the scene, keeps a constant radiance. This results in a relationship between local speed vectors and the local derivatives of function $I(x, y, t)$ [HW83]:

$$\frac{\partial I}{\partial x} V_x + \frac{\partial I}{\partial y} V_y + \frac{\partial I}{\partial t} = 0$$

However, to solve this equation and compute velocity fields, it is necessary to make some regularity assumptions, as local derivatives at a given pixel do only give one equation for two unknowns V_x and V_y. Major Optical Flow resolution algorithms are described and compared for example in [BFB94]. They use elaborate techniques to exploit assumptions on topological properties of the physical objects represented (connexity...), without which the above equation would be completely unconstrained.

Once the velocity field calculated, it would be theoretically possible to use it to determine rotations in the image, in order to extract solid rotating regions; but in practice the speed components are too noisy to be able to differentiate them safely.

With the same Optical Flow hypothesis on the conservation of pixels' radiance along time, it may be shown from geometrical considerations [BL95], that if a region is moving with a solid apparent 2D motion, then:

$$-\omega\frac{\partial I}{\partial x}(y - \eta) + \omega\frac{\partial I}{\partial y}(x - \xi) + \frac{\partial I}{\partial t} = 0$$

where:

ξ, η are the coordinates of the Instantaneous Rotation Centre (IRC);

ω is the rotation speed.

This formula ("IRC equation") does not involve second derivatives, but gives a relationship between easily computed values at the current pixel (pixel's coordinates and first-order derivatives) and the three rotation parameters (ω, ξ, η). It is rather similar to the Optical Flow equation, in that it links several second-order motion descriptors with the local first-order derivatives of function $I(x, y, t)$.

Let us now focus on how to use it to detect *solid motion* in the 2D image sequence. All the pixels in a solid region will share a common IRC (ξ, η) and a common rotation speed ω. Moreover, the equation above states that for each pixel (x, y) and each value of ω, all corresponding potential rotation centres are on one straight line. Therefore if the scene contains a sufficiently large solid region, and if we suppose that we know the rotation speed ω, then these straight lines will normally converge onto the rotation centre. The first resolution method (§ 2.2) uses an accumulation method to find a single solid motion; an evolutionary preprocessing algorithm is then introduced in § 2.3 to find multiple motion primitives and exploit the spatial coherence of moving regions.

2.2 An accumulation algorithm to find a single IRC

The following algorithm is based on the fact that resolving the IRC equation does not require solving explicitly the Optical Flow equation. For each value of ω and each image pixel (x, y), the equation gives a straight line in the (ξ, η) domain; we know that the local IRC belongs to this line. The idea is that a "good" IRC will belong to many such lines. We implemented an algorithm based on a vote technique inspired by the Hough method [BFB94]. The main steps of the algorithm are:

- define an *accumulation space* in the domain (ξ, η), i.e. a 2-dimensional buffer initialised at 0;
- for each value of ω in a given interval:
  - for each pixel (x, y), calculate the local derivatives and increment the values

in the accumulation space domain (ξ, η) along the line defined by the IRC equation

$$-\omega\frac{\partial I}{\partial x}(y - \eta) + \omega\frac{\partial I}{\partial y}(x - \xi) + \frac{\partial I}{\partial t} = 0$$

- detect the position and value of the maximum peak, thus defining a function $F: \omega \rightarrow (\xi, \eta, \text{ peak value})$
- for the value ω_i detected, $F(\omega_i)$ gives the coordinates (ξ_i, η_i) of the IRC.;
- the corresponding region is the set of pixels which verify the IRC equation.

Experimental results

We tested the algorithm on a natural image sequence from a 2D scene. The scene consists here in a C-shaped object moving on a textured background. Two consecutive 256×256 frames are shown. The rotation centre is a screw near the centre of the image, with coordinates (69,140).

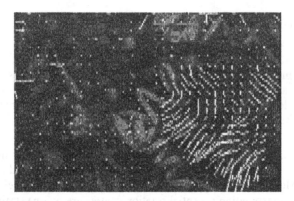

Fig. 1 - A frame from the sequence, with velocity field calculated using a classical method.

We only used image pixels with a sufficient gradient norm value to increment the accumulation space..

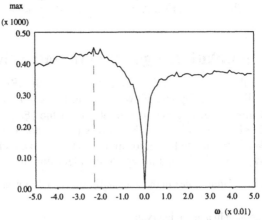

Fig. 2 - Accumulation space maxima vs. ω

Exploring values of ω between -.05 and .05 radians per frame, gives the best peak for $\omega = -.023$ (fig. 2) and the corresponding accumulation space (fig. 3), giving a good estimate of the coordinates of the IRC: $\xi = 72, \eta = 142$

Fig. 3 - The Hough accumulation space for ω_{max}

Conversely, the image pixels corresponding to the rotating object should verify the IRC equation with $\omega = -0.023$, $\xi = 72, \eta = 142$. Figure 4 shows brighter intensities for pixels verifying the IRC equation with parameters values as found (image pixels with insufficient space derivatives are eliminated first).

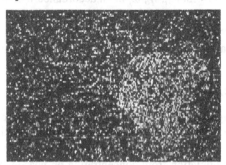

Fig. 4 - Blob detection: brighter pixels verify the equation of motion detected.

2.3 Finding multiple IRCs: Rigid Motion Segmentation

2.3.1 Outline of the algorithm

The algorithm described above, uses a single accumulation space to detect the rotation centre. Therefore, it cannot detect multiple motion as occurring in the general case. Moreover, it makes no assumption on the topology (connectivity) of solid moving regions. The values in the accumulation space are cluttered because of the small size of the objects actually moving compared to the full image size. To solve simultaneously these problems, we devised a new algorithm in three steps:

The first step consists in a primary, raw detection of multiple rotation components. We chose to use an evolutionary algorithm with sharing, to give multiple concurrent solutions.

The next step consists in refining the shapes of these components in order to get a better coverage of the apparent individual rigid objects. We used an optimisation method based on active contours (snakes, [IP94]).

The third step consists in refining the rotation characteristics of these shapes by using an accumulation method as in section 2.1, but restricted to each of the refined shapes determined above.

2.3.2 An evolutionary algorithm to combine local and topological constraints

In order to detect multiple rotation components, we define a population consisting in rotation descriptors, and make it evolve through an evolutionary algorithm. The basic idea is to *introduce the topological constraint into the population individuals themselves*. An individual will therefore consist in the association of a rotation, with a circular neighbourhood of a pixel: the cost function of an individual will be low if the pixel values inside the neighbourhood are consistent with the rotation.

Coding

Each individual i is an ordered sequence $(x_i, y_i, r_i, \xi_i, \eta_i, \omega_i)$ of real numbers, where: x_i, y_i are the coordinates of the centre of a disc; r_i the radius of the disc; ξ_i, η_i the coordinates of a rotation centre; ω_i the rotation value.

The aim of the algorithm is to make the population converge to a final population in which the discs have a fair repartition on the image area, do not overlap, and their corresponding rotation descriptors ξ_i, η_i, ω_i be in accordance with the local motion inside the disc.

Cost function and sharing

The cost function is based on a normalized version of the IRC equation:

$$Cost(i) = \alpha \sum_{(x-x_i)^2+(y-y_i)^2 < r_i^2} \left[\frac{-\frac{\partial I}{\partial x}(y-\eta_i) + \frac{\partial I}{\partial y}(x-\xi_i) + \frac{1}{\omega_i}\frac{\partial I}{\partial t}}{\left(\frac{\partial I}{\partial x}\right)^2 + \left(\frac{\partial I}{\partial y}\right)^2} \right] + \beta \cdot \frac{R}{R+r_i}$$

The "over-normalizing" term in the denominator, helps preventing the discs from spreading over image regions with low gradient values. An additional term gives a slight advantage to larger discs. R, α, β are parameters.

Sharing prevents the population from concentrating on near-identical individuals, and is implemented through the addition of an extra cost to individuals with overlapping discs: the cost function is first calculated for each individual, and the individuals are sorted accordingly. The shared cost function is initialised as equal to the cost function. Then, all pairs of individuals are examined in turn: for each overlapping pair, the shared cost function of the individual with higher (old) cost function is incremented proportionally to the overlapping surface, while the shared cost function of the other individual is kept unchanged.

Selection

At each generation, the individuals are sorted again, according to their shared cost function values. The selection process is controlled by the rank rather than by the shared cost function value. The 20% most performing individuals are then kept, the remaining 80% are deleted and replaced by new individuals created by the mutation process.

Mutations

At each generation, four different mutation processes are applied in parallel. No crossover process is applied.

The first one consists in creating totally new individuals through random functions.

The other ones consist in creating three slightly altered copies of each of the 20% best individuals selected. The two first ones are obtained through applying a random noise to the disc's position and radius, and the rotation speed ω_i. The last one consists in projecting (with a damping coefficient) the rotation centre (ξ_i, η_i) onto the straight lines defined by equation [3]. The projection is repeated for each pixel in the disc.

Initialising

The population is initialised with random values for x, y, r, ξ, η, ω (between given bounds).

Results

With a typical population of 100 individuals, a reasonable stability is obtained after 50 to 100 generations. The final 20% best individuals are retained.

The image below shows the results of primary rotation segmentation after 50 generations on a population of 150 individuals. The circles represent neighbourhood discs. Small squares are the rotation centres found. The straight line from the rotation centre to the corresponding disc, is omitted when the rotation centre is out of the picture. The shorter straight lines from the discs centres represent the local estimated velocities.

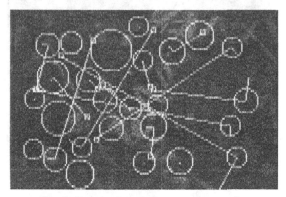

Fig. 5 - Result of evolutionary preprocessing

2.3.3 Refining shapes using snakes

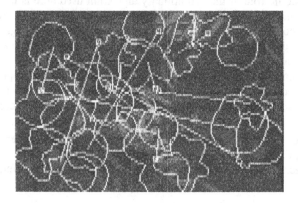

Fig. 6 - Refining the discs' shapes using snakes.

This step consists in refining the shapes found, and deform the circular discs to get more complex shapes by using a snake (active contour) technique, where regions are defined by their radial equation. The regions' cost incorporates a stiffness term to avoid erratic contours; a simple algorithm inflates the contour and retain it when it gives a better energy value. Stabilisation is obtained after typically 5 to 20 iterations (figure 6).

2.3.4 Determining rotations

This last step consists now in refining the rotation characteristics of each region. We apply the algorithm of section 2.2, exploring values of ω around the initial value ω_i found by the evolutionary preprocessing. The essential difference is that values in the domain (ξ_i, η_i) are incremented only for the pixels (x, y) belonging to the refined region (fig. 6) determined above, thus allowing both to reduce noise and potentially detect as many solid motion elements as the number of discs detected in 3.2.1. Experimental results show that the IRCs corresponding to a single solid region do concentrate fairly well into a single point, as shown in figure 7.

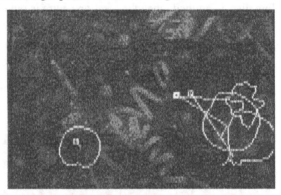

Fig. 7 - Refining the rotations detected.

2.4 Conclusion

The method above allows to detect regions with solid 2D motion in an image sequence. The specific role of the evolutionary algorithm used here is to allow simultaneous determination of unknown motion of unknown image regions, involving several independent constraints: a connexity constraint on the shape of regions, and a motion consistency constraint on grey levels derivatives. The evolutionary preprocessing algorithm gives also a good answer to the need of finding several simultaneous solutions to a problem, as it is often needed in other image processing applications. It is rather fast (a fraction of the time taken e.g. by snake optimisation) and gives good first-approximation solutions which can then be refined using mode conventional approaches.

Here, we do not use any crossover mechanism: this would be no benefit, because a combination of e.g. the disc of one individual with the rotation of another individual, would be physically meaningless and have no reason to yield a new individual with a lower cost.

3 Identifying internal parameters of a cloth animation model

3.1 Physical model identification

An interesting property of the algorithm described above was the ability of an evolutionary approach to find easily multiple approximate solutions to an image processing problem, and to exploit the topological consistency of images by creating individuals strongly related to topological constraints.

Another useful property in image applications corresponds to the current trend towards model-based image interpretation. Fitting a model to given image data requires in particular to be able to optimise functions depending on large numbers of parameters. We described in [L94] an evolutionary algorithm to identify the internal parameters (spring lengths and stiffness...) of a physical animation model to fit given motion data. An original feature was there the use of multiple cost functions, again to exploit more efficiently the 3-D topology of the object to be identified, resulting in the algorithm's convergence in an average number of generations independent of the number of particles (and therefore the number of parameters) involved.

However, in many cases, the same bond may be replicated a great number of times with identical parameters, in a single object. This is the case in cloth models or in several molecular modelling problems. The model parameters are then independent of the location of the bond considered.

This section describes such an evolutionary identification approach, using a cloth animation physical model developed by X.Provot [P95], and giving a visual counterpart to the quality of parameters' identification.

3.2 A Mass-Spring Cloth Model

The elastic model is a periodic mesh of masses, each one linked to its neighbours by massless springs of non-zero natural length at rest. The links between neighbours are:

- springs between masses [i, j] and [i+1, j], or between masses [i, j] and [i, j+1]: "structural springs";
- diagonal springs between masses [i, j] and [i+1, j+1], or [i+1, j] and [i, j+1]: "shear springs" ;
- double-length springs between masses [i, j] and [i+2, j], or [i, j] and [i, j+2]: "flexion springs".

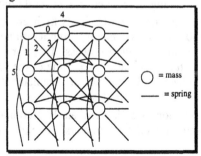

Fig. 8 - The periodic mesh of masses and springs used in our model.

We chose not to introduce any "ternary" bond. In practice, if not strictly equivalent, a flexural stiffness effect can be achieved by giving planar internal stress, e.g. by increasing the lengths of type 2 and 3 springs while reducing the lengths of types 0 and 1.

Motion is obtained by calculating at each step: mutual distances between particles, resulting forces on particles, particles' accelerations, new speeds and new positions.

To take into account the non-linear elasticity of cloth, an `inverse dynamics' procedure is applied if necessary to the two ends of the spring, so that its deformation rate cannot be greater than τ_c.

3.3 Identifying mechanical parameters of cloth

The fabric model above uses (in the case of homogeneous fabrics) equal masses, and 6 different spring types. Each spring being described by three parameters, the fabric is completely described by 18 parameters. In a first approach to the identification problem, in order to reduce the algorithmic cost of analysis, we have partly simplified this general model, making the assumption the fabric is isotropic and all the springs share a common stiffness value. The simplified model contains 5 parameters:

- the springs stiffness
- the elongation rate;
- lengths at rest of springs 0 and 1;
- lengths at rest of springs 2 and 3;
- lengths at rest of springs 4 and 5.

The cost function is based on the difference between the predicted and actual behaviours. One formulation could be:

$$f(parameters) = \sum_{time\ steps} \sum_{i,j} [(x_p - x_r)^2 + (y_p - y_r)^2 + (z_p - z_r)^2 + \Delta t^2 ((vx_p - vx_r)^2 + (vy_p - v$$

where

x_r, y_r, z_r, vx_r, xy_r, vz_r are the actual (recorded) positions and velocities of particles;

x_p, y_p, z_p, vx_p, xy_p, vz_p are the positions and velocities of particles as predicted by the model. Δt is an arbitrary coefficient.

In order to prevent a quick divergence between actual and predicted coordinates, which would result in the cost function being too sensitive to the parameters values, positions and velocities are predicted using actual positions and velocities at the preceding time step, rather than values predicted over several time steps.

The major problem with this cost function is its high computational cost (about 1 minute on a Sparc 4 for 50 time steps), remembering it will have to be calculated several thousands times in the identification process. Therefore we have defined a new "small" cost function, which only involves one frame of the sequence. It proved to be sufficient in practice to obtain good estimates of parameters (see below) if the frame chosen is far enough from the initial conditions.

To optimise this cost function, we devised an Evolutionary Algorithm. The *individuals* are tentative sets of model parameters. An individual is a n-uple of real numbers.

The population is randomly *initialised*, then evolves using three random basic processes controlled by the cost function values: a *selection* process, a *mutation* process and a *crossover* process.

Selection

At each generation, individuals are sorted according to their cost function values. The selection process is guided by *ranking* rather than by the cost function value in itself. We chose to keep the 50% most performing individuals unchanged, the remaining 50% are deleted and replaced by new individuals created by the mutation (20%) and crossover (30%) processes.

Mutations

At each generation, 20% new individuals are created through mutations of randomly chosen parameters among the 80% most performing individuals.

Crossover

At each generation, 30% new individuals are created through uniform crossover. For each new individual to be created, we choose a number of parents equal to the number of model parameters, among the 50% most performing individuals .

Initialising

The population is initialised with random values for all the parameters. However, natural spring length values are chosen around the average length values observed on the reference data.

3.4 Experimental results

We tested the algorithm in the case of a cloth hanging from two corners, with different parameter values. In order to check the suitability of the cost function, we calculated its theoretical values using the simplified 5-parameter model in a neighbourhood of the reference values. The following diagrams show variations of theoretical cost functions, in function of two parameters: the spring stiffness K and elongation rate τ, for two parameter sets. The cost function has been calculated on the domain $\tau \in [0.995, 1.01]$, $K \in [2.0, 4.0]$. Spring natural lengths are fixed to their reference values.The accuracy of the new cost function appears good in the absence of noise, and very sensitive to the elongation rate: the last image in the animation corresponds to a state where the elongation rate plays a dominant role.

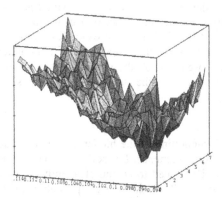

Fig. 9. Cost function for a reference trajectory computed with the fixed parameters $K = 4$ and $\tau_c = 0.1$.

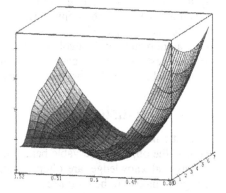

Fig. 10. Cost function for a reference trajectory computed with the fixed parameters $K = 4$ and $\tau_c = 0.5$.

Convergence is generally obtained after 50 to 100 generations and gives results in good accordance with the reference. In some cases, the cost function suffers from numerical stability problems, as it can be seen above on the left, but this only results in a less accurate identification. The listings below shows the parameters values for the 5 best individuals in a 100 individual population, after 49 generations, in typical good and bad situations, with a 17×17 mesh. The first line shows the reference values.

```
              /   101   /   123   /   145   /   K    /   tau   /   cost  /
reference       0.093750 0.132583 0.187500 4.000000 0.500000 0.000000
candidate  4:0.093984 0.133019 0.187106 3.938169 0.497108 0.024129
candidate 26:0.093887 0.133019 0.187111 3.906064 0.497108 0.022811
candidate 57:0.093960 0.133019 0.187107 3.926550 0.497108 0.022180
candidate 39:0.093954 0.133019 0.187108 3.927238 0.497108 0.021845
candidate 88:0.093984 0.132705 0.187083 4.084211 0.497108 0.014619
```

In the second situation:

```
reference       0.093750 0.132583 0.187500 4.000000 0.500000 0.000000
candidate 73:0.086170 0.132667 0.199857 0.394229 0.630850 0.645847
candidate  7:0.086170 0.132667 0.198607 0.394229 0.630850 0.645059
candidate  1:0.086170 0.132670 0.199827 0.400431 0.630850 0.644962
candidate 40:0.086170 0.132668 0.199821 0.406031 0.630850 0.644202
candidate 51:0.086170 0.132671 0.200991 0.435233 0.630850 0.642728
```

In the first case, the cost of 0.014 corresponds to an average error per frame of about one twentieth of the springs' natural length and gives very small visual differences with the original sequence (Figure 11, middle column).

In the second case, the cost function has been trapped into the valley visible on the left side of figure 10 and the number of generations was not sufficient to reach the global optimum. The cost of 0.64 corresponds to a quadratic average error per frame on particles' positions about one half the structural springs' natural length.

Figure 11 shows in the left column, the original image sequence generated using arbitrary parameters. The centre column shows a reconstruction of the sequence using only the parameters resulting from the first identification case above (with final cost function 0.014619), the column right shows another reconstruction using the set of parameters obtained in the second case (cost 0.642726). These examples show how premature results of the evolutionary algorithm affect the visual results in animation, and suggest that in this case, the criterion used to stop the evolutionary algorithm should be based on the cost function values rather than the number of generations.

4 Conclusion

Realistic image animation requires explicit modelling of the hidden processes which control the objects' observable behaviours in the scene. Such models, based on general physical knowledge, are often complex. They need to be run efficiently to create images, but also to be *built accurately from real life data*. The examples above, show two applications of Evolutionary Algorithms to the resolution of problems on the border between Computer Graphics and Image Processing, in both kinematic animation and physical animation modelling. They suggest that Evolutionary techniques are likely to play a significant role in the field of model-driven image analysis, where statistical methods alone, like Hough's accumulation algorithms, are not able to cope with the large numbers of parameters often involved in physical or even phenomeno-logical models.

Moreover, they illustrate the strong link between the *semantics* of the function being optimised and the corresponding *evolutionary scheme*: in particular, crossover was omitted in our first example because of its physical meaninglessness; in the second example, the use of multiple cost functions, shown in [L94] to be essential in the parameter identification of complex mass-spring objects, is left aside in the special case of fabrics modelling where the same set of parameters is used throughout the object.

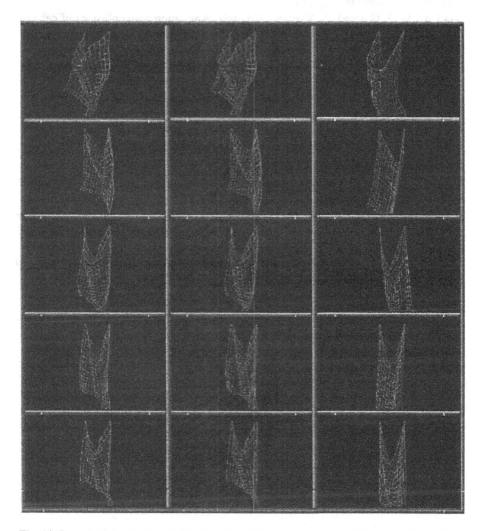

Fig. 11. The original animation (left column), and two reconstructions of the animation using different identification results (centre column: cost = 0.014; right column: cost = 0.642).

References

[BHW94] D.Breen, D. House, M. Wozny, "Predicting the drape of woven cloth using interacting particles", Proc. Siggraph 94, Comp. Graph. Proc., 1994, pp. 365-372.

[BL95] M.Boccara, J.Louchet "Recherche de points d'articulation dans les séquences d'images", internal report, Ecole Nationale Supérieure de Techniques Avancées, 1995.

[G89] D.A.Goldberg, "Genetic Algorithms in Search, Optimization and Machine Learning", Addison-Wesley 1989.

[GVP90] M.-P. Gascuel, A. Verroust, C.Puech "Animation with collisions of deformable articulated bodies", Eurographics Workshop on Animation & Simulation, Sep. 1990.

[HW83] B.K.P.Horn, E.J.Weldon Jr., "Determining Optical Flow", Artificial Intelligence 17: 185-204, 1981.

[IP94) Jim Ivins, John Porrill, *Statistical Snakes: Active Region Models,* British Machine Vision Conference, York, Sep. 1994.

[K92] John Koza, *Genetic Programming*, MIT Press, 1992.

[L94] J. Louchet, "An Evolutionary Algorithm for Physical Motion Analysis", British Machine Vision Conference, York, Sep. 1994.

[L94a] J.Louchet, "Identification évolutive de modèles physiques d'animation", Journées Evolution Artificielle 94, Toulouse Sep. 1994.

[LJFCR91] A. Luciani, S. Jimenez. J.L. Florens, C. Cadoz, O. Raoult, "Computational Physics: a Modeller Simulator for Animated Physical Objects", Proc. Eurographics Conference, Wien, Sep. 1991, Elsevier.

[P95] X.Provot, "Deformation Constraints in a Mass-Spring Model to describe Rigid Cloth behavior", Graphics Interface 1995, Québec, April 1995.

A Genetic Algorithm with Sharing for the Detection of 2D Geometric Primitives in Images

Evelyne LUTTON, Patrice MARTINEZ

INRIA - Rocquencourt
B.P. 105, 78153 LE CHESNAY Cedex, France
Tel : 33 1 39 63 55 23 - Fax : 33 1 39 63 53 30
email : evelyne.lutton@inria.fr

Abstract. We investigate the use of genetic algorithms (GAs) in the framework of image primitives extraction (such as segments, circles, ellipses or quadrilaterals). This approach completes the well-known Hough Transform, in the sense that GAs are efficient when the Hough approach becomes too expensive in memory, i.e. when we search for complex primitives having more than 3 or 4 parameters.

Indeed, a GA is a stochastic technique, relatively slow, but which provides with an efficient tool to search in a high dimensional space. The philosophy of the method is very similar to the Hough Transform, which is to search an optimum in a parameter space. However, we will see that the implementation is different.

The idea of using a GA for that purpose is not new, Roth and Levine [18] have proposed a method for 2D and 3D primitives in 1992. For the detection of 2D primitives, we re-implement that method and improve it mainly in three ways :

- by using distance images instead of directly using contour images, which tends to smoothen the function to optimize,
- by using a GA-sharing technique, to detect several image primitives in the same step,
- by applying some recent theoretical results on GAs (about mutation probabilities) to reduce convergence time.

Keywords : Genetic Algorithms, Image Primitive extraction, Sharing, Hough Transform.

1 Introduction

Geometric Primitives extraction is an important task in image analysis. For example, it is used in camera calibration, tridimensional stereo reconstruction, or pattern recognition. It is important especially in the case of indoor vision, where most of the objects to be analysed are manufactured. The description of such objects with the help of bidimensional or tridimensional geometric primitives is well adapted.

Our aim is to present an alternative to the well-known Hough transform [11], widely used for the primitive extraction problem. The Hough transform is a very efficient method for lines of simple primitives detection (see for example [13, 19, 14]), but reaches its limits when we try to extract complex primitives.

The method consists in the searching of maxima within the space of parameters which describe the primitive. Let us consider for example the search of lines on a binary image where black points represent some contour points. Image lines can be represented with an equation $cos\theta x + sin\theta y = p$ (polar representation of a line). A line is thus defined with help of two parameters : θ and p. Each image point can belong to several lines, and if we consider the parameters space (θ, p), these particular lines are represented by a curve (a sinusoid). We thus build a so-called accumulator, wich represents the parameters space (θ, p), discretized into "cells". For each image point, cells which represent lines that contain it are updated. Cells are thus a sort of counter. A search for the maximal cells provides finally the lines that are effectively represented in the image.

The Hough Transform constructs explicitly the function to optimize, which is represented by the accumulator. This accumulator can be filled in with two equivalent techniques :

— *the 1 to m technique*, where for one point of the image, we draw a curve (or update the cells) in the parameter space, which represents the parameters of all the primitives which the considered image point may belong to, This is the technique we have just described in the example of lines detection,

— *the m to 1 technique*, also called *randomized* or *combinatorial Hough Transform*, where for all possible m-uple of image points (couple of points for the line detection), we draw a point in the parameter space, which represents the unique primitive that can pass trough the considered m-uple of image points.

The effective detection of primitives is thus done by a rough sequential search on the accumulator. If we search for circles, we have to build an accumulator of dimension 3 (center coordinates and radius), for ellipses, an accumulator of dimension 5. It is thus obvious that this technique becomes rapidly very expensive in memory, with regard to the complexity of the primitives we wish to extract

To summarize, the Hough Transform is a very quick and precise technique for simple geometrical primitives, but it becomes rapidly untractable to store an accumulator and detect optima on it when the number of parameters to estimate increases.

This is why we have to think about efficient optimization techniques to solve the problem for complex geometric primitives. As we have seen, it can be easily formulated as an optimization problem : optimizing the position and size of a geometric primitive (or equivalently the values of parameters), knowing the edges detected on an image. The function optimized in the Hough Transform is a function of the parameters, that is the total number of image points which coincide with the trace of the primitive defined by these parameters.

Another problem is that, when the dimension of the space to search is large, the function to be optimized can be very irregular.

When a function has a certain type of regularity, a number of optimization methods exists, mostly based on gradient or generalized gradient computations (see for instance [4]).

Generalized gradient methods work well when :

- some sort of gradient can be defined and computed at any point of the space of solutions (for instance, directional derivatives),
- the function does not have too many local minima, or the value taken by the function at these minima is significantly greater than the value at the absolute minimum.

For very irregular functions, different methods have to be used for optimization. Most of them are based on stochastic schemes.

One of the most known stochastic algorithms is Simulated Annealing. It is a powerful technique for finding the global minimum of a function when a great number of parameters have to be taken into account. It is based on an analogy with the annealing of solids, where a material is heated to a high temperature, and then very slowly cooled in order to let the system reach its ground energy. The delicate point is not to lower the temperature T too rapidly, thus avoiding local minima. Application to other optimization problems is done by generalizing the states of the physical system to some defined states of the system being optimized, and generalizing the temperature to a control parameter for the optimization process ; most of the time, the Metropolis algorithm is used : at "temperature" T, the jump from a state of energy E to a state of energy E' is made with probability of one if E' is lower than E and with a probability proportional to $e^{(E-E')/T}$ otherwise ([1, 16]).

The main drawback of Simulated Annealing is the computational time : the optimal solution is guaranteed only if the temperature is lowered at a logarithmic rate([6]), implying a huge number of iterations. Most of the time, a linear rate is used to obtain affordable converging times, but, for certain very wild functions, the logarithmic rate has to be used.

In this work, we investigate the use of another recently introduced method for stochastic optimization, namely Genetic Algorithms [9, 18, 17]. In section 2, we present the characteristics of the genetic algorithms we have developped for the detection of image primitives. We then introduce in section 3 the Sharing method, which enables an improvement of the efficiency of our method, and sum up our results in section 4, proposing various desirable extensions.

2 The GA for images primitives extraction

The benefit of using Genetic Algorithms (GA) to optimize irregular functions is that they perform a stochastic search over a large search space, by making a population of solutions evolve together, instead of using a single solution as in the Simulated Annealing scheme.

So for, as well as stochastic optimization methods as under the form of classifier systems or genetic programming, GA have not been largely used in robot vision and image analysis applications.

The GA we use here is a classical one, in the sense that it makes evolve (via elitist selection with scaling, crossover and mutation) solutions *encoded* as chromosomes. The discrete representation is generally convenient for image analysis applications, where initial data are pixels matrices. In the following, we detail the modification we have implemented, in comparison to the Goldberg scheme [7], and the specific components (solutions encoding and fitness computation) of our application.

2.1 The chromosomal representation: primitives encoding

For the purely optimization applications of GA, the chromosome is simply a concatenation of the binary codes of all the components of the vector of the space to be searched. This space must therefore be bounded. If there are real values, they are sampled with a certain precision, for the case of natural numbers the binary code is straightforward. Notice also that the choice of sampling rates for real values can be problematic [20], and must be handled with care.

In the approach of Roth and Levine [17], the chromosome representing a primitive is the coding of the points needed to define that primitive. These points are contour points of the primitive. The representation of a primitive with a minimal set of points makes the extraction process less noise-sensitive and the chromosomal coding trivial : a chromosome is the concatenation of the codes of its minimal set of points. But that representation has a main drawback, the redundancy : the same primitive can be represented by several differents chromosomes, one for each possible permutation of the points of the minimal set.

In our application, we propose a different coding, which insures the unicity of the representation :

- **Segment** : 2 points of the image $I = [0..x_{max}, 0..y_{max}]$, with integer coordinates, representing the vertices of the segment,
- **Circle** : 1 point of I for the center of the circle, and a positive integer of $[0.. \max(x_{max}, y_{max})]$ for its radius,
- **Ellipse** : 2 point of I, the center O and the point P, a positive real a, between 0 and $\frac{\pi}{2}$ (sampled on 8 bits), representing the rotation angle of the ellipse. The characteristics of this coding are classical for computer graphics and are detailed in [12], see figure 1,
- **Rectangle** : 2 points of I, coordinates of the top-left and bottom-right vertices. This rectangle is parallel to the axes of the image. For different orientations, we add a positive real, between 0 and $\frac{\pi}{2}$, for the rotation angle.
- **Quadrilateral** : 4 points of I, for the 4 vertices.

2.2 Genetic Operators

The creation of a new individual is performed first by *selection* (with scaling) of two "parents", the selection criterion is given accordingly to the fitness function, then by application of the genetic operators: *crossover* and *mutation*. These

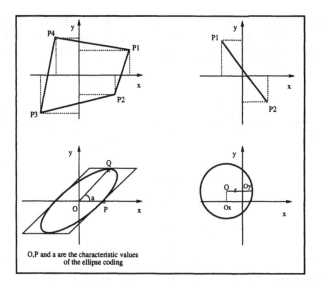

O,P and a are the characteristic values
of the ellipse coding

Fig. 1. Graphical coding of a segment, a circle, a quadrilateral, and of an ellipse

operators are randomly applied. The crossover probability is fixed between 0.7 and 0.9. We have tested classical crossovers: one point crossover and uniform crossover (they provide comparable results in our experiments).

The mutation probability is classically fixed at a very low value along the evolution of generations. Recent results [5] on the convergence of finite population GA with fixed crossover probability, insure a theoretical convergence onto the global optimum if the mutation probability $p_m(k)$ decreases at each generation k with respect to a monotonic lower bound :

$$p_m(k) = \frac{1}{2} * k^{-\frac{1}{M \cdot L}}$$

M is the population size and L is the length of the chromosomes.

Of course such a decreasing rate is very slow (see figure 2), and needs an infinite number of generations to make the GA to converge. When we practically implement a Simulated Annealing algorithm, we use a faster decreasing rate of the temperature than the theoretical one. Here we propose a decreasing rate which is given by the formula :

$$p_m(k) = p_m(0) * \exp(-\frac{k}{\alpha})$$

$p_m(0)$ is the initial mutation probability, α is computed to yield a very low mutation probability (namely 10^{-4}) at the end of the evolution :

$$\alpha = \frac{\text{Max Nb of Generations}}{\ln \frac{p_m(0)}{10^{-4}}}$$

The continuous curve of figure 2 is drawn for a Maximal number of Generation of 100 and an initial mutation probability of 0.25. The theoretical curve (dotted) corresponding to a length of chromosome of 32 bits is drawn for comparison with 1000 generations.

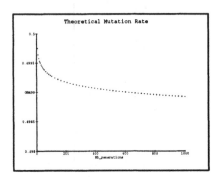 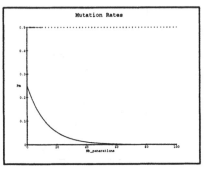

Fig. 2. Left, the theoretical curve of Davis (dotted) *on 1000 generations*, Right, comparison with the one we adopt (continuous) *on 100 generations*

The decreasing rate that we propose allows to perform a large exploration of the search space at the beginning, and then a rapid convergence at the end of the evolution of the GA. This rather rough decreasing rate is of course dependent on the form of the function to optimize. If this function is too irregular, some slower decreasing rate is necessary, just as in the case of Simulated Annealing.

2.3 Computation of the fitness function

For the computation of fitness function we prefer to use distance images, instead of simple contour images. Indeed, if we use contour images, the fitness function is a measure of the contour points in coincidence with the trace of the primitive. See figure 3 a) for the example of segment fitness computation. To tolerate small errors, it is often necessary to make a computation on a strip centered on the primitive, which increases the computational time of an evaluation of the fitness function, see figure 3.

Moreover, the form of the fitness function in that case is very irregular, and a primitive near a real contour has no more information than a primitive which is far away from it. The convergence of a GA in that case can be very slow, especially when the contours are sparse in the image.

We use a well-known tool of mathematical morphology, which furnish distance images, i.e. grey-level images computed from contour images, where each pixels gives the value of its distance to the nearest contour point. These images are easily created by application of two masks on a contour image, see [3]. The distances computed are parameterized by $d1$ and $d2$, see figure 4, which represent

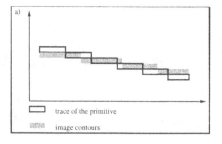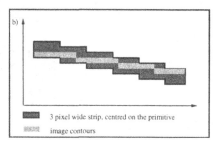

Fig. 3. Fitness computation on contour images

the two elementar distances on vertical/horizontal and diagonal directions. We use Chamfer distance ($d1 = 3$ and $d2 = 4$), or more "abrupt" distances ($d1 = 10$ and $d2 = 14$), in our application, see figure 5. The tuning of $d1$ and $d2$ depends on the frequency of the contour points on the image, and we can think of an "adaptative" tuning of these parameters (we have not implemented it yet).

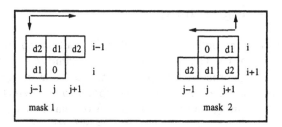

Fig. 4. Distance Masks

The benefit of using such distances images is double : first, the fitness function is more rapid to compute (using the trace of the primitive is largely sufficient), and secondly, the tolerance to small errors is improved. The fitness function takes into account the mean intensity of the pixels of the distances image in coincidence with the trace of the primitive (to position the primitive), plus a counting term of effective contour pixels on the trace (to favor bigger primitives).

2.4 Analysis of the final population : solutions extraction

An interesting characteristic of the convergence of GA is that they "visit" several good local optima, before converging to the global one. A finer analysis than simply finding the solution with best fitness of the final population, can give some interesting information when the GA is stopped before complete convergence.

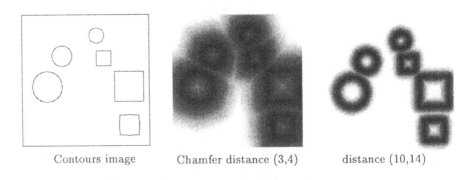

Contours image   Chamfer distance (3,4)   distance (10,14)

Fig. 5. Example of distance images on a synthetic image

This can be important, especially in our case, where we search for several local - or global - optima. Thus, we have developed a simple clustering technique, which permits to locate several local optima present in the final population. Such convergence behaviors (in section 4) can be favored using sharing techniques. We describe this method and our particular implementation in the following section.

3   Use of a sharing technique

The primitive extraction method described before allows to detect only one primitive at each run of the GA. To detect all the primitives of the image it is necessary to iterate the process, by updating the contour image (just by "removing" the contours on the image that correspond to the detected primitive), regenerate the contour image, and re-run the GA on that new environment. We stop the process when there are no more contours in the image, or when the best detected primitive by the GA has a fitness under a fixed threshold.

The interest of detecting several primitives in the same GA-run is evident. For that purpose, we propose to use a sharing technique, followed by the simple clustering we have described in section 2.4.

Sahring techniques [7, 8] simulate the natural phenomena of populations "niching", and subspecies creation. Individuals of a same subpopulation have to share the local resources. Due to overcrowding, the local resources decrease, and individuals tend to search other places, thus creating new subpopulations

In GAs several solutions have been proposed, based on explicit or implicit creation of niches. More precisely, we can divide these approaches in two classes. The first one represents techniques to maintain the diversity in the population along the GA evolution, thus in a certain measure it favors the creation of separate subpopulations. The second one uses a modification of the fitness function to simulate the sharing of local resources in the population. (Goldberg and Richardson [8]). This approach is based on the notion of distance between individuals: the fitness of an individual is reduced according to the number of its neighbours, with the help of a so-called sharing function.

These methods have been carefully studied these last five years, and the ability of the sharing technique has been theoretically demonstrated to find multiple, good solutions. For example using the finite Markov Chain Analysis as Horn in 1993 [10]. A niched GA tell us more about the fitness landscape than what the best solution is : each niche, representing a "good solution" has a subpopulation proportional to its fitness.

3.1 The sharing function

The sharing function is a way of determining the degradation of an individual's fitness due to a neighbor at some distance. Of course, we have to define a distance on our search space. It can be computed on chromosomes (genotypic distance), as Hamming distances between strings, or in the search space itself (phenotypic distance).

In our application, we have preferred to use the phenotypic distance between primitives. Indeed phenotypic distances, when we can use them, have been demonstrated as being more powerful [8].

The neighborhood notion is a fuzzy one, we define a fuzzy neighborhood from a membership function $sh()$, which is a function of the chosen distance d. The membership function that we use, proposed by Goldberg and Richardson [8], depends on two constants : σ_{share} which commands the extent of the neighborhood, and α for its "shape", see figure 6

$$Sh(d) = \begin{cases} 1 - (\frac{d}{\sigma_{share}})^\alpha & \text{if } d < \sigma_{share} \\ 0 & \text{else} \end{cases}$$

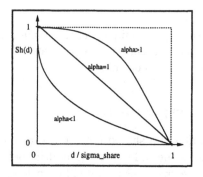

Fig. 6. Neighborhood shape according to α

Sharing is implemented by changing the fitness function to a shared fitness, which is the fitness divided by its niche count m_i :

$$\text{NewFitness}(i) = \frac{\text{Oldfitness}(i)}{m_i}$$

$$m_i = \sum_{k=1}^{N} Sh(d_{ik})$$

The effect of sharing is to separate the population in subpopulations of sizes proportional to the height of the optima. Goldberg and Richardson [8] have explained that a sharing is stabilized when $\frac{f_h}{m_h} = \frac{f_k}{m_k}$ ($\forall h, k$ with h and k different local optima), see figure 7.

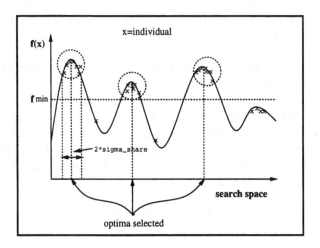

Fig. 7. Repartition of the population after a sharing

The results of the sharing technique depend mainly on the tuning of the parameter σ_{share}, which is a measure of the separation we accept between two detected peaks of the function to optimize.

3.2 Sharing with mating restriction

If we examine the evolution of a shared GA, on a simple multi modal function, we notice that there are some individuals that drift between two peaks. This is due to the fact that the crossover of individual belonging to different niches rarely results in an individual near these optima. Booker [2] has proposed to limit the crossover of different "species". He modified the selection technique in the following way :

- choice of the first parent (selection on the classical fitness function),
- scanning of the population to find the subpopulation of individuals that have a distance with it less that σ_{cross} (σ_{cross} is currently take equal to σ_{share}),

– if the size of this subpopulation is bigger that 1, application of the classical elitist selection in that subpopulation, otherwise, the other parent is chosen in the whole population.

Goldberg and Richardson [8] have proven on simple multi modal functions that the mating restriction scheme improves the efficiency of the sharing. We experimentally verified that fact in our application.

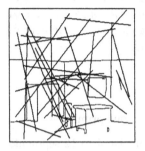 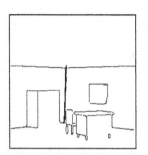 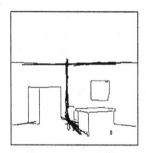

Fig. 8. Initial random population (segments) **Fig. 9.** Final classical convergence of the population (after 80 generations) **Fig. 10.** Convergence with sharing

3.3 A modified sharing technique

We propose another scheme for the sharing, based on the relative importance of the individuals in a neighborhood. An individual with a low fitness in a niche will have a few influence on crossovers in that particular sub-population, and will surely disappear to the benefit of better individuals. Goldberg and Richardson just take into account the number of neighbors to share the fitness. We prefer to take also into account the "force" of the neighbors. The new sharing scheme is thus :

$$\text{NewFitness(i)} = \frac{\text{OldFitness(i)}}{\mu_i}$$

$$\mu_i = \sum_{k=1}^{N}(\text{OldFitness}(k) * Sh(d_{ik}))$$

μ_i represents a fuzzy mean fitness (a sort of density) in the neighborhood of the individual i. Thus $\frac{fitness(i)}{\mu_i}$ is a measure of the relative importance of the individual with respect to his neighborhood. Following [8], we can tell also that the evolution process is stabilized when :

$$\frac{fitness(peak_h)}{\mu_{peak_h}} = \frac{fitness(peak_k)}{\mu_{peak_k}}$$

It tends to equilibrate the subpopulations of the peaks, independently of their height, since they are bigger than a certain threshold. This particular fact permits to "inhabit" more peaks with the same population size, than with the classical sharing technique, where the best peaks attract much more individuals, and thus reduces the number of individuals which can occupy other peaks.

This comportment is very interesting in our case because the function we have to optimize is strongly multi modal, and we have noticed an important improvement in the performances in using our sharing scheme, in comparison with the classical sharing. For example, for a population of 100 individuals (see figures 8, 9, and 10), we can detect 4 to 7 optima in the same run. Of course the shared GA takes more CPU time to converge but we noticed a global improvement of the computational time in comparison with the simple GA.

4 Results

We present here results on synthetic and real images, for four primitives :

- **segments** : on figures 13 and 21, a GA run (wich furnish 4 to 12 segments at the same time) takes 10 to 15 seconds on a Sparc II station,
- **circles** : on figure 14, it takes 70 to 80 seconds for circles,
- **ellipses** : on figures 22 and 23,
- **rectangles** : on figures 15 and 18.

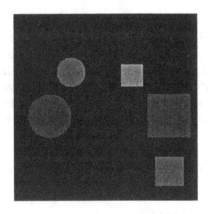

Fig. 11. Synthetic image

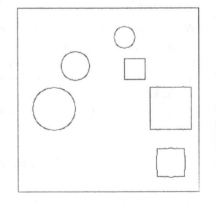

Fig. 12. Contours

5 Conclusion

The method we have presented, is complementary to the Hough Transform, in the sense that for simple primitives (less than 4 parameters), GA is not efficient,

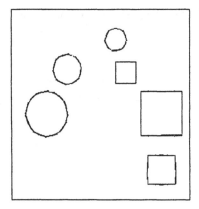

Fig. 13. Segments detected (continuous) and contours (dotted)

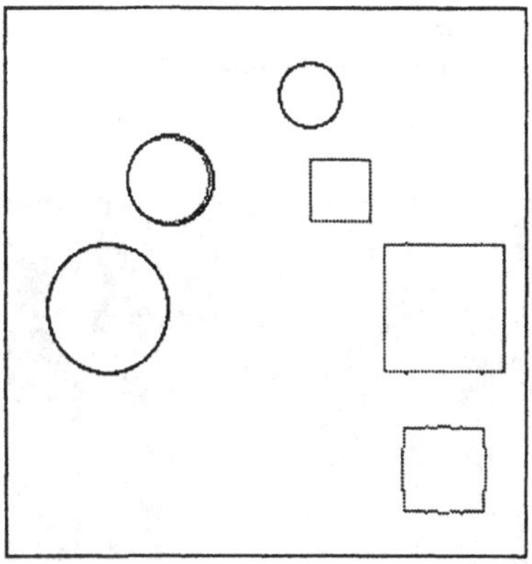

Fig. 14. Circles detected (continuous) and contours (dotted)

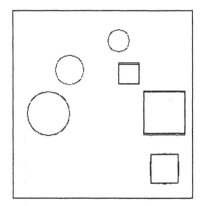

Fig. 15. Rectangles detected (continuous) and contours (dotted)

Hough Transform must be used. For the segments detection task, the GA application we have presented (in a parallel implementation), could be concurrent to the Hough Transform. GA is much more interesting for complex primitives (circles, ellipses, quadrilaterals, etc ...). The application to the detection of another type of primitive is easily done by updating the fitness function, and the distance function.

We can also think about some applications of the Hough Transform, called Generalized Hough Transform, where we have a reference form (non parameterized) and search the presence of that form in an image. This is done through the construction of an accumulator of possible translations and rotations of the form. Once again the Hough Transform is limited, and the search is mostly done on a few parameters (translation *or* rotation *or* dilatation). The GA approach can also furnish a tool to do such a search with displacement, and deformation parameters together.

Fig. 16. Synthetic image

Fig. 17. Contours

Fig. 18. Rectangles detected (continuous) and contours (dotted)

The particular formulation of GA approach permits to easily use tools as distance images, which smoothen the function to optimize, and decreases the convergence time. We have also exploited the sharing scheme to improve the efficiency of the search on multi modal fitness functions.

The main problem of such an approach remains the parameters tuning, because it severely influences the convergence speed, and the quality of results. Except for the mutation probability where we could use some theoretical results,

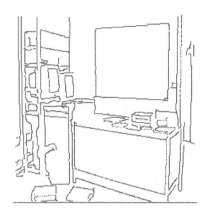

Fig. 19. Real image

Fig. 20. Contours

Fig. 21. Segments detected (continuous) and contours (dotted)

this tuning is now experimentally done, and varies for each type of primitives.

To conclude, we can tell that theoretical researches on GAs are directed towards the problem of judicious choice of parameters, see for example the recent results we use for the mutation probability variation ([5] in 1991). Another point to mention is that GAs can be very easily parallelized, and some authors claim that it permits to divide the computational time almost by the number of processors used.

We hope to have pointed out in this paper the interest of considering GA approaches for complex optimization problems involved in image processing and robot vision tasks. We do not claim that GA can replace some well-known techniques, but we think that they can be considered as a complementary approach for some problems which are untractable with classical techniques.

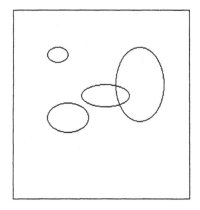

Fig. 22. Synthetic contours

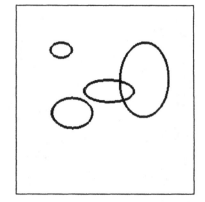

Fig. 23. Ellipses detected

References

1. E. Aarts and P. Van Laarhoven. Simulated annealing : a pedestrian review of the theory and some applications. *AI Series* F30, NATO.
2. L. B. Booker. Intelligent behavior as an adaptatition to the task environment. *PhD thesis, University of Michigan,* Logic of Computers Group, 1982.
3. Gunilla Borgefors. Distance transformations in arbitrary dimensions. *Computer Vision, Graphics, and Image Processing,* 27:321–345, 1984.
4. A. Bihian C. Lemarechal, J.J. Strodiot. *On a bundle algorithm for nonsmooth optimization,* pages 245–282. Academic Press, 1881. Non-Linear Programming 4, Mangasarian, Meyer, Robinson, Editeurs.
5. T. E. Davis and J. C. Principe. A Simulated Annealing Like Convergence Theory for the Simple Genetic Algorithm . In *Proceedings of the Fourth International Conference on Genetic Algorithm,* pages 174–182, 1991. 13-16 July.
6. S. Geman and D. Geman. Stochastic relaxation, gibbs distributions, and the bayesian restoration of images. *IEEE Trans. on Pattern Analysis and Machine Intelligence,* 6(6):712–741, November 1984.
7. D. A. Goldberg. Genetic Algorithms in Search, Optimization, and Machine Learning . *Addison-Wesley,* January 1989.
8. David E. Goldberg and J. Richardson. Genetic algorithms with sharing for multimodal function optimization. In J. J. Grefenstette, editor, *Genetic Algorithms and their Applications,* pages 41–49, Hillsdale, New Jersey, 1987. Lawrence Erlbaum Associates.
9. J. H. Holland. Adaptation in Natural and Artificial System . *Ann Arbor, University of Michigan Press,* 1975.
10. J. Horn. Finite Markov Chain Analysis of Genetic Algorithms with Niching . *IlliGAL Report 93002, University of Illinois at Urbana Champaign,* February 1993.
11. P. V. C. Hough. A new method and means for recognizing complex pattern, 1962. *U. S. Patent 3,0690,654.*
12. A. Van Dam J.D. Foley. Computer graphics - principles and practise.
13. V. F. Leavers. The dynamic generalized hough transform for the concurrent detection of circles and ellipses. In *Progress in Image Analysis and Processing II.* 6th International Conference on IASP, 1991.

14. Evelyne Lutton, Henri Maitre, and Jaime Lopez-Krahe. Determination of vanishing points using hough transform. *PAMI*, 16(4):430–438, April 1994.
15. Evelyne Lutton and Patrice Martinez. A genetic algorithm for the detection of 2d geometric primitives in images. *Research Report 2110, INRIA*, November 1993.
16. R. Otten and L. Van Ginneken. Annealing : the algorithm . *Technical Report RC 10861*, March 1984.
17. G. Roth and M. D. Levine. Geometric Primitive Extraction Using a Genetic Algorithm . In *IEEE Computer Society Conference on CV and PR*, pages 640–644, 1992.
18. Gerhard Roth and Martin D. Levine. Extracting geometric primitives. *CVGIP: Image Understanding*, 58(1):1–22, July 1993.
19. Frank Tong and Ze-Nian Li. On improving the accuracy of line extraction in hough space. *International journal Of Pattern Recognition and Artificial Intelligence*, 6(5):831–847, 1992.
20. J. Lévy Véhel and E. Lutton. Optimization of fractal functions using genetic algorithms. In *Fractal 93*, 1993. London.

Applications

A Genetic Algorithm to Improve an Othello Program

Jean-Marc Alliot, Nicolas Durand

ENAC*, CENA**

Introduction

Let a computer program learn to play a game has been for long a subject of studies. It probably began in 1959 with A. Samuel's program [Sam59], and similar methods are still used on the most advanced computer programs, such as Deep Thought ; however, in chess, the different methods used are mainly linear regression on parameters of the evaluation function.

Othello was, from the start, a game where learning proved to be very useful. The BILL program used Bayesian learning [LM90] to improve parameters of the evaluation function. Similar methods are still applied for the best Othello programs (such as LOGISTELLO[Bur94]). However, these methods require large databases of games, and are very efficient only on Othello game playing, because the game always terminates in a fixed number of moves, with perfect end-games up to 15 moves. Such methods would be very difficult to use on chess playing algorithms.

Other techniques used include co-evolution [SG94], use of neural networks trained by GA to focus minimax-search [MM94], evolving strategies with GA [MM93]. However, [SG94] did not lead to world class Othello program, [MM94] was unable to improve search as soon as the evaluation function was good enough to correctly predict the move, and [MM93] was apparently not able to improve the classical (positional+mobility) strategy.

In this article, we show how genetic algorithms can be used to evolve the parameters of the evaluation function of an Othello program. We must stress that our method can be used on any algorithm using an evaluation function, for any two players game. In the first part of the article, we explain the structure of the Othello program. In the second part, we explain the structure of our genetic algorithm, the operators (crossover, mutation) used, and our method to compute fitness. In the third part, results are presented and, in the last part, possible improvements and generalization of this method are discussed.

1 The Othello program

This work is based on an Othello program developed by the author a few years ago. This program was tested against other public domain programs and by a

* Ecole Nationale de l'Aviation Civile, 7 Avenue Edouard Belin, 31055 Toulouse CEDEX, France. e-mail : alliot@dgac.fr
** Centre d'Etudes de la Navigation Aérienne, e-mail : durand@cenatls.cena.dgac.fr

good French player (rated 10th in France). The program performed very correctly against these opponents.

The main advantage of this program is its very simple structure. The evaluation function has very few parameters. First, a static evaluation is obtained by using a static valuation of each part of the Othello board. For each disc on one square of the board, the value shown on table 1 is added to the current evaluation value if the disc belongs to the computer or subtracted if the disc is owned by its opponent.

500	-150	30	10	10	30	-150	500
-150	-250	0	0	0	0	-250	-150
30	0	1	2	2	1	0	30
10	0	2	16	16	2	0	10
10	0	2	16	16	2	0	10
30	0	1	2	2	1	0	30
-150	-250	0	0	0	0	-250	-150
500	-150	30	10	10	30	-150	500

Table 1. Static square values

Of course, this evaluation function has to be slightly enhanced. First, when one corner of the board is already occupied, the values of the three squares next to it are set to 0. On figure 1, the values of square G8, G7 and H7 are set to 0, instead of -150, -250, -150. Second, also when one corner is occupied, all squares on the edge which are connected to this corner and of the same color get a bonus (called thereafter the *bonus value*). On figure 1, three white discs get a bonus.

Then a second important factor has to be taken into account : the *liberty score*. The number of liberties of a disc is the number of empty squares close to the disc. The number of liberties of a player is the sum of all the liberties of all his discs. For example, on figure 1, black discs have 8 liberties. The *liberty score* is the difference between the liberties of the computer and the liberties of its opponent. This number is multiplied by a *penalty coefficient* and the result is subtracted to the evaluation function computed above.

So the evaluation function has three terms : the static evaluation of the position, a bonus given for connected discs on the edge and a penalty value for the liberties of the discs. Given a board position, the function can be computed with only 12 parameters : 10 (given the symmetries) for the static evaluation of the board, one for the value of the bonus coefficient for each connected disc on the border, and one for the value of the penalty coefficient for each liberty of each disc. On the original program, the values of the 10 coefficients where [500, -150, -250, 30, 0, 1, 10, 0, 2, 16] (that can be easily deduced from table 1). The bonus value was set to 30 and the liberty penalty set to 8.

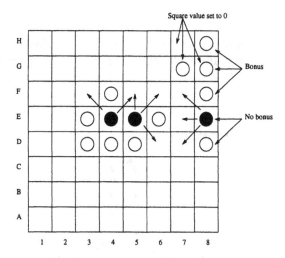

Fig. 1. Evaluation function principles

The rest of the program was a classical α-β iterative deepening algorithm, with an exhaustive search 11 to 14 plies before the end of the game.

As simple as it might look, the program was quite efficient, as the evaluation function was fast to compute and quite good. It was tested against different opponents, such as other public domain Othello programs, and human players, and always obtained good results. This is not very surprising, as almost all good Othello playing programs use such functions, starting with Rosenbloom's IAGO program [Ros82].

2 Optimizing the evaluation function parameters

Tuning the parameters of the evaluation function is a problem very difficult to solve. The values shown above are the results of a few interviews of good players which were also programmers, with no other justification. Using local climbing hill methods (modifying slightly one coefficient, while freezing the others) never gave conclusive results. The process was unstable : no stable maximum was ever reached, and coefficients were modified in one direction, then in the other. Classical optimization methods fail to solve this kind of problem.

Moreover, it is always difficult to evaluate if a program is performing well or badly. Even if the modified program wins 3 games in a row against the original one, this does not mean it is better. It could just mean it was lucky (of course, if it loses 3 in a row, it could also mean it was unlucky). To evaluate if a program is better or worst than another, a more complex methodology must be designed.

2.1 Principles

Classical Genetic Algorithms and Evolutionary Computation principles such as those described in the literature [Gol89, Mic92] are used; Figure 2 describe the main steps of GAs.

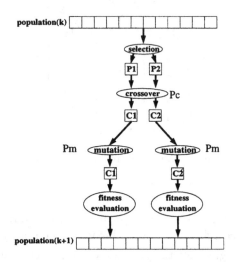

Fig. 2. GA principle

First a population of points in the state space is randomly generated. Then, the value of the function to optimize, called *fitness*, is computed for each population element. In a second step, elements are selected[3] according to their fitness. Afterward, classical operators of crossover and mutation are applied to diversify the population (they are applied with respective probabilities P_c and P_m). At this step a new population has been created and the process is iterated.

This GA has been improved by including different enhancements, described in different papers [DASF94, DAAS94]. We shortly present them again here (a full description is available in [All94]).

First a Simulated Annealing process is used after applying the operators [MG92]. For example, after applying the crossover operator, there are four individuals (two parents $P1,P2$ and two children $C1,C2$) with their respective fitness. Afterward, those four individuals compete in a tournament. The two winners are then inserted in the next generation. The selection process of the winners is the following: if $C1$ is better than $P1$ then $C1$ is selected. Else $C1$ will be selected according to a probability which decreases with the generation number. At the beginning of the simulation, $C1$ has a probability of 0.5 to be selected even if its

---

[3] Selection aims at reproducing better individuals according to their fitness. We tried two kinds of selection process, Roulette Wheel Selection" and "Stochastic Remainder Without Replacement Selection", the last one always gave better results.

fitness is worse than the fitness of $P1$ and this probability decreases to 0.01 at the end of the process. A description of this algorithm is given on figure 3.

Tournament selection brings some convergence theorems from the Simulated Annealing theory. On the other hand, as for Simulated Annealing, the (stochastic) convergence is ensured only when the fitness probability distribution law is stationary in each state point [AK89].

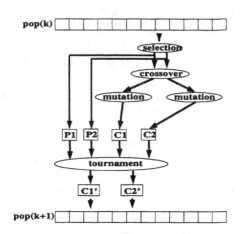

Fig. 3. GA and SA mixed up

Another enhancement was to use sharing as described in [YG]. This prevents the GA to fall too early in local minima. The method described in [YG] has the advantage of having a complexity in $n \log n$ instead of n^2 but is not perfect, as it does not always compute real clusters, but only approximates them. The algorithm has been modified in order to adapt dynamically the number of clusters during computation.

Last, parallelism is used to reduce time in computation. This method is mandatory on this problem as the time needed to compute one fitness is very long. The model of parallelism used is "island parallelism" : on a network of machines, each machine runs one genetic algorithm with its own population (one island). From time to time (every 10 generations for example), each island sends "missionaries" (some elements of its population, around 10%) to another island. Elements are picked at random in the origin population and replace elements taken also at random in the destination population (an elitist strategy is used to prevent destruction of best element, or best clusters element).

2.2 Chromosome structure

The chromosome is a very simple one. It contains 12 integers defining the evaluation function of the Othello program. It contains also the fitness computed at the previous generation, as it will be needed to compute the fitness at the next generation.

2.3 Computing fitness

To compute fitness, the idea is to let one population element play against a reference program (the original one) and use the result as fitness. However, this is not satisfactory. First of all, results are probabilistic. To have a good estimation, a large number of games would be required. This is not possible because of the time required. It takes 0.080s to play one game with a depth of 0 ply (only using the evaluation function), 0.190s to play one game at depth 1, 0.807s at depth 2 and 3.56s at depth 3.

An other problem is determinism: a computer program playing against another computer program at the same depth will always play the same moves. It is useless to play many games with the same depth or the same starting position. These problems are classical when trying to evaluate two computer programs playing against each other, and some of the solutions presented here were already proposed in [LM90].

The following methodology is used; for each element of the population:

- 4 different starting positions are randomly generated. In each of these positions there are at least 4 discs (the 4 center discs) and at most 14 (there can be 10 more discs randomly placed in the inner 4×4 board).
- the new program then plays for each of these positions once with white and once with black. For each of this two games, the number $s = nw - nb$ is computed where nw stands for the number of white discs at the end of the game, and nb the number of black discs; then the difference $s1 - s2$ is computed. If the result is positive, we say that it is a victory for the new program, if the result is negative we say it is a loss, if the result is zero, it is a draw. For example, if the new program plays with white and wins $50 - 14$, then plays with black on the same position and loses $40 - 24$, we add $50 - 14 = 36$ to $24 - 40 = -16$, which gives 20, a victory for the new program.
- This operation is repeated with 3 different depths of evaluation, 0, 1 and 2.
- As 4 different positions are played at 3 different depths, the number of victories v and the number of draw d are computed. The number of discs of the program minus the number of disk of its opponent at the end of each game are also computed: we will call it c. Then we the current fitness f_c is :

$$f_c = v + d/2 + c/1000$$

- The real fitness is f computed with :

$$f = (f_c + f_p)/2$$

where f_c is the current fitness and f_p is the fitness computed at the previous generation. This way, a program, if it is not modified, can accumulate results as time goes by. Of course, mutation and crossover must be correctly handled. This will be discussed in the next section.

This methodology was not chosen at random. In the first time, only one depth was used for searching. The new program was very efficient at this given depth, but was easily defeated at all other depths. With only one starting position (always the same) instead of five different randomly chosen at each generation, the program was very efficient for that given position, but played poorly with other starting positions. Last, having a fitness which uses past experience helped a lot to have stable results.

However, the results of the program were disappointing; it is easy to understand why: a program winning 12 games out of 12 has a better adaptation than a program winning 46 games out of 48. However, it is clear that, with "luck", it is "easier" to win 12/12 than 46/48. It is necessary to introduce also this factor when computing the fitness.

From a mathematical point of view, the probability of winning m games out of n, if the probability of winning one game is p, is given by the following formula:

$$P(p, m, n) = \binom{n}{m} p^m (1 - p)^{n-m}$$

If n games are played, there are $n + 1$ different possible results (only wins or losses are considered). They have all the same probability. So, we have:

$$\int_0^1 P(m, n, p) dp = 1/(n + 1)$$

Let's consider the following function : $h(p, m, n) = (n+1)P(p, m, n)$. The fitness $f(m, n)$ of an element which has won m points over n games played will be given by the following implicit equation (we choose a 95% confidence value):

$$\int_{f(m,n)}^1 h(p, m, n) dp = 0.95$$

This seriously modifies fitness values. For example $f(12, 12) = 0.79$ while $f(23, 24) = 0.83$, $f(44, 48) = 0.82$ and $f(65, 72) = 0.83)$. All tables for n=12,24,36,... were pre-computed with Maple and included in the program.

2.4 The crossover operator

Different operators to cross programs were tried. A stochastic barycentric crossover was finally chosen.

When we cross 2 elements, a number α is picked randomly in the range $[-0.5, 1.5]$ and a number i in the range $[1, 11]$. Then the ith element of each of the two chromosomes is crossed. The new ith component of each chromosome is:

$$c'_{1i} = \alpha c_{1i} + (1 - \alpha)c_{2i}$$

$$c'_{2i} = (1 - \alpha)c_{1i} + \alpha c_{2i}$$

There is a 13th element in a chromosome, which is the fitness computed at the last generation. The fitness of the new elements is the mean of the fitness of the two parents. This is of course not perfect, but gives quite good results.

2.5 The mutation operator

A number i is picked in the range $[1, 11]$. Then a Gaussian noise centered on 0 is added, with a deviation of 50 if it is one of the 10 first components (static values of squares), and 20 if it is the 11th or the 12th.

The fitness of the new element is the same as the fitness of the parent.

2.6 Computing a distance for the sharing operator

To use sharing, distance between two elements has to be defined. A classical Euclidean distance could be used: $\sum_{i=1}^{12}(c_{1i} - c_{2i})^2$. But, here, it is not a very good idea. As the evaluation function is partly linear regarding its parameters, multiplying the tenth first parameters by 2 gives a program which will play exactly as the original one ; however, regarding Euclidean distance, they are very far from each other... To solve this problem, the value of the first element of the evaluation function is set to 500, and the GA is not allowed modify this value on any element of the population. This is not completely a neutral choice. By choosing this value we prevent the GA to evolve programs that might have a negative value for the first coefficient (all programs with a positive value for the first coefficient are equivalent). Half of the search space disappears. However, we strongly believe that corners must have a positive value...

2.7 Parameters settings for the GA

It is very long to run a genetic algorithm on such a problem. Evaluating one element of population takes more than 10s. Running a GA with 100 elements and 100 generations would require 30 hours. That's why a parallel GA was used. Each machine ran a GA with 30 elements of populations for 150 generations. Due to workload and speed differences between the 30 workstations (HP-720 and HP-730), the test ran for 12 hours instead of theoretically 10 hours.

We used a crossover probability of 60% and a mutation probability of 15%. Sharing, evolutive fitness, simulated annealing and an elitist strategy were also used.

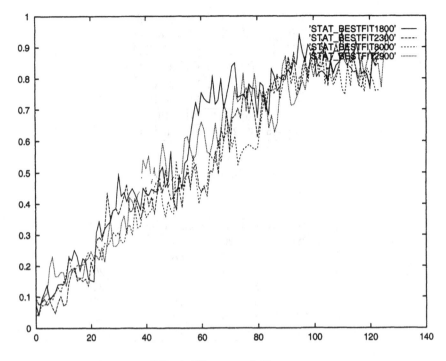

Fig. 4. Fitness evolution

3 Results

The evolution of fitness on the 8 best machines (those which gave the best results in the end) is presented on figure 4.

Fitness evolves and can decrease, as it is not a static value but depends on the starting positions which are generated randomly at each turn. However, having a value which is always above 0.8 is good.

We chose the program which looked best. The static values are shown in table 2. The bonus value was set to 26 and the liberty penalty set to 104.

The program was intensively tested against the reference program. the new program was created by running some sequences of test, with 4 positions played and a depth going from 0 to 2. So it was tested at depths ranging from 0 to 6, by playing 400 positions at each depth, except at depth 6 where we only played 100 positions. Results are given in table 3. The program wins 68% to 77% of points (1 point is given for a win one half point for a draw and zero for a loss), which is a definite improvement. At depths 3, 4, 5 and 6 which were not used to "train" the program, results are even better than at depth 0, 1 and 2.

We might fear that the new program is very efficient against our test program, but will be easily defeated against other programs or human players. The new program (called the stage-1 program) was tested against human players, but it was very inconclusive, as both program always won against all the players

500	-86	96	26	26	96	-86	500
-86	-1219	-6	0	0	-6	-1219	-86
96	-6	52	15	15	52	-6	96
26	0	15	-17	-17	15	0	26
26	0	15	-17	-17	15	0	26
96	-6	52	15	15	52	-6	96
-86	-1219	-6	0	0	-6	-1219	-86
500	-86	96	26	26	96	-86	500

Table 2. Static square values

Depth	Wins	Losses	Draw	% points	Disc difference
0	282	98	20	73	+28
1	252	101	47	68	+24
2	278	78	44	75	+20
3	281	75	44	75	+22
4	280	75	45	75	+22
5	286	68	46	77	+25
6	72	20	8	76	+23

Table 3. Results of the stage-1 program against the original

we could find. So, the stage-1 program was used as the reference program to evolve a stage-2 program. Again, the GA was able to improve the program; The static values are shown in table 4. The bonus value was set to 18 and the liberty penalty set to 312.

500	-240	85	69	69	85	-240	500
-240	-130	49	23	23	49	-130	-240
85	49	1	9	9	1	49	85
69	23	9	32	32	9	23	69
69	23	9	32	32	9	23	69
85	49	1	9	9	1	49	85
-240	-130	49	23	23	49	-130	-240
500	-240	85	69	69	85	-240	500

Table 4. Static square values

Then, similar tests were ran again, having the stage-2 program playing against the stage-1 program. Only tests for depths 0,1,2,3 and 4 were ran, due to lack of time. Results are very good (see table 5), except at depth 0. We can only guess that the stage-1 program is already excellent at that depth and very difficult to improve. At depth 1 and 2, which were also used to train the program, results are above 75%, and are still excellent at depth 3 and 4.

Depth	Wins	Losses	Draw	% points	Disc difference
0	188	194	18	49	-3
1	285	82	33	75	+27
2	303	55	42	81	+28
3	298	50	52	78	+37
4	320	46	34	85	+36
5	329	40	31	86	+38

Table 5. Results of the stage-2 program against the stage-1 program

Is the stage-2 program still better that the original one? The same tests (see table 6) show that results are excellent at every depth, as good as the results of the stage-1 program.

Depth	Wins	Losses	Draw	% points	Disc difference
0	274	107	19	71	+26
1	266	93	41	72	+21
2	243	120	37	65	+20
3	292	61	47	79	+27
4	277	82	41	74	+22

Table 6. Results of the stage-2 program against the original program

4 Remaining problems and improvements

It would of course be interesting to build a stage-3, a stage-4 program, etc, and see how long the program can be improved, or if, in the long time, the resulting program becomes less efficient against the original program, etc. This will be done in the near future.

We have stressed the fact that our methodology could be extended to almost any two-players game which rely on an evaluation function. The only problem would be the time required to play the game. It would be very interesting to try to evolve parameters of a chess program with our algorithm. However, this would require a tremendous power in calculation: it is much longer to evaluate a chess position and the branching factor in chess is higher than in Othello. Moreover, an evaluation function in chess needs much more parameters. It would probably require weeks to get results, even with many workstations working together.

Building the evaluation function by evolving the structure of the evaluation function itself with Genetic Programming, instead of evolving only the parameters was inconclusive. The problem was to find the right terminal functions. Using functions too basic gave no result at all, and using too macroscopic functions (already including most of the structure) was not very different from evolving only the parameters of the evaluation function.

We had previously worked on evolving neural nets with Genetic Algorithms. This gives very good results on some test problems such as car parking [SR92], or aircraft conflict resolution. These examples are quite simple as there are few inputs (6 for the second example) to the network, and the GA has only a hundred parameters to evolve. We are currently working on the following program: the base idea is to have an evaluation function computed by a neural network, and to build this neural network with a GA. However, a direct approach with random initialization of all weights in the network gave no result. The GA was never able to build a network that could even defeat the reference program once in 20 games. So, we now train a very simple neural network (64 inputs, 8 neuronal units in the hidden layer, and one output) with our reference program. We then have a program playing Othello with a neural network instead of a classical evaluation function. This program plays quite good Othello, and sometimes defeat the reference program. Now, we are going to evolve the neural net with a GA, starting with this NN as the base for all elements. This may give interesting results.

It must be noted that this approach is very slow, as the neural networks we used are based on classical floating-point functions. Results might be obtained by using ALN networks [ALLR91] instead of classical feed forward networks : in Othello, inputs of the network are almost binaries, and ALN are much faster and easy to evolve.

References

[AK89]   Emile Aarts and Jan Korst. *Simulated annealing and Boltzmann machines.* Wiley and sons, 1989. ISBN: 0-471-92146-7.

[All94]   Jean-Marc Alliot. *A parametrable parallel genetic algorithm with simulated annealing and adaptive sharing.* Technical report, Ecole Nationale de l'Aviation Civile, 1994.

[ALLR91]   William Armstrong, Jiandong Liang, Dekang Lin, and Scott Reynolds. Experience using adaptative logic networks. In *Proceedings of the IASTED International Symposium on Computers, electronics, communication and control,* 1991.

[Bur94]   Michael Buro. *Techniken fur die bewertung von Speilsituationen anhand von beispielen.* PhD thesis, Universitat GH Paderborn, 1994.

[DAAS94] Nicolas Durand, Nicolas Alech, Jean-Marc Alliot, and Marc Schoenauer. Genetic algorithms for conflict resolution in air traffic. In *Proceedings of the Second Singapore Conference on Intelligent Systems*. SPICIS, 1994.

[DASF94] Daniel Delahaye, Jean-Marc Alliot, Marc Schoenauer, and Jean-Loup Farges. Genetic algorithms for air traffic assignment. In *Proceedings of the EuropeanConference on Artificial Intelligence*. ECAI, 1994.

[Gol89] David Goldberg. *Genetic Algorithms*. Addison Wesley, 1989. ISBN: 0-201-15767-5.

[LM90] Kai-Fu Lee and Sanjoy Mahajan. The development of a world class othello program. *Artificial Intelligence*, 43, 1990.

[MG92] Samir W. Mahfoud and David E. Goldberg. Parallel recombinative simulated annealing: a genetic algorithm. IlliGAL Report 92002, University of Illinois at Urbana-Champaign, 104 South Mathews Avenue Urbana IL 61801, April 1992.

[Mic92] Zbigniew Michalewicz. *Genetic algorithms+data structures=evolution programs*. Springer-Verlag, 1992. ISBN: 0-387-55387-.

[MM93] David Moriarty and Risto Miikulainen. Evolving complex othello startegies using marker-based genetic encoding of neural networks. Technical Report AI93-206, University of Texas, Austin, TX 78712-1188, September 1993.

[MM94] David Moriarty and Risto Miikulainen. Evolving neural networks to focus minimax search. In *Proceedings of the Twelfth National Conference on Artificial Intelligence AAAI-94*, Seattle, WA, 1994.

[Ros82] P. Rosenbloom. A world championship-level othello program. *Artificial Intelligence*, 19:279–320, 1982.

[Sam59] Arthur Samuel. Some studies in machine learning using the game of checkers. *IBM journal of research development*, 3(3):210–229, 1959.

[SG94] Robert Smith and Brian Gray. Co-adaptive genetic algorithms: An example in othello strategy. In *Proceedings of the Florida Artificial Intelligence Symposium*, 1994.

[SR92] Marc Schoenauer and Edmund Ronald. Evolving neural nets for control. Technical report, Centre de mathématiques appliquées de l'Ecole Polytechnique, 1992.

[YG] X. Yin and N. Germay. A fast genetic algorithm with sharing scheme using cluster analysis methods in multimodal function optimization. Technical report, Laboratoire d'Electronique et d'Instrumentation, Catholic University of Louvain.

A Comparative Study of a Penalty Function, a Repair Heuristic, and Stochastic Operators with the Set-Covering Problem

Thomas Bäck[1], Martin Schütz[2], and Sami Khuri[3]

[1] Informatik Centrum Dortmund,
Joseph-von-Fraunhofer-Straße 20, D–44227 Dortmund
E-mail: baeck@ls11.informatik.uni-dortmund.de

[2] Universität Dortmund
Fachbereich Informatik, LS XI, D–44221 Dortmund
E-mail: schuetz@ls11.informatik.uni-dortmund.de

[3] Department of Mathematics & Computer Science,
San José State University, One Washington Square,
San José, CA 95192-0103, U.S.A.
E-mail: khuri@sjsumcs.sjsu.edu

Abstract. In this paper we compare the effects of using various stochastic operators with the non-unicost set-covering problem. Four different crossover operators are compared to a repair heuristic which consists in transforming infeasible strings into feasible ones. These stochastic operators are incorporated in GENEsYs, the genetic algorithm we apply to problem instances of the set-covering problem we draw from well known test problems. GENEsYs uses a simple fitness function that has a graded penalty term to penalize infeasibly bred strings. The results are compared to a non GA-based algorithm based on the greedy technique. Our computational results are then compared, shedding some light on the effects of using different operators, a penalty function, and a repair heuristic on a highly constrained combinatorial optimization problem.

1   Introduction

The set-covering problem is a combinatorial optimization problem that can be used as a model for many practical problems. Applications can be mainly found in resource-selection and committee forming tasks. The problem consists of a finite set $E = \{e_1, e_2, \ldots, e_m\}$ and a family F of subsets S_i of E, $F = \{S_1, S_2, \ldots, S_n\}$, such that every element of E belongs to at least one subset in F, i.e., $\cup_{i=1}^{n} S_i = E$. The unicost minimum set-covering problem consists in finding a minimum-size subset $C \subseteq F$ such that $\cup_{S_i \in C} S_i = E$. Any set $Y \subseteq F$ with $\cup_{S_j \in Y} = E$ is said to cover E. The problem thus consists in finding a cover of E with minimal cardinality. The set-covering problem is NP-complete as can be shown by polynomially transforming the vertex cover problem into it (see [1], p. 392) or by the transformation from the exact cover by 3-sets [14].

In many applications, the subsets have different costs. The non-unicost set covering problem, denoted by mscp, consists in finding a cover of E with cheapest overall cost. In other words, if $c_1, c_2, \ldots, c_n$ denote the costs of $S_1, S_2, \ldots, S_n$, then it is required to find a cover C of E such that $\sum_{S_i \in C} c_i$ has the smallest possible value.

In our work, we study the effects of using different stochastic operators with a penalty-based fitness function and a repair heuristic for infeasible solutions on the mscp. More precisely, the traditional one- and two-point crossover, uniform crossover, and the recently suggested fusion crossover [5] are incorporated in the genetic algorithm GENEsYs. This package is based on Grefenstette's popular software package GENESIS [11]. The fitness function uses a graded penalty term to penalize infeasibly bred strings. Unlike other works, such as Beasley et al. [7], that use a tailor-made fitness function, a variable mutation rate, or Huang et al. [12], whose custom made algorithm targets specific instances of the mcsp, GENEsYs is not augmented with any a priori knowledge of the problem except for our fitness function. We investigate a repair method (following earlier work by Orvosh and Davis [19]) which guarantees feasibility of solutions by changing the genotype of strings that represent infeasible solutions.

Following the formal introduction of the mscp, a greedy heuristic for that problem is introduced. The paper then introduces the fitness function that is used in the subsequent sections and the repair heuristic we use to transform infeasible strings into feasible ones. In order to compare the effects of different genetic operators and the penalty approach vs. the repair method, the paper then focuses on the experimental results. The last section gives the conclusions obtained after the recording of the performance of the different stochastic operators and the penalty function with and without using the repair technique.

2 The Minimum Set-Covering Problem

As mentioned in the introduction, applications of the set-covering problem can be found in resource allocation and committee forming endeavours.

Practical applications of the mscp range from assignment applications such as assigning flying crews to particular flights [21] and fire companies to fire houses [25], to the simplification of boolean expressions [8].

As is the case with many NP-hard problems, one of the only two algorithms, short of exhaustive enumeration, that delivers optimum solutions is based on the branch-and-bound technique (the other being dynamic programming). Approximation algorithms include greedy techniques [9], simulated annealing [22], Lagrangian relaxation and sub-gradient optimization techniques [5] and GA-based heuristics [17]. The interested reader can find more references to the unicost and mscp including more applications in [4].

The mscp can be recast by using matrix notation. Elements of E and subsets of F can be represented by an m-row, n-column, zero-one matrix A, with elements a_{ij}, $1 \leq i \leq m$ and $1 \leq j \leq n$. The columns of A represent the subsets while the rows represent the elements of E. Thus, $a_{ij} = 1$ means that

e_i is covered by S_j while $a_{ij} = 0$ signifies that e_i is not covered by S_j. The columns have costs $c_1, c_2, \ldots, c_n$. A solution to the mscp is represented by a vector $\mathbf{x} = (x_1, x_2, \ldots, x_n)$ in which $x_j = 1$ means that S_j (i.e., column j) is chosen while $x_j = 0$ signals the omission of S_j in the solution. The objective of the mscp is to cover the rows of A by a subset of columns at minimal cost.

The following is a formal definition of the mscp in which we make use of Stinson's terminology for combinatorial optimization problems [23].

Problem instance:

A set of objects $E = \{e_1, \ldots, e_m\}$ and a family of subsets of E, $F = \{S_1, \ldots, S_n\}$, such that $\cup_{i=1}^{n} S_i = E$. Each S_i costs c_i where $c_i > 0$ for $i = 1, \ldots, n$. Or equivalently, an $m \times n$ matrix

$$A = \begin{pmatrix} a_{11} & \cdots & a_{1n} \\ & \cdot & \\ \cdot & & \cdot \\ & \cdot & \\ a_{m1} & \cdots & a_{mn} \end{pmatrix}$$

such that $\sum_{j=1}^{n} a_{ij} \geq 1$ for $i = 1, 2, \ldots, m$.

Feasible solution:

A vector $\mathbf{x} = (x_1, x_2, \ldots, x_n)$ where $x_i \in \{0, 1\}$, such that

$$\sum_{j=1}^{n} a_{ij} x_j \geq 1$$

for $i = 1, 2, \ldots, m$.

Objective function:

A cost function $P(\mathbf{x}) = \sum_{j=1}^{n} c_j x_j$ where $\mathbf{x} = (x_1, x_2, \ldots, x_n)$ is a feasible solution.

Optimal solution:

A feasible vector that gives the minimum cost, i.e., a vector that minimizes the objective function.

We proceed by considering the non-GA based greedy heuristic, which we call "greedy-mscp."

In the process of building the cover C of E, we need to devise a scheme for greedily choosing the next subset (or column), among the remaining competing subsets, to cover uncovered elements (or rows).

The greedy step can be implemented by choosing the least expensive subset that contains uncovered elements of E, or the subset that contains the maximum number of uncovered elements.

The greedy strategy we choose for "greedy-mscp" makes use of the cost-ratio $c_j/|S_j'|$ where $S_j' \subseteq S_j$ is the set of elements of S_j that are still uncovered. The algorithm chooses among the subsets that are still in contention, the one that yields the smallest cost-ratio.

The procedure has to keep updating the subsets that are still competing by removing from each one of them the elements that have already been covered by previously selected subsets. It is thus very possible to have subsets drop out of contention. If U denotes the set of uncovered elements of E, then S_j ceases being a candidate for inclusion in C if $S_j \cap U = \emptyset$ or equivalently $S'_j = \emptyset$ with $j \in J$, where J is the index set of subsets still competing for inclusion in C.

procedure *greedy-mscp*

$U := E$;	(* *set of uncovered elements* *)
$C := \emptyset$;	(* *cover to be constructed* *)
$J := \{1, 2, \ldots, n\}$;	(* *index set of unused subsets* *)
for all $i \in \{1, \ldots, n\}$	
$\quad S'_i := S_i$;	(* S'_i *contains uncovered elements of* S_i *)
$K := \emptyset$;	(* *index set of subsets in* C *)
while $U \neq \emptyset$	
$\quad$ *choose* $S_j \in F$ *where* $j \in J$ *such that*	
$\qquad \frac{c_j}{|S'_j|} = \min_{i \in J} \left\{ \frac{c_i}{|S'_i|} \right\}$;	
$\quad C := C \cup \{S_j\}$;	(* *include* S_j *in the cover set* *)
$\quad K := K \cup \{j\}$;	
$\quad U := U - S_j$;	(* *remove elements of* E *covered by* S_j *)
$\quad J := J - \{j\}$;	(* S_j *has been chosen* *)
$\quad$ **for all** $i \in J$	
$\qquad S'_i := S'_i - S_j$;	(* *remove from* S'_i *elements covered by* S_j *)
$\qquad$ **if** $S'_i = \emptyset$ **then**	
$\qquad\qquad J := J - \{i\}$;	(* *elements of* S'_i *are already covered* *)
$\quad$ **for all** $i \in K$	
$\qquad S'_i = \cup_{t \in K, t \neq i} S_t$	(* S'_i *contains the covered elements* $\notin S_i$ *)
$\qquad$ **if** $S'_i = \bar{U}$ **then**	(* S_i *is a redundant subset* *)
$\qquad\qquad C := C - \{S_i\}$;	
$\qquad\qquad K := K - \{i\}$;	

return C

It is very possible to have a subset in C, $S_k \in C$ for instance, have all its elements covered by the union of other subsets in C. The subset is then redundant and should be removed from C. In other words, if $S_k = \cup_{i \in K, i \neq k} S_i$ where K is the index set for subsets that are in the cover C, then S_k is redundant and should be dropped out of C. Thus, each time a subset S_j with $j \in J$ is chosen to join C, every subset in the cover should be checked to see whether its elements are covered by the union of the other subsets in the cover. We point out that procedure *greedy-mscp* also checks the newly added subset for redundancy, which is not necessary. We did not delete this step from the algorithm for the sake of clarity. We note that when S_j with $j \in J$ is chosen to join C, its corresponding prime set, S'_j, is not needed anymore. Rather than discarding the prime sets, the heuristic uses them for detecting redundant subsets. More precisely, each prime set corresponding to a subset in the cover will contain the union of all other subsets in C excluding its own elements, i.e. $S_i' = \cup_{t \in K, t \neq i} S_t$ for every $i \in K$. The prime sets thus play a dual role; S_i' with $i \in J$ is the set of elements in S_i that are still uncovered by C. On the other hand, S_i' with $i \in K$ is the union of all other subsets in the cover.

As mentioned earlier, the procedure uses two index sets, J and K. The first contains the indices of the subsets that are still competing for inclusion in C, while K contains the indices of the subsets that are already in C. It is important to realize that due to the dropping of subsets from contention, and due to redundant subsets that are removed from the cover C, the union of J and K rarely yields $\{1, \ldots, n\}$.

To get a better understanding of the procedure, we make use of the problem instance of the unicost set-cover problem found in "Introduction to Algorithms" by Cormen et al. [9] (pp. 974–975). The problem instance will also be used in the next section to demonstrate the workings of the graded penalty we use in the fitness function. The example is first converted to a non-unicost problem instance by adding costs to the various subsets, and we then recast it by using the matrix notation.

Example: Consider the following problem instance with 12 elements and 6 subsets: $E = \{e_1, e_2, \ldots, e_{12}\}$ and $F = \{S_1, \ldots, S_6\}$, where $S_1 = \{e_1, e_2, e_5, e_6, e_9, e_{10}\}$, $S_2 = \{e_6, e_7, e_{10}, e_{11}\}$, $S_3 = \{e_1, e_2, e_3, e_4\}$, $S_4 = \{e_3, e_5, e_6, e_7, e_8\}$, $S_5 = \{e_9, e_{10}, e_{11}, e_{12}\}$ and $S_6 = \{e_4, e_8\}$. Suppose that $c_1 = 10$, $c_2 = 5$, $c_3 = 11$, $c_4 = 10$, $c_5 = 8$ and $c_6 = 2$ are the costs we assign to the subset in order to convert the original instance [9] into a non-unicost problem instance. The following 12×6 matrix represents the problem, where the first column costs 10, the second 5, and so on.

$$A^T = \begin{pmatrix} 1 & 1 & 0 & 0 & 1 & 1 & 0 & 0 & 1 & 1 & 0 & 0 \\ 0 & 0 & 0 & 0 & 0 & 1 & 1 & 0 & 0 & 1 & 1 & 0 \\ 1 & 1 & 1 & 1 & 0 & 0 & 0 & 0 & 0 & 0 & 0 & 0 \\ 0 & 0 & 1 & 0 & 1 & 1 & 1 & 1 & 0 & 0 & 0 & 0 \\ 0 & 0 & 0 & 0 & 0 & 0 & 0 & 0 & 1 & 1 & 1 & 1 \\ 0 & 0 & 0 & 1 & 0 & 0 & 0 & 1 & 0 & 0 & 0 & 0 \end{pmatrix}$$

After the initialization phase of the procedure, S_6 is the first chosen subset since:

$$2/2 = \min\{10/6, 5/4, 11/4, 10/5, 8/4, 2/2\}.$$

At the end of the updating remaining steps in the while loop of the procedure we have:

$$S_1' = \{e_1, e_2, e_5, e_6, e_9, e_{10}\}, \ S_2' = \{e_6, e_7, e_{10}, e_{11}\}, \ S_3' = \{e_1, e_2, e_3\}, \ S_4' = \{e_3, e_5, e_6, e_7\}, \ S_5' = \{e_9, e_{10}, e_{11}, e_{12}\}, \text{ and } C = \{S_6\}.$$ Note that $S_1' = S_1$, $S_2' = S_2$, and $S_5' = S_5$ since none of these subsets contain either e_4 or e_8, and $S_6' = \emptyset$.

The second pass through the while loop will choose S_2 since

$$5/4 = \min\{10/6, 5/4, 11/3, 10/4, 8/4\}$$

Thus, at the end of the of the second pass of the while loop we have:

$S'_1 = \{e_1, e_2, e_5, e_9\}$, $S'_3 = \{e_1, e_2, e_3\}$, $S'_4 = \{e_3, e_5\}$, $S'_5 = \{e_9, e_{12}\}$, and $U = \{e_1, e_2, e_3, e_5, e_9, e_{12}\}$. The cover consists of $C = \{S_2, S_6\}$ and $S'_2 = \{e_4, e_8\}$ and $S'_6 = \{e_6, e_7, e_{10}, e_{11}\}$.

The procedure then proceeds by including S_1, S_5 and S_4 in C, in that order. Each inclusion is followed by updating S'_i, for $i = 1, \ldots, 6$, U, C, and the index sets J and K. The inclusion of S_4 in C in the last time through the while loop triggers the removal of e_3 from S'_3, and e_7 from S'_2. But since these were the only elements remaining in S'_3 and S'_2, respectively, both sets are now empty. S_3 drops out of contention and is not any more considered for inclusion in the cover, while S_2 becomes a redundant subset since the other elements of C, namely S_1, S_4, S_5 and S_6, cover all the elements in S_2. Thus, S_2 is removed from the cover and the procedure returns $C = \{S_1, S_4, S_5, S_6\}$ which has a cost of 30 (notice that this is not the optimum because the cost of the cover $\{S_3, S_4, S_5\}$ is 29).

In the next section we discuss the various stochastic operators including the fitness function we use in our experimental runs.

3 Fitness Functions and Stochastic Operators

Prior to the application of the genetic algorithm to problem instances of the mscp, one needs to encode the problem. Recall that a solution to the mscp can be represented by $\mathbf{x} = (x_1, x_2, \ldots, x_n)$ in which $x_j = 1$ means that S_j is in the cover, and $x_j = 0$ signals the omission of S_j from the cover. Each binary string of length n in the population can be thought of as being the components of a vector $\mathbf{x}$.

3.1 Fitness Functions

The fitness function we use in our comparisons makes use of graded penalty terms to penalize infeasibly bred strings. We remark that a string $\mathbf{x} = (x_1, x_2, \ldots, x_n)$ might represent an infeasible solution, i.e., a collection of subsets that does not cover all elements of E. Rather than discarding the latter, we follow Richardson et al.'s conclusions [20], and allow infeasible strings to be part of the population but at a certain price. The infeasible string's strength relative to the other strings in the population is reduced by adding a penalty term to its fitness. The farther away from feasibility the string is, the higher its penalty term is. We also make sure that the best infeasible string is never superior to the weakest feasible one. We note that in the worst case, we need all the subsets to cover the elements of E. In other words, $B = \sum_{i=1}^{n} c_i$ is the least upper bound on all possible costs. Hence, B is added to any infeasible string's fitness. By using B as an offset term invoked for infeasible strings, we are sure that infeasible strings will never outperform feasible ones. We would like to point out that this approach has worked quite well with other combinatorial optimization problems [3, 15].

In our fitness function, U is the set of elements from E for which the subsets represented by $\mathbf{x}$ have failed to cover. Obviously, U is empty when $\mathbf{x}$ is feasible.

The function considers the minimum cost required to transform the infeasible string into a feasible one:

$$f(\mathbf{x}) = \sum_{i=1}^{n} c_i x_i + s(B + \sum_{i=1}^{m} c_i^*) \tag{1}$$

where
$$s = \begin{cases} 0 & , \quad \text{if } \mathbf{x} \text{ is feasible} \\ 1 & , \quad \text{if } \mathbf{x} \text{ is infeasible} \end{cases}$$

and
$$c_i^* = \begin{cases} \min_{e_i \in U}\{c_j \mid e_i \in S_j\} & , \quad \text{if } e_i \in U \\ 0 & , \quad \text{if } e_i \notin U \end{cases}$$

This fitness function, which is based on the cheapest cost of covering every single uncovered element of E, is more graded than the one which uses the naive penalty term $s(B + c_{max}|U|)$, where $c_{max} = \max\{c_1, c_2, \ldots, c_n\}$ and $|U|$ is the cardinality of E (our first experiments were based on this simpler fitness function and demonstrated the need for a better penalty term).

While evaluating the graded penalty term, it is easy to transform an infeasible string into a feasible one by changing $x_j = 0$ to $x_j = 1$ when $c_i^* = \min_{e_i \in U}\{c_j \mid e_i \in S_j\}$ is calculated. This can be interpreted as a kind of repair heuristic and is also compared here with the greedy heuristic and the fitness function given above. Since the repaired string is always feasible, the fitness function in this case is simply $f'(\mathbf{x}) = \sum_{i=1}^{n} c_i x_i$.

Example: Consider the previous problem instance with $\mathbf{x} = (1, 0, 1, 0, 1, 0)$.

$$A \cdot (1, 0, 1, 0, 1, 0)^T = (2, 2, 1, 1, 1, 1, 0, 0, 2, 2, 1, 1)^T \quad .$$

So $\mathbf{x}$ is an infeasible string since e_7 and e_8 are not covered by $\{S_1, S_3, S_5\}$ and $U = \{e_7, e_8\}$. Now $e_7 \in S_2$ and $e_7 \in S_4$, so $c_7^* = \min\{c_2, c_4\} = 5$. Similarly, $e_8 \in S_4$ and $e_8 \in S_6$, i.e., $c_8^* = \min\{c_4, c_6\} = 2$. We compute $B = \sum_{i=1}^{6} c_i = 46$. Thus $f(\mathbf{x}) = 29 + 46 + (5 + 2) = 82$, while the naive penalty term $s(B + c_{max}|U|)$ would yield $29 + 46 + 2 \cdot 11 = 97$. The repair heuristic would incorporate S_2 and S_6 into the solution and thus change $\mathbf{x}$ into $\mathbf{x}' = (1, 1, 1, 0, 1, 1)$, with $f'(\mathbf{x}) = 36$.

Next we consider the crossover operator we make use in our experimental runs.

3.2 Stochastic Operators

The first three crossover operators we use, one- and two-point [10, 13] and uniform crossover [24], need no introduction. The fusion crossover introduced by Beasley [7] was designed to generate an offspring which represented a *newer* solution when compared to one or two-point crossover. It is based on the "differences" of the parents and as pointed out by Beasley, is advantageous especially when both parents are quite similar. If $\mathbf{a} = (a_1, a_2, \ldots, a_n)$ and $\mathbf{b} = (b_1, b_2, \ldots, b_n)$ are the parent strings with fitness $f_\mathbf{a}$ and $f_\mathbf{b}$, then $\mathbf{c} = (c_1, c_2, \ldots, c_n)$ is the offspring where

for all $i \in \{1, \ldots, n\}$
  if $a_i = b_i$
  then $c_i := a_i$;
  else $c_i := \begin{cases} a_i \text{ with probability } p = \frac{f_b}{f_a + f_b} \\ b_i \text{ with probability } 1 - p \end{cases}$

We note that when $a_i \neq b_i$, the offspring inherits the bit from the better of the parents with higher probability (because smaller fitness values correspond with better solutions).

4 Experimental Results

The genetic algorithm is used with the one-point, two-point, uniform, and fusion crossover operators and a crossover rate of 0.6. As suggested independently by Mühlenbein [18] and Bäck [2], the mutation rate $p_m = 1/n$ is determined by the problem dimension n. A population size $\lambda = 100$ and proportional selection with linear dynamic scaling (and a scaling window of five generations) complete the parameter settings of this standard genetic algorithm.

Problem	m	n	Density (%)
4.1–4.10	200	1000	2
5.1–5.10	200	2000	2
6.1–6.5	200	1000	5
a.1–a.5	300	3000	2
b.1–b.5	300	3000	5
c.1–c.5	400	4000	2
d.1–d.5	400	4000	5
e.1–e.5	50	500	20

Table 1. Characteristics of test problems

We apply the algorithm with the penalty-based fitness function (1) and the repair method to 50 problem instances from Beasley's OR-library [6]. The characteristics (problem sizes, matrix densities: percentage of ones in the matrix) of these problems are summarized in table 1.

Independently of the problem size, we applied each genetic algorithm configuration (crossover operator and fitness function) a total of ten experiments to each single problem instance, and each experiment was allowed to run for $5 \cdot 10^5$ function evaluations (i.e., 5,000 generations). The best final objective function value found within these ten experiments is reported in tables 2 (fitness function f without repairing) and 3 (fitness function f' with repairing). The results in these tables are also compared to the solution quality as obtained by our greedy heuristic as well as the best solution value known so far (columns 6 and 7 of the tables). From the experimental results, the following observations can be made:

- The fitness function f without repairing infeasible strings always yields a feasible solution to the set covering problem and thus confirms the usefulness of the graded penalty function approach. The final solution quality, however, always remains far from the best known objective function value and can neither compete with the solution quality reached by genetic heuristics using more problem-specific operators [7, 16] nor with the greedy heuristic.
- Repairing infeasible strings according to the minimum cost principle improves the results remarkably, such that final objective function values close to the best known solution are obtained. The repair procedure competes well with the greedy heuristic and yields better solutions in most cases
- The choice of the crossover operator has no significant impact on the quality of the results.
- Our simple greedy heuristic always outperforms the genetic algorithm without repairing.

The multimodal fitness function of the set covering problem includes many local optima, which are characterized by a relatively small number of ones distributed over a long bit string. As table 1 shows, the densities of most problem instances are very low. Different local optima are separated by a large Hamming distance and a variety of infeasible strings (i.e., the feasible region is not connected), and the objective function values of local optima might differ a lot. Consequently, the genetic algorithm sooner or later concentrates its search in the region of attraction of a local optimum and is not able to leave this region by means of the mutation operator. This might give the reason for which the genetic algorithm without repairing quickly gets stuck in relatively bad local optima.

In combination with a repair heuristic, which certainly represents a greedy technique, the volume-oriented search principle of a genetic algorithm is able to identify much better local optima than without repairing. This combination of global search and locally optimal repair performs even better than our greedy technique.

5 Discussion

The set covering problem certainly represents a very complicated, highly multimodal optimization problem with a complex, non-connected feasible region. Genetic algorithms normally require the incorporation of certain problem-specific heuristics to yield solutions of acceptable quality; see e.g. [7, 12, 17].

The simple penalty function approach to the set covering problem as presented in this paper confirmed this general impression by yielding solutions which, although being feasible in all cases, are far from the solution quality which is easily obtained by our greedy heuristic. A variation of the main search operator of genetic algorithms, the crossover operator, did not produce substantial improvements.

Incorporating a simple repair method into the evaluation of solutions, however, improved the solution quality remarkably and caused the genetic algorithm to encounter better solutions than the greedy heuristic in most cases. Although repairing infeasible solutions is almost a trivial operator, this technique received some attention in genetic algorithm research only recently [19]. Little is known about the frequency of repairing and further research is required to determine which portion of infeasible strings should be repaired.

Acknowledgements

The first two authors gratefully acknowledge support by the BMFT under project EVOALG, grant 01 IB 403 A (Thomas Bäck) and the DFG, grant Schw 361/5-2 (Martin Schütz). The third author would like to thank the California State University system for their research award GS850-851 that allowed him to conduct some preliminary work on set-covering problems. We are also very grateful for the help of Jörg Heitkötter, who organized the data files from OR-library and helped us with formatting the article.

References

1. A. V. Aho, J. E. Hopcroft, and J. D. Ullman. *The Design and Analysis of Computer Algorithms*. Addison Wesley, Reading, MA, 1974.
2. Th. Bäck. The interaction of mutation rate, selection, and self-adaptation within a genetic algorithm. In R. Männer and B. Manderick, editors, *Parallel Problem Solving from Nature 2*, pages 85–94. Elsevier, Amsterdam, 1992.
3. Th. Bäck and S. Khuri. An evolutionary heuristic for the maximum independent set problem. In *Proceedings of the First IEEE Conference on Evolutionary Computation*, pages 531–535. IEEE Press, 1994.
4. E. Balas and S. M. Ng. On the set covering polytype: I. All the facets with coefficients in $\{0, 1, 2\}$. *Mathematical Programming*, 43:57–69, 1989.
5. J. E. Beasley. A lagrangian heuristic for set-covering problems. *Naval Research Logistics*, 37:151–164, 1990.
6. J. E. Beasley. OR-Library: Distributing test problems by electronic mail. *Journal of the Operational Research Society*, 41(11):1069–1072, 1990.
7. J. E. Beasley and P. C. Chu. A genetic algorithm for the set covering problem. Submitted to *European Journal of Operational Research* for publication, 1994.
8. M. A. Breuer. Simplification of the covering problem with application ot boolean expressions. *Journal of the Association of Computing Machinery*, 17:166–181, 1970.
9. T. H. Cormen, C. E. Leiserson, and R. L. Rivest. *Introduction to Algorithms*. The MIT Press, Cambridge, MA, 1990.
10. D. E. Goldberg. *Genetic algorithms in search, optimization and machine learning*. Addison Wesley, Reading, MA, 1989.
11. J. J. Grefenstette. GENESIS: A system for using genetic search procedures. In *Proceedings of the 1984 Conference on Intelligent Systems and Machines*, pages 161–165, 1984.

12. Wen-Chih Huang, Cheng-Yan Kao, and Jorng-Tzong Horng. A genetic algorithm approach for set covering problems. In *Proceedings of the First IEEE Conference on Evolutionary Computation*, pages 569–574. IEEE Press, 1994.

13. K. A. De Jong. *An analysis of the behaviour of a class of genetic adaptive systems*. PhD thesis, University of Michigan, 1975. Diss. Abstr. Int. 36(10), 5140B, University Microfilms No. 76–9381.

14. R. M. Karp. Reducibility among combinatorial problems. In R. E. Miller and J. W. Thatcher, editors, *Complexity of Computer Computations*, pages 85–104. Plenum Press, New York, 1972.

15. S. Khuri, Th. Bäck, and J. Heitkötter. An evolutionary approach to combinatorial optimization problems. In D. Cizmar, editor, *Proceedings of the 22nd Annual ACM Computer Science Conference*, pages 66–73. ACM Press, New York, 1994.

16. D. M. Levine. A genetic algorithm for the set partitioning problem. In S. Forrest, editor, *Proceedings of the 5th International Conference on Genetic Algorithms*, pages 481–487. Morgan Kaufmann Publishers, San Mateo, CA, 1993.

17. G. E. Liepins, M. R. Hilliard, J. Richardson, and M. Palmer. Genetic algorithms applications to set covering and traveling salesman problems. In Donald E. Brown and Chelsea C. White III., editors, *Operations Research and Artificial Intelligence: The Integration of Problem-Solving Strategies*, pages 29–57. Kluwer Academic Publishers, 1990.

18. H. Mühlenbein. How genetic algorithms really work: I. mutation and hillclimbing. In R. Männer and B. Manderick, editors, *Parallel Problem Solving from Nature 2*, pages 15–25. Elsevier, Amsterdam, 1992.

19. D. Orvosh and L. Davis. Shall we repair ? Genetic algorithms, combinatorial optimization, and feasibility constraints. In S. Forrest, editor, *Proceedings of the 5th International Conference on Genetic Algorithms*, page 650. Morgan Kaufmann Publishers, San Mateo, CA, 1993.

20. J. T. Richardson, M. R. Palmer, G. Liepins, and M. Hilliard. Some guidelines for genetic algorithms with penalty functions. In J. D. Schaffer, editor, *Proceedings of the 3rd International Conference on Genetic Algorithms*, pages 191–197. Morgan Kaufmann Publishers, San Mateo, CA, 1989.

21. J. Rubin. A technique for the solution of massive set-covering problems with applications to airline crew scheduling. *Transportation Science*, 7:34–48, 1973.

22. S. Sen. Minimal cost set covering using probabilistic methods. In *Proceedings 1993 ACM/SIGAPP Symposium on Applied Computing*, pages 157–164, 1993.

23. D. R. Stinson. *An Introduction to the Design and Analysis of Algorithms*. The Charles Babbage Research Center, Winnipeg, Manitoba, Canada, 2nd edition, 1987.

24. G. Syswerda. Uniform crossover in genetic algorithms. In J. D. Schaffer, editor, *Proceedings of the 3rd International Conference on Genetic Algorithms*, pages 2–9. Morgan Kaufmann Publishers, San Mateo, CA, 1989.

25. W. Walker. Using the set-covering problem to assign fire companies to fire houses. *Operations Research*, 22:275–277, 1974.

Problem	1-point	2-point	uniform	fusion	greedy	Best known
4.1	1,030	1,030	1,003	829	436	429
4.2	1,208	1,152	1,008	880	533	512
4.3	1,135	1,079	1,025	962	540	516
4.4	1,042	1,064	981	1,003	512	494
4.5	1,006	1,058	1,034	1,018	520	512
4.6	1,130	1,224	993	1,106	605	560
4.7	912	943	847	1,026	447	430
4.8	1,159	1,090	861	1,002	525	492
4.9	1,404	1,246	1,257	1,263	671	641
4.10	1,046	1,153	1,079	1,083	533	514
5.1	714	757	630	656	269	253
5.2	827	809	663	696	330	302
5.3	741	638	606	540	232	226
5.4	806	664	717	585	250	242
5.5	850	784	600	637	218	211
5.6	717	743	673	663	227	213
5.7	786	836	710	845	310	293
5.8	806	799	680	756	311	288
5.9	776	805	686	822	292	279
5.10	912	777	723	835	278	265
6.1	317	338	310	324	144	138
6.2	294	309	299	351	157	146
6.3	323	365	268	291	157	145
6.4	344	302	298	303	141	131
6.5	383	376	355	355	186	161
a.1	897	955	760	886	264	253
a.2	842	868	745	786	272	252
a.3	787	823	681	730	245	232
a.4	848	704	694	821	251	234
a.5	773	746	705	875	248	236
b.1	342	253	245	290	74	69
b.2	262	278	253	267	78	76
b.3	291	272	282	275	85	80
b.4	303	324	290	262	83	79
b.5	311	296	259	279	75	72
c.1	1054	927	650	782	239	227
c.2	809	854	729	829	225	219
c.3	885	896	767	841	259	243
c.4	839	861	772	703	240	219
c.5	892	845	735	834	219	215
d.1	310	294	325	280	68	60
d.2	285	319	293	296	71	66
d.3	298	246	281	286	80	72
d.4	280	305	256	284	67	62
d.5	318	340	308	331	66	61
e.1	6	7	7	7	5	5
e.2	7	7	7	7	5	5
e.3	7	6	7	6	5	5
e.4	7	7	7	7	6	5
e.5	8	7	7	7	5	5

Table 2. Experimental results for the penalty-based fitness function and the greedy heuristic.

Problem	1-point	2-point	uniform	fusion	greedy	Best known
4.1	439	440	440	438	436	429
4.2	536	551	525	556	533	512
4.3	529	532	532	532	540	516
4.4	528	517	508	512	512	494
4.5	522	526	518	527	520	512
4.6	568	579	572	568	605	560
4.7	444	437	437	437	447	430
4.8	504	504	509	499	525	492
4.9	678	689	686	675	671	641
4.10	549	538	553	553	533	514
5.1	274	271	270	274	269	253
5.2	318	319	319	320	330	302
5.3	238	235	235	239	232	226
5.4	247	245	245	244	250	242
5.5	219	219	219	219	218	211
5.6	229	236	238	232	227	213
5.7	313	310	309	309	310	293
5.8	302	309	304	306	311	288
5.9	301	304	298	292	292	279
5.10	275	276	276	277	278	265
6.1	146	145	147	147	144	138
6.2	154	154	153	154	157	146
6.3	154	150	151	151	157	145
6.4	140	133	136	133	141	131
6.5	180	180	177	179	186	161
a.1	255	261	260	260	264	253
a.2	270	268	267	267	272	252
a.3	247	247	244	245	245	232
a.4	244	246	249	247	251	234
a.5	240	239	239	239	248	236
b.1	77	77	79	75	74	69
b.2	87	83	84	81	78	76
b.3	84	85	83	83	85	80
b.4	84	82	84	84	83	79
b.5	72	75	72	77	75	72
c.1	234	235	234	235	239	227
c.2	231	232	226	233	225	219
c.3	266	261	263	264	259	243
c.4	237	234	238	237	240	219
c.5	220	219	222	219	219	215
d.1	62	61	61	62	68	60
d.2	72	72	70	70	71	66
d.3	77	79	79	79	80	72
d.4	66	64	66	66	67	62
d.5	64	65	65	64	66	61
e.1	8	7	7	8	5	5
e.2	7	6	6	7	5	5
e.3	7	7	8	6	5	5
e.4	6	7	7	7	6	5
e.5	7	8	7	7	5	5

Table 3. Experimental results for the repair heuristic and the greedy heuristic.

Study of Genetic Search for the Frequency Assignment Problem[*]

Jin-Kao Hao and Raphaël Dorne

LGI2P
EMA-EERIE
Parc Scientifique Georges Besse
F-30000 Nîmes
France
email: {hao, dorne}@eerie.fr

Abstract. The goal of this paper is twofold. First, we present an evolutionary approach to a real world application: the Frequency Assignment Problem (FAP) in Cellular Radio Networks. Second, we present an empirical study on the effectiveness of crossover for solving this problem. Experiments carried out on a set of real-size FAP instances (up to 300 cells, 30 frequencies and 30,000 interference constraints) show the interest of EAs. At the same time, empirical evidence suggests that the contribution of crossover is marginal for this application.

1 Introduction

The Frequency Assignment Problem (FAP) in cellular radio networks is a very complex application in the field of telecommunications. Although different versions can be defined for FAP, the main goal consists in assigning one or more frequencies, a very limited resource, to each radio cell in a cellular radio network while minimizing electromagnetic interferences due to the reuse of frequencies by adjacent cells. The difficulty of this application comes from the fact that an acceptable solution of FAP must satisfy a set of multiple constraints, some of these constraints being orthogonal. The most severe constraint concerns a very limited radio spectrum consisting of a small number of frequencies (or channels). Indeed, telecommunications operators such as France Telecom must cope with only up to 60 frequencies for their networks whatever the traffic volume to cover may be, and this, in agreement with national and international regulations. In addition to this frequency constraint, two other types of constraints must be satisfied to insure good communication quality:

1. the *traffic constraint* for each cell corresponding to the minimum number of frequencies required by the cell to cover the communications of the cell.

---

[*] Supported by the CNET (French National Research Center for Telecommunications) under the grant No.940B006-01.

2. two categories of frequency *interference constraints*:
   - *Co-cell constraints*: any pair of frequencies assigned to a radio cell must have a certain distance between them in the frequency domain.
   - *Adjacent-cell constraints*: the frequencies assigned to two adjacent cells[2] must be sufficiently separated in the frequency domain.

The basic FAP can be shown to be NP-complete because, in its simplest form, it is reduced to the graph coloring problem [9]. So far, many methods have been proposed to solve FAP, including 1) classic methods: graph coloring algorithms [9, 8] and integer programming; 2) heuristic methods: neural networks [12, 7], genetic algorithms (GAs) [4, 11], local search such as simulated annealing (SA) [6, 1] and Tabu search (TS) [10], and constraint programming (CP) [3].

The goal of this paper is twofold. First, we present various (hybrid) evolutionary algorithms (EAs) and their performances on real-size FAP instances (up to 300 cells, 30,000 interference constraints with only 30 frequencies). Second, we investigate empirically the effects of crossover for this application.

The paper is organized as follows. In Section 2, FAP is modelled as an optimization problem. In Section 3, the different components of our EAs are summarized. In Section 4, experimental results are presented and compared. Conclusions and perspectives are given in the last section.

2 Modelling the Frequency Assignment Problem

2.1 Notations and Model

In this paper, the following notations will be used to simplify the presentation.

N the number of cells
NF the number of available frequencies in the spectrum
NCI the number of interference constraints defined for adjacent cells
NC the number of constraints ($NC=NCI+N$)
$f_{i,m}$ the mth frequency assigned to the cell C_i

FAP can be modelled with a quadruple FAP $= <X,D,C,F>$ representing an optimization problem with:

$X = \{C_i \mid C_i$ is a cell of the network, $i \in [1..N]\}$.
$D = \{F_i \mid F_i$ is an available frequency of the spectrum, $i \in [1..NF]\}$.
$C = T \cup I$

$T = \{T_i \mid T_i$ minimal number of frequencies necessary for C_i, $i \in [1..N]\}$.
$I = \{I_i \mid I_i$ interference constraints of frequencies, $i \in [1..NC]\}$.

f = cost function of a frequency assignment.

Finding a solution to FAP $= <X,D,C,F>$ means assigning one or more frequencies in D to each cell C_i of X in such a way that the constraints of C are simultaneously satisfied and the cost f is minimized.

---

[2] Two cells are adjacent if they emit within a common area even if they are not geographically adjacent.

2.2 Constraints

There are essentially two big families of constraints: *traffic constraints* and *interference constraints*. The traffic constraint for each cell C_i is represented by an integer T_i coding the minimum number of frequencies necessary for C_i to cover its maximum traffic volume. In practice, this maximum traffic value is defined by an estimation of the maximum number of communications which can simultaneously arise within this cell. The interference constraints over the network are represented by a symmetric compatibility matrix M[N,N] defined by:

- M[i,j] with $i \neq j$ represents the minimum frequency separation required to satisfy the adjacent-cell constraints between the cells C_i and C_j.
  $\forall n \in [1..T_i], \forall m \in [1..T_j], |f_{i,n} - f_{j,m}| \geq M[i,j]$
- M[i,i] represents the minimum frequency separation necessary to satisfy the co-cell constraints:
  $\forall n, m \in [1..T_i], n \neq m, |f_{i,n} - f_{i,m}| \geq M[i,i]$
- M[i,j]=0 means there is no constraint between the cells C_i and C_j.

Several optimization problems can be defined for FAP. For example, given a traffic volume in the network, we may minimize the frequency interference and the number of frequencies used. Otherwise, given a number of available frequencies, we may maximize the traffic volume (assigning more frequencies to a cell) while minimizing the frequency interference. In the latter case, the two optimization objectives are orthogonal. Indeed, requiring a higher degree of reuse of frequencies implies a higher risk of frequency interference. In this paper, we deal with the first problem, i.e. we try to mimimize the frequency interference using a minimum number of frequencies for a given traffic volume.

3 EAs for FAP

The main obstacle of applying EAs to a new application lies essentially in two difficulties. First, we must define an efficient encoding for the given application and a set of tailored genetic operators. Second, we must find an efficient combination of these operators working with the encoding. The first point is related to the application at hand and can be solved by integrating, at different levels of an EA, specific knowledge about the application and various efficient local search techniques. On the contrary, the second point is a methodological issue and must be dealt with independently of applications. Indeed, until now, neither the roles of genetic operators nor their combinations are well understood. From an application point of view, the simultaneous presence of all the genetic components makes it difficult to build efficient EAs and to evaluate the efficiency of each component. For these reasons, to tackle our application with EAs, we take a step-by-step approach which is based on the bottom-up principle.

Instead of integrating all the genetic components in an EA, we begin with a very simple EA which uses only mutation and selection. Using this largely simplified EA, many alternative mutation and selection operators can be experimented with and evaluated. Another interesting point of this simple EA is that it may work with a singleton population. In this case, the EA can be used to simulate numerous (stochastic) hill-climbers. Once the most efficient selection and mutation operators are identified, crossover may be added to the initial EA if this proves to be necessary for the application.

Problem Encoding & Solution Space

Given a problem $FAP = <X, D, C, F>$, a feasible solution $s = <C_1, C_2, ..., C_N>$ corresponds to a complete assignment of frequencies to cells where

- $|C_i| = T_i$, i.e. each C_i is composed of T_i frequencies (genes) assigned to it.

An individual or a chromosome I simply represents a feasible solution. A population of chromosomes represents a part of the total search space composed of all the feasible solutions of FAP noted by $S = \{s \mid s$ is a feasible solution of FAP$\}$. Note that this encoding satisfies implicitly the traffic constraint.

Other encodings are possible. For instance, one may use a $NF * \sum_{i=1}^{T_i}$ boolean matrix, each element of the matrix indicating if a frequency is assigned to a cell. One may also use a heterogeneous 2 dimensional encoding whereby a number of frequencies equal to its traffic corresponds to each cell. However, the chosen encoding has some desirable properties compared with others. First, the number of genes in a chromosome is minimized. Second, mutation can be directly applied. Third, crossover can also be directly applied or be applied with a minimum constraint of choosing crossover points at the first gene of each cell.

Fitness Function

Two functions are used to evaluate an individual's fitness: the first is for evaluating the initial population and offspring produced by crossover and the second is for offspring produced by mutation.

For any individual I, one natural way to calculate its fitness is to apply the following function $f: S \rightarrow NC$ (the number of interference constraints).

$$f(I) = \sum_{i=1}^{N} \sum_{j=i+1}^{N} \sum_{k=1}^{T_i} \sum_{p=1}^{T_j} CI(i,j,k,p) + \sum_{i=1}^{N} \sum_{k=1}^{T_i} \sum_{p=k+1}^{T_i} CO(i,k,p) \quad (1)$$

$$CI(i,j,k,p) = 1 \text{ if } |f_{i,k} - f_{j,p}| < M[i,j]$$
$$= 0 \text{ otherwise}$$
$$CO(i,k,p) = 1 \text{ if } |f_{i,k} - f_{i,p}| < M[i,i]$$
$$= 0 \text{ otherwise}$$

It is easy to see that this function examines in pairs all the cells of the networks in order to count the total number of interference constraints violated by I. Given that T_i ($i \in [1..N]$) are bounded by NF, the time complexity of the fitness function (1) is $\Theta(N^2 * NF^2)$.

When mutation is applied to an individual I producing I', only one cell in I is affected. More precisely, only one frequency of the cell is changed. Thus, to calculate the fitness of I', only cells adjacent to this modified cell need to be re-examined to calculate the fitness difference between I and I'. This can be realized by using the following formula which counts the number of violated constraints induced by the frequency $f_{i,k}$ of the cell C_i.

$$f(Ci, k) = \sum_{j=1, j \neq i}^{N} \sum_{p=1}^{T_j} CI(i, j, k, p) + \sum_{p=1, p \neq k}^{T_i} CO(i, k, p) \quad (2)$$

CI and CO are the same as defined in (1). Since the traffic of the cells is bounded by the number of frequencies NF, the time complexity of this function is $\Theta(N * NF)$. Note finally that, in practice, the complexity is much lower since the number of adjacent–cells of C_i is usually much smaller than N-1.

Selection

For the purpose of investigating the effects of crossover for FAP, a single selection mechanism, Baker's SUS (Stochastic Universal Sampling) selection algorithm [2] is used for all our EAs (with and without crossover). Similar to the standard *spinning wheel* method with one pointer, SUS can be considered as a spinning wheel method with K equally spaced pointers (K being the size of population). Hence, all K samples will be achieved in a single spin. Like other probabilistic selection methods, well-fit individuals will have more chances to be selected by SUS.

Compared with other selection methods, SUS has some desirable qualities such as simplexity and efficiency (complexity of $\Theta(K)$), and accuracy and precision (zero bias and minimum spread). Moreover, according to our experiments, SUS gives good results on average.

Crossover

Three crossover operators are studied: two standard crossovers, i.e. one-point and uniform, and one specialized crossover called conflict-based.

- *one-point*: a crossover point is randomly chosen with equal probability and the right portions of the two parents are exchanged, producing two children.
- *uniform*: each gene of the two children receives the corresponding gene from either parent one or parent two according to a given probability [13].
- *conflict-based*: the idea of this specialized crossover is to use specific knowledge about the application. It tries to pass on good genes from the parents to their children. More precisely, frequencies (genes) free of conflict are directly copied and frequencies in conflict are stochastically or deterministically chosen. Figure 1 gives an example whereby deterministic choices are made for genes in conflict.

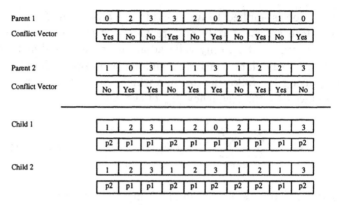

Fig. 1. Conflict-based crossover

These crossover operators can be considered to be sufficiently representative for studying the effects of crossover for FAP. In fact, one-point and uniform crossover are well known and have some complementary properties. One-point crossover exhibits the minimum distributional bias and the maximum positional bias. On the contrary, uniform crossover exhibits the maximum distributional bias and the minimum positional bias. Moreover, uniform crossover is known to be more disruptive, which is particularly suitable for small populations to sustain a highly explorative and diversifying search. On the other hand, specialized crossovers use application knowledge to insure a better transition of useful information from parents to offspring.

Mutation

For the same reason we fix the selection operator, we have chosen to fix the mutation operator to be used in our experiments in order to isolate the effects of crossover. More precisely, a special mutation operator which is based on frequency conflict and defined in [5] is used by all our EAs. This conflict-based mutation (CBM) operator is composed of three choices based on specific knowledge of the application: 1) random selection of a cell among the cells in conflict, 2) random selection of a frequency from this cell, and 3) the best frequency value which is different from the current value.

Compared with other mutation operators we have tested, including classic random and specialized mutations of the same class defined in [5], the chosen mutation proves to be the most efficient on average for this application.

4 Experimentation and Results

4.1 Tests

In this section, we present our experimental results for the various EAs on two sets of FAP instances provided by the French National Research Center for Telecommunications. These instances are produced by a generator in such a way that they correspond to real situations encountered in real networks or sub-networks in France. Consequently, some instances have a very big size in terms of the number of cells (variables) and of inteference constraints. The biggest instance has 300 cells (600 genes) and more than 30,000 constraints. Moreover, each gene may take up to 30 possible values (frequencies).

Test Set No.1 (Traffic=1)

The first set of FAP instances has the following characteristics.
- *traffic constraints*: $T_i = 1$ (i $\in$ [1..N]); i.e. each cell is assigned one frequency. Consequently, a co-cell constraint does not exist (M[i,i] = 0 for i $\in$ [1..N]).
- *adjacent constraints*: $|f_i - f_j| \geq$ M[i,j] = 1 (i, j $\in$ [1..N]) for two adjacent cells C_i and C_j; i.e. C_i and C_j must be assigned different frequency values.

It is easy to see that these instances correspond to the classic graph coloring problem. In fact, it will be sufficient to replace frequencies by colors, cells by n-odes and adjacent constraints by edges. Finding an optimal frequency assignment is equivalent to coloring a graph using only a minimum number (the chromatic number) of necessary colors. For these instances, the chromatic number corresponds to the minimum (optimal) number of frequencies necessary for having a frequency assignment without interference (*optimal solution*). The names of these instances consist of three integer parameters *nf.nc.d* where

nf : the optimal number of frequencies needed for an optimal solution.
nc : the number of cells in the network.
d : the density of interference constraints defined as a percentage of all the possible constraints over the network.

For example, the instance 8.150.30 defines a problem composed of 150 cells with 8 available frequencies and 30% of 150*(150-1)/2 total constraints. For large problems having a high density, we obtain up to 13,000 constraints.

Test Set No.2 (Traffic=2)

The second set of FAP instances has the following characteristics.
- *traffic constraints*: $T_i = 2$ (i $\in$ [1..N]); i.e. each cell is assigned two frequencies.
- *co-cell constraints*: $|f_{i,n} - f_{i,m}| \geq$ M[i,i] = 3 (i $\in$ [1..N]); i.e. two frequencies assigned to the same cell must have a minimum distance of 3.
- *adjacent constraints*: $|f_{i,n} - f_{j,m}| \geq$ M[i,j] $\in$ [1,2] (i, j $\in$ [1..N]) if C_i and C_j are adjacent cells; i.e. two frequencies assigned to two adjacent cells must have a minimum distance of 1 or 2 according to the cells.

The names of the instances consist of four integer parameters *nf.nc.d.p*. The first three parameters have similar meanings to those presented before. The fourth parameter indicates the average degree for the nodes of the graph (the average number of constraints associated with a cell). Due to the way the instances are generated, the exact number of frequencies for an optimal solution is no longer known in advance. However, this optimal number is bounded by a value given by the generator.

Compared with the first set, these instances are naturally much harder due to the doubled traffic and more constraints. For large instances having a high density, we obtain up to 300 cells (therefore 600 genes) and 30,000 constraints[3].

4.2 Measure Criteria

Two criteria are used: *excess of frequencies* and *number of fitness evaluations*.

Excess: the number of frequencies added to the minimum of the optimal solution. For instance, for the problem 8.150.20 which requires 8 frequencies, an excess of 2 of a method means that the method can only find an optimal solution by adding 2 extra frequencies. This criterion is essential because adding even one frequency may make the initial problem much easier to solve. This criterion reflects the *quality* of a solution found by an algorithm.

Nb_evaluation: the evaluation number needed to obtain an optimal solution, corresponding to the exact number of points in the search space visited by an algorithm. This criterion reflects the *speed* of an algorithm and is the most objective for measuring an algorithm's performance.

The criterion of generation number is not used since two algorithms may have completely different complexities for a generation.

4.3 Results

This section describes two (classes of) EAs and their results on the two sets of FAP instances. The first EA called SM is a simple EA which uses only selection (S) and mutation (M). The second EA called SCM is SM augmented by crossover (C). The set up of SM and SCM is defined as follows.

- *SM*: SM uses the SUS selection and the conflict-based mutation defined in §3. Beginning with an initial population, SM selects first some individuals using SUS favoring well-fit individuals. Then the conflict-based mutation is applied to certain individuals according to the mutation rate.
- *SCM*: SCM uses the same selection and mutation operators as SM, but before mutation is activated, crossover (one of those described in §3) is applied to certain individuals according to the crossover rate.
- *Population size*: the population size is fixed at 50 for all of our experiments.
- *Initial population*: it is generated randomly, i.e. for each gene of an individual, a random value is taken from among all the possible frequency values.
- P_c (the crossover rate): it defines the percentage of the selected individuals that will receive a crossover operation, producing two offspring. According to our experiments, P_c is fixed at 40% in this paper.
- P_m (the mutation rate): it defines the percentage of individuals that will receive a mutation operation, i.e. one frequency (gene) of a cell will be changed. According to our experiments, P_m is fixed at 10% in this paper.

---

[3] The difficulty of a problem depends not only on the number of variables and the number of possible values for each variable, but also on the number of constraints.

– *Generation*: The maximum generation for each try is fixed between 10,000-50,000 according to the difficulty of the instances.

Remember that the main objective to reach in the application is to minimize the frequency interference with a minimum number of frequencies. In practice, our EAs begin with a certain number K of frequencies (a bigger value than the optimal). If an optimal solution is found within a certain number of tries (fixed at 5 in this paper) with K frequencies, then the procedure will try to solve the problem with K-1 frequencies and so on. This process continues until the procedure can no longer find an optimal solution (an interference-free assignment).

Table 1 gives the results of SM and SCM on the first set of FAP instances.

problems	Opt.	SM Excess /nb_eval.	SCM(O) Excess /nb_eval.	SCM(U) Excess /nb_eval.	SCM(CB) Excess /nb_eval.
8.150.10	8	+0/14280	+0/25510	+0/21500	+0/3000
8.150.20	8	+2/68900	+2/116250	+2/291300	+2/130200
8.150.30	8	+0/605620	+5/478170	+0/351900	+0/437500
15.150.10	15	+0/13400	+0/19700	+0/19170	+0/1630
15.150.20	15	+0/32985	+0/44952	+0/42939	+0/5484
15.150.30	15	+0/101170	+0/427240	+0/554170	+0/465900
15.300.10	15	+0/58890	+0/73280	+0/62470	+0/8540
15.300.20	15	+2/742300	+2/1107000	+2/1034000	+3/1985000
15.300.30	15	+8/756930	+8/1594558	+9/893348	+9/781000
30.300.10	30	+0/78390	+0/71360	+0/59030	+0/4250
30.300.20	30	+0/130110	+0/122900	+0/110350	+0/14150
30.300.30	30	+0/237480	+0/312470	+0/256180	+0/65370

Table 1. Comparative results on test No. 1

SCM(X) indicates the crossover operator used, O, U and CB represent one-point, uniform and conflict-based crossover respectively.

From the data in the table, two remarks can be made. First, although all of our EAs manage to solve most of the instances, they have serious difficulties with 3 instances (in bold) which are intrinsically hard. Second, according to the quality criterion, all the EAs behave similarly. However, according to the speed criterion, SCM(CB) is much more efficient. This seems logical given that the CB crossover operator directly passes on good frequencies (genes) to future generations, which favors the convergence of the algorithm.

Note also that the number of evaluations used in the table does not distinguish Nb_evaluation caused by crossover from that caused by mutation. However, as discussed in §3, evaluating an individual produced by crossover is much more expensive than evaluating an individual produced by mutation ($\Theta(N^2 * NF^2)$ v.s $\Theta(N * NF)$). The same remark remains true for Table 2. Thus, in practice, crossover can justify its utility only if it produces solutions of better quality.

Table 2 gives the results of SM and SCM on the second set of FAP instances. Since now there are two genes for each cell (the traffic is of 2), one-point and uniform are adapted to take this fact into account. We use O2 (U2) to designate the adapted one-point (uniform) crossover and O1 (U1) the original one. Now the crossover point is limited to the first gene of each cell. In particular, for uniform crossover U2, the two genes of each cell will be simultaneously exchanged in the two parents. It should be pointed out that these adapted crossovers can be also considered to be specialized operators.

problems	Opt.	SM Excess /nb eval.	SCM(O1) Excess /nb eval.	SCM(U1) Excess /nb eval.	SCM(O2) Excess /nb eval.	SCM(U2) Excess /nb eval.	SCM(CB) Excess /nb eval.
4.75.05.10	11	+1/104529	+2/98943	+2/125057	+1/260428	+2/359733	+2/14945
4.75.05.30	11	+1/243279	+1/262432	+2/193739	+1/118270	+2/97548	+2/114257
4.75.15.10	11	+6/87593	+5/724826	+6/92105	+6/457314	+5/401066	+7/17880
4.75.15.30	11	+3/507024	+5/274746	+5/957438	+5/112524	+5/425902	+6/300893
8.75.05.10	16	+0/33920	+0/33622	+0/46166	+0/33246	+0/38660	+0/4382
8.75.05.30	16	+0/32928	+0/36514	+0/43146	+0/33946	+0/40214	+0/4642
8.75.25.10	16	+7/165876	+7/130708	+5/1736100	+5/401363	+5/169621	+7/31892
8.75.25.30	16	+3/603915	+3/537415	+4/352140	+3/1017383	+4/131194	+7/22933
8.150.05.30	16	+0/191264	+0/154114	+0/445288	+0/198962	+0/151854	+0/55000
8.150.05.60	16	+0/153216	+0/118812	+0/184286	+0/140346	+0/129712	+0/31000
8.150.15.30	16	+5/1044435	+7/409000	+6/2491000	+6/1945000	+7/2374000	+8/834000
15.300.05.60	30	+0/370000	+0/221000	+0/194000	+0/196000	+0/203000	+0/8000
30.300.05.60	60	+0/413000	+0/205000	+0/193000	+0/224000	+0/179000	+0/6000

Table 2. Comparative results on test No. 2

The second set of FAP instances is, in general, much harder than the first set since the search space is much bigger: the number of genes is doubled and the number of interference constraints increases notably. This can be seen easily from the table. In fact, the EAs have serious difficulties on more instances and several extra frequencies are often needed to find a conflict-free assignment.

As was the case for Table 1, similar observations can be made for Table 2. In terms of solution quality, there is no significant difference between the three crossover operators on the one hand, and no significant difference between SM and SCM on the other. Note also that SCM(CB) is very efficient in terms of evaluation number, though its solution quality is sometime worse than others.

Finally, the resolution time varies greatly following instances. Easy instances, often having a low density of constraints, can be solved instantaneously while the solving of hard instances is, in general, very time consuming and may require several hours CPU time on a SPARC-10 station.

4.4 Discussion

In terms of solution quality, the results of SM and SCM presented in Tables 1&2 are better than those of two complete methods: graph coloring algorithms and constraint programming (CP) presented in [1, 3]. In terms of resolution speed, CP is much faster for easy instances than other methods. Compared with two of the most efficient incomplete methods SA [1] and TS [10], the results of SM and SCM are worse especially in terms of quality. For example, both SA and TS find optimal solutions for all the instances of the first set and require less extra frequencies than SM and SCM for the second set.

However, it should be noted that an efficient EA without crossover, called M&S, exists [5]. Compared with SM, there are two important differences. First, M&S uses the elitist strategy instead of SUS for selection. Second, like a stochastic hill-climber, M&S uses the conflict-based mutation augmented by a deterioration control probability, i.e. non-improved offspring produced by mutation are only accepted stochastically. These two points result in M&S and SM having different behavior and, consequently, producing different results. Indeed, M&S gives competitive or better results compared with SA and TS cited above.

5   Conclusions & Future Work

This study shows the interest of the evolutionary approach for FAP. Meanwhile, empirical evidence suggests that the contribution of the tested crossovers is marginal both in terms of solution quality and resolution speed. This is especially true for hard instances. Whether this conclusion can be generalized is not clear. However, considering its higher complexity for implementation and for fitness evaluation with respect to mutation, we can say that crossover is useful only if it proves to significantly improve the search in terms of solution quality. In the case where low quality solutions are sufficient for practical need, specialized crossovers may be developed.

At the same time, these remarks should be interpreted carefully for several reasons. First, we cannot exclude the possibility of existence of other specialized and efficient crossovers. Second, in this paper, crossover is studied in the context of traditional genetic search, i.e. crossover is used as the main search operator. Other possibilities of using crossover are worthy of investigation. For example, in mutation-oriented algorithms, any specialized efficient mutation operator may be used as the main search mechanism to reach rapidly local optima and crossover is occasionally activated to create new promising individuals in order to help the search to escape from the local optima.

Finally, the EAs presented here can be improved in several respects. First, it is evident that more efficient, application knowledge-based genetic operators may be researched. Second, a more technical and interesting improvement is possible concerning the constraint handling technique. Indeed, the current encoding takes into account only traffic constraints but not co-cell constraints. Consequently, two frequencies may be assigned to one cell during the search even if they violate the co-cell constraint. This conflict situation is only repaired (hopefully) by the search operators and may re-appear later. One way to solve this problem is to include co-cell constraints directly in the solution encoding, i.e. we make sure from the beginning of the search process that conflict frequencies will not be assigned to the same cell. In this way, the search space will be greatly reduced since both the number of constraints to be checked explicitly and the possible combinations of frequencies to be assigned to cells are reduced. Preliminary study on this issue shows very promising performance and a complete study of this technique will be reported in the near future.

Acknowledgements

We would like to thank the referees for their useful comments which helped to improve this paper and the CNET which supported this work. Special thanks go to A. Caminada and R. Mignone from the CNET for their various assistance.

References

1. A. Akrout. Problèmes d'affectation de fréquences: méthodes basées sur le recuit simulé. Technical Report RP/PAB/SRM/RRM/4123, CNET, 1994.
2. J.E. Baker. Reducing bias and inefficiency in the selection algorithm. In *Proc. of Intl. Conf. on Genetic Algorithms (ICGA'87)*, 1987.
3. A. Caminada. Résolution du problème de l'affectation des fréquences par programmation par contraintes. Technical Report FT.CNET/CNET BEL/POH/CDI/71-95/CA, CNET, 1995.
4. W. Crompton, S. Hurley, and N.M. Stephen. A parallel genetic algorithms for frequency assignment problems. In *Proc. of IMACS SPRANN'94*, pages 81–84, 1994.
5. R. Dorne and J.K. Hao. An evolutionary approach for frequency assignment in cellular radio networks. In *Proc. of IEEE Intl. Conf. on Evolutionary Computation (ICEC'95)*, Perth, Australia, 1995.
6. M. Duque-Anton, D. Kunz, and B. Rüber. Channel assignment for cellular radio using simulated annealing. In *IEEE Trans. on Vehicular Technololy*, volume 42, pages 14–21, 1993.
7. N. Funakini and Y. Takefuji. A neural network parallel algorithm for channel assignment problems in cellular radio network. In *IEEE Trans. Vehicular Technology*, volume 41, pages 430–437, 1992.
8. A. Gamst. Some lower bounds for a class of frequency assignment problems. In *IEEE Trans. on Vehicular Technololy*, volume 35, pages 8–14, 1986.
9. A. Gamst and W. Rave. On the frequency assignment in mobile automatic telephone systems. In *Proc. of GLOBECOM 82*, pages 309–315, 1982.
10. J.K. Hao and L. Perrier. Tabu search for channel assignment problems. submitted for publication, 1995.
11. A. Kapsalis, V.J. Rayward-Smith, and G.D. Smith. Using genetic algorithms to solve the radio link frequency assignment problem. In *Proc. of Intl. Conf. on ANN and GAs (ICANNGA'95)*, pages 37–40, Alès, France, 1995.
12. D. Kunz. Channel assignment for cellular radio using neural networks. In *IEEE Trans. on Vehicular Technology*, volume 40, pages 188–193, 1991.
13. G. Syswerda. Uniform crossover in genetic algorithms. In *Proc. of Intl. Conf. on Genetic Algorithms (ICGA'89)*, pages 2–9, 1989.

An Application of Evolutionary Algorithms to the Scheduling of Robotic Operations

Vittorio Gorrini and Marco Dorigo
IRIDIA - CP 194/6
Université Libre de Bruxelles
50, av. Franklin Roosevelt
1050 Bruxelles - Belgium
gorrini@ulb.ac.be, mdorigo@ulb.ac.be
Tel.: +32/2/650.27.29, Fax: +32/2/650.27.15

Abstract. The application problem we discuss in this paper is how to translate schedules for the activity of a robot whose task is to feed some machines with cylinders of raw material into loading and routing instructions. The application can be modeled as the interplay of two combinatorial optimization problems: a particular kind of knapsack and a routing problem. The knapsack problem consists of placing the largest possible number of cylinders of raw material on the robotic platform, respecting some physical constraints. To do this we apply an evolutionary strategy algorithm to individuals which code the disposition of cylinders of raw material. The coding is obtained by a set of real values which are used by an algorithm which simulates a gravitational process to produce possible solutions. After a feasible and possibly good solution is found, this is given as input to a routing optimization routine. The routine is implemented by means of a classical genetic algorithm. The main contribution of this paper is to show how evolutionary computation can be applied to solve a relatively complex real problem. Results have shown this to be a viable approach whose main merit is to be simple and effective.

1 Introduction

In this paper we consider the problem of translating a high level description (a schedule) of the activities of an autonomous robot which moves in a given environment into low level plans. In our application the robot task is to load cylinders of raw material (cylinders, hereafter) on its carrying platform, and to bring them to some machines which need the raw material to function. The schedule here is a list of cylinders C_i with a delivery time limit T_{li}. We want to translate this schedule into a loading plan (where to place the cylinders) and a routing plan (which path to follow in the magazine). The problem is rather difficult, due to its "double" combinatorial nature and to the numerous constraints which make most of the possible solutions unfeasible.

The double combinatorial nature is due to the fact that the complete scheduling problem is composed of (i) a kind of knapsack problem [Garey and Johnson, 1979] (that is, how to saturate the robot platform), and (ii) a path minimization problem (that is, how to minimize the route length in the store).

Moreover, the problem is constrained: (i) the machines which consume the raw material should never stop, (ii) due to the robot geometry, the order in which cylinders are loaded must be the inverse of the order in which they are unloaded, and (iii) again due to the geometry of the robot, cylinders should be positioned on the robot platform in such a way that the unloading of cylinders that must be unloaded first is not impeded by the presence of cylinders which must be unloaded later.

In this work we are interested in proposing a "natural algorithm" approach to the solution of the scheduling problem in which all algorithms comprising the solution system are algorithms which were inspired by the observation of some natural phenomena.

In the following we describe, see Section 2, how we solved the optimization of the loading phase, and in Section 3 the optimization of the routing problem. To solve the loading problem we combined an evolutionary strategy [Schwefel, 1975] with a simulation of the gravitational fall of cylinders in a vertical plane. The routing problem was solved by a rather straightforward application of the genetic algorithm [Holland, 1975]. In section 4 we give the complete algorithm and present results. We draw some conclusions in Section 5.

2 Optimization of Loading

The optimization of loading presents two difficulties. First, the circular shape of cylinders requires the definition of an appropriate algorithm to build solutions to the problem; we designed an algorithm called *gravitational fall* to solve this problem. Second, choosing the position of cylinders on the platform is NP-hard; this problem was attacked by means of an evolutionary approach. The problem is defined as follows.

A robot receives as input a list of cylinders to load; this list (an example is reported in Table 1), is ordered by machines and by increasing delivery time (the ordering by machines was decided to simplify the problem: once the robot is unloading cylinders for a given machine it will unload all of them before moving to the next one). The goal of the algorithm is to find a good feasible solutions for the disposition of the cylinders on the carrying platform of the robot. The determination of a feasible solution is particularly difficult due to the combinatorial nature of the problem, and to presence of two constraints: the fixed dimension of the carrying surface, and the fixed unloading order.

Figure 1 shows an example of solution for the problem reported in Table 1: the seventeen cylinders are placed on the carrying surface and can be unloaded following the ordering 1 → 17. Finding a good solution to the loading problem requires to be able to build a solution, evaluate it, and improve it.

2.1 The gravitational fall algorithm

The unusual geometry of elements required us to design an algorithm which builds solutions which are feasible from the unloading order point of view. This algorithm, which we call *gravitational fall*, simulates a gravitational field: cylinders are seen as heavy circles which fall in a vertical plane within a region with the same dimensions as the robot carrying surface (see Figure 2). Cylinders, which are ordered as in the input list, fall one after the other in the direction given by the arrow g in Figure 2; therefore, once loaded they can be unloaded in the reverse order using the opposite of direction g as unloading direction. This process alone does not guarantee the feasibility of solutions, given that situations like the one presented in Figure 2 are highly probable. Given that the order in which cylinders fall is fixed by the input list, the resulting solution depends only on the choice of so-called *dropping points*. A dropping point is the abscissa x_c of the center of the cylinder C when it starts its fall (that is, when its $y_c=L$).

Table 1. Example of list of cylinders C_i given as input to the robot

	Radius	Max time	Machine
C_1	300	65	1
C_2	280	67	1
C_3	110	68	1
C_4	280	69	1
C_5	100	72	2
C_6	120	73	2
C_7	80	74	2
C_8	140	80	5
C_9	120	83	5
C_{10}	100	85	5
C_{11}	140	89	7
C_{12}	320	91	7
C_{13}	100	96	7
C_{14}	140	99	7
C_{15}	80	97	9
C_{16}	110	99	9
C_{17}	120	101	9

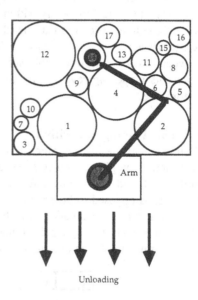

Fig. 1. Example of configuration

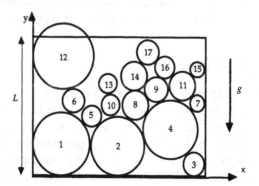

Fig. 2. Example of unfeasible configuration

2.2 The fitness function

The fitness function is defined as

$$FF = \begin{cases} \sum_{i \in B} R_i^2 & \text{if } B \neq T \quad (1) \\ \sum_{i \in T} R_i^2 + \sum_{i \in T} R_i(L - Y_i) & \text{if } B = T \quad (2) \end{cases}$$

where L is the height of the carrying surface (see Figure 2), B is the set of cylinders which are completely within the robot carrying surface boundaries, and T is the set of all the cylinders in the input list. FF is such that its value raises with the number of cylinders which are completely within the carrying surface boundaries (formula (1)). When the solution becomes feasible (that is $B=T$), then the second term in formula (2) determines a considerable increase in the FF which makes conspicuous the difference between feasible and unfeasible solutions. The second term of formula (2) also represents a packing reward, given that it increases as cylinders are closer to the bottom line.

2.3 The loading algorithm

To implement the loading algorithm we used an evolutionary algorithm based on evolution strategies [Schwefel, 1975]. The main difference between evolution strategies based algorithms and genetic algorithms is that the first work on real numbers, as opposed to binary strings, and that they use an adaptive, and therefore variable at run time, mutation probability [Bäck, Hoffmeister and Schwefel, 1991]. Mutation is controlled by a pair of genes (Figure 3): the mutation probability P_m and the maximum amplitude of mutation Δ_m.

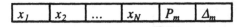

Fig. 3. Example of chromosome

The evolutionary algorithm is applied iteratively until the FF remains stationary for a fixed number of iterations. A solution is therefore represented by a string (chromosome) whose elements (genes) are the dropping points of the cylinders, ordered as in the input list. The quality of solutions is evaluated by computing the fitness function of the solution obtained applying the gravitational algorithm to the string.

3 Routing Optimization

Once the gravitational algorithm has produced a good feasible solution (in Section 4 we will see what can be done in case the algorithm was not able to produce a feasible solution), the routing optimization algorithm receives the list of Table 1 as input.

The problem consists in determining a tour, possibly the shortest, in the store which allows the robot to load all the cylinders in the order specified in the input list and within the given time limit T_l. The store used in our example is represented in Figure 4: it is a grid in which each cell contains cylinders of a given radius (in cells there are two numbers: the first is the name of the cell, the second the radius of cylinders stored in the cell). Cylinders with a same radius can be stored in many different cells, and some cells can be empty.

IN									
1	7	13	19	25	31	37	43	49	55
60	120	110	80	140	100	320	140	empty	empty
2	8	14	20	26	32	38	44	50	56
120	220	empty	80	empty	200	140	140	90	100
3	9	15	21	27	33	39	45	51	57
110	250	140	200	empty	210	100	100	60	100
4	10	16	22	28	34	40	46	52	58
100	280	160	100	160	120	120	empty	250	empty
5	11	17	23	29	35	41	47	53	59
empty	110	160	110	empty	empty	140	110	320	220
6	12	18	24	30	36	42	48	54	60
300	280	110	280	100	120	80	empty	280	260
OUT									

Fig. 4. Example of magazine

To solve this problem we used a genetic algorithm which works on a population of individuals representing the path followed by the robot. The value of each gene is the name of a specific location in the magazine. An example of individual is shown in Figure 5. A feasible solution must be both *spatially* and *temporally* feasible. A spatially feasible solution is a solution which respects the input list ordering, while a solution is temporally feasible if it satisfies the following inequality:

$$T_i = T_c + T_v + T_{di} \le T_{li} \tag{3}$$

where T_i represents the delivery time of C_i to the machine, T_c the time required to load all the cylinders (time which the routing algorithm is in charge to minimize), T_v the time required for the robot to move from the magazine to the production area, T_{di} the time required to unload cylinder C_i, and T_{li} the deadline.

The genetic algorithm duty is to search for good *temporally feasible* solutions, given that spatial feasibility is assured by construction. In fact, we constrain the value of each gene to belong to the set of locations in which cells contain cylinders with the same radius as the one required for that gene by the input list. In other words, if the input list contains for example 2 cylinders with radius 10 and 20, and if cells in the magazine containing cylinders of radius 10 and 20 are respectively $C_{10}=\{1,4,6\}$ and $C_{20}=\{2,3,5\}$, then the first gene can get any value belonging to C_{10}, and the second gene any value belonging to C_{20}.

The initial spatial feasibility of individuals is maintained by the GA through the application of non disruptive operators: mutation is defined to cause changes in a gene value that are within the set of locations which contain cylinders with the same radius as that in the mutated gene, and standard one-point crossover cannot introduce unfeasible individuals.

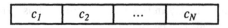

c_1	c_2	...	c_N

Fig. 5. Example of individual

The fitness function used is a function of the traveling time within the magazine; that is, the time T_c required to load all the cylinders. The fitness function was defined to be $FF = e^{-kT_c}$, with k a time constant which in our experiments was set to 100.

4 Implementation and Results

The flow chart of the whole algorithm is shown in Figure 6. First, before the algorithm is started, a schedule is generated. That is, a decision is taken about which are the cylinders that the robot should deliver and what are the delivery deadlines. The schedule, which is the input list to the loading algorithm, is called *critical list*, in that the cylinders in the list must be delivered within the time given by the deadline. An example of critical list was reported in Table 1. A second list, called *non critical list* and also ordered by deliver deadline, is composed of those cylinders which can wait to be delivered later on.

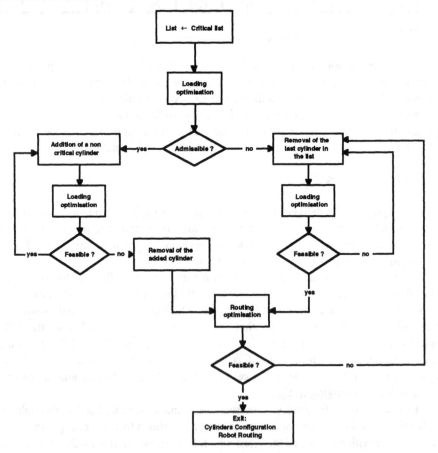

Fig. 6. The whole system

The algorithm tries first to find a good feasible solution for the critical list. If successful, it tries to ameliorate the solution adding one or more non critical cylinders. In case the algorithm cannot find a feasible solution for the critical list, than the last cylinder in the list is removed and the algorithm applied again to the reduced list (the removed cylinder becomes the first element of the non critical list). Once an element is removed from the critical list the algorithm does not try any longer to add non critical cylinders.

Starting from the solution provided by the loading algorithm, that is the original critical list augmented by non critical cylinders or reduced by critical cylinders which could not fit on the robot carrying surface, the routing optimization algorithm tries to generate a good final solution to the scheduling problem. In case it cannot produce a solution which meets the deadline, it is necessary to reduce again the cylinders list by removing the last cylinder in the list (again, this cylinder is added to the first position of the non critical list), and to reapply the whole algorithm to the new list (that is, to apply again the loading algorithm and then the routing one).

Experiments have been run with a robot with two carrying surfaces (see Figure 7). This configuration does not require any change to the previously introduced loading algorithm: the two surfaces are independent of each other and before choosing the dropping point it must be decided in which surface the cylinder must fall.

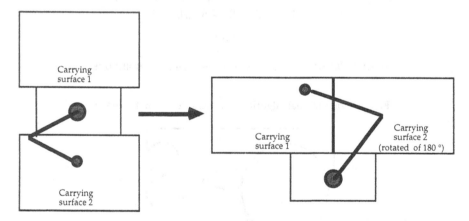

Fig. 7. The robot with two carrying surfaces

We now present the results obtained on the test problem represented by the critical list of Table 1. In Table 2 we give the values we used for the GA parameters and the time constants which characterize the actions of the robot.

The graph in Figure 8 shows the behavior of the fitness function for the loading algorithm. During the first six generations (dotted line) the best found solution is unfeasible; at generation 7 the GA finds a feasible solution, which makes the best solution fitness jump to a much higher value (for this reason we use two different scales on the same graph: on the left side we report the fitness for unfeasible solutions, on the right side for feasible solutions). A very good solution, shown in Figure 9) was found after 16 generations.

Table 2. Optimization parameters

	Optimization of loading	Optimization of routing
Crossover probability	0.725	0.725
Mutation probability	(variable)	0.070
Number of generations	40	250
Population dimension	20	50
Time constant k	—	100
Time required to move from a cell to the cell adjacent	—	1
Time to load/unload a cylinder	—	2

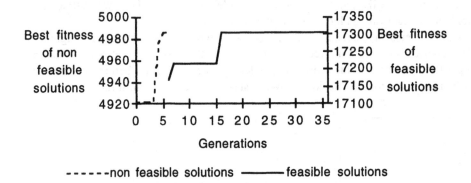

non feasible solutions ------ **feasible solutions** ——

Fig. 8. Gravitational algorithm: fitness function behavior

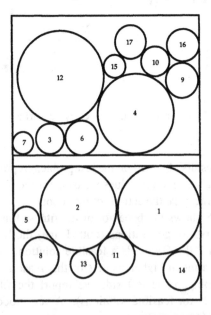

Fig. 9. A solution found by the loading algorithm

In order to provide a better visualization of the results of the routing algorithm, the cylinders have been placed in the magazine in such a way that the optimal routing is along the gray path in Figure 10. This path was always found by the algorithm; on the average it was found after 235 iterations. The behavior of the fitness function is shown in Figure 11.

IN									
1	7	13	19	25	31	37	43	49	55
60	120	110	80	140	100	320	140	empty	empty
2	8	14	20	26	32	38	44	50	56
120	220	empty	80	empty	200	140	140	90	100
3	9	15	21	27	33	39	45	51	57
110	250	140	200	empty	210	100	100	60	100
4	10	16	22	28	34	40	46	52	58
100	280	160	100	160	120	120	empty	250	empty
5	11	17	23	29	35	41	47	53	59
empty	110	160	110	empty	empty	140	110	320	220
6	12	18	24	30	36	42	48	54	60
300	280	110	280	100	120	80	empty	280	260
OUT									

Fig. 10. Best solution found by the routing algorithm

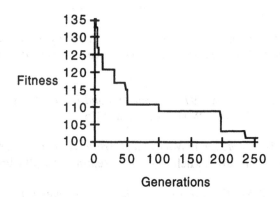

Fig. 11. Routing algorithm: fitness function behavior

5 Conclusions

In this article we have presented an approach to the scheduling of the operations of an autonomous robot completely based on algorithms which have a "natural" inspiration. We have applied twice the evolutionary computation approach to two combinatorial problems. In the first application the problem is a kind of knapsack, that is a NP-hard problem, made more difficult by the geometry of the elements (cylinders) to be packed. We call this the circle-knapsack problem. Given that, at the authors knowledge, there is no standard approach to solve knapsack problems with non rectangular shapes, we have devised a constructive algorithm which produces possible solutions to the problem by simulating a gravitational field. The interplay of this gravitational algorithm with an evolution strategy was able to produce with reasonable efficiency good solution to the circle-knapsack problem. The second application was to find a good routing in a magazine in which the robot goes to load the cylinders. Also in this case the algorithm was able to quickly find good solutions.

The two algorithms have been integrated so that whenever it is impossible to produce a feasible solution within the deadline, the algorithm restarts with a shorter input list. Also, when there is some time left, the algorithm tries to improve the use of the robot by adding a non critical cylinder to the initial list.

Our application shows that evolutionary computation can be successfully used to solve real factory problems. In particular, we believe that it can be most useful for the solution of those problems which, due to their peculiarities, cannot be solved by the straightforward application of any specialized algorithm. In our application this is certainly the case for the loading problem.

Acknowledgments

This work has been partially supported by a fellowship granted to Vittorio Gorrini for the years 1994-95 by the Université Libre de Bruxelles and by an Individual CEC Human Capital and Mobility Programme Fellowship to Marco Dorigo for the years 1994-1996.

References

Bäck T., F. Hoffmeister and H.P. Schwefel (1991). A survey of evolution strategies. *Proceedings of the Fourth international conference on genetic algorithms*, Morgan Kaufmann, 2–9.

Garey M.R. and D.S.Johnson (1979). *Computers and Intractability: A Guide to the theory of NP-Completeness*. W.H. Freeman and Co.

Holland J.H. (1975). *Adaptation in Natural and Artificial Systems*. The University of Michigan Press, Ann Arbor, Michigan, 1975. Reprinted by MIT Press, 1992.

Schwefel H.P. (1975). *Evolutionsstrategie und Numerische Optimierung*. Ph.D.Thesis, Technische Universität Berlin, 1975. Also available as *Numerical Optimization of Computer Models*, J.Wiley, 1981.

Genetic Operators for Two-Dimentional Shape Optimization

Couro Kane and Marc Schoenauer

Centre de Mathématiques Appliquées de l'Ecole Polytechnique

91128 Palaiseau Cedex - France

E-mail: {Couro.Kane,Marc.Schoenauer}@polytechnique.fr

Abstract. This paper presents a Genetic Algorithm approach to two-dimensional shape optimization. Shapes are represented as arrays of boolean pixels (material/void), or bit-arrays. The inadequacy of the (one-dimensional) bitstring representation is emphasized, both *a priori* and experimentally. This leads to the design of crossover operators adapted to the two-dimensional representation. Similarly, some non standard mutation operators are introduced and studied. A strategy involving evolutionary choice among these different operators is finally proposed. All experiments are performed on a simple test-problem of Optimum Design, as the computational cost of real-world problems forbids extensive experimental tests.

1 Introduction

The choice of a representation space is a major issue in Genetic Algorithms. Since the pioneering work of Holland [Hol75], the binary encoding constraint has been relaxed, either in general [Ant89], or for specific domains of applications, e.g. for the TSP [Gre87, WSF89, RS94], or for handling real numbers [JM91, Rad91]. But even within the binary frame, different encodings can produce different behaviors of the Genetic Algorithm on the same problem: The now widely used Gray-coding for ordinal numbers does produce better results, as stated experimentally in [CS88].

This paper addresses the problem of the choice of the right binary representation for another domain of application: The background here is Topological Shape Optimization and the chosen representation is the bit-array representation of a shape in a regular grid. First introduced in [Jen92] and [CSJ94], on the standard Optimum Design problem of the cantilever plate, the bit-array representation for topological shape optimization has been extended to a more general mechanical model and compared to the deterministic method of homogenization in a previous work of the authors [KJS95].

In this paper, we investigate new recombination and mutation operators that take into account the specificity of the bit-array representation with regards to the standard bitstring representation; these operators are experimented and compared with the classical operators used in [KJS95]. The test problem used to perform these experiments is the optimization of the section of a beam in order to maximize its moment of inertia. The low computing cost of the fitness function, compared to the cantilever plates problems, allows for systematic experimental settings of the different parameters that govern the application of the genetic operators. The framework of Topological Shape Optimization, together with the test problem, are presented in section 2, defining the phenotype space. The representation of shapes on a regular grid is introduced in the same section, defining the genotype space, or search space for the GAs. Section 4 focuses on crossover operators: the poor performances of standard bitstring crossover lead to define specific two-dimensional crossover operators. Two kinds of mutation operators are similarly introduced in section 5. Both sections present experimental comparisons of these operators alone, while some results concerning their interactions are given in section 6. A mixed strategy involving a change of operators during the evolution is also proposed and experimented: The clear increase of performances resulting from that mixed strategy unfortunately vanishes as generations pass, and no overall best strategy emerges. Finally, some concluding remarks, together with possible future directions for using evolutionary computation in the domain of topological shape optimization are sketched in the last section.

2 Background

The aim of shape optimization is to find a geometrical description of an object in two or three dimensions such that this object optimizes a given function (e.g. minimizes the weight of the structure having that shape) while satisfying some constraints (e.g. resisting a given force applied at a given point without breaking). Standard methods of Structural Mechanics (sensitivity analysis) and Applied Mathematics (domain variation) tackle this objective by continuously modifying a given initial shape, but the result heavily depends on this initial shape: these methods cannot modify the *topology* of the shape (e.g. add a hole).

2.1 Topological shape optimization

Topological shape optimization deals with situations where the topology of the solution is unknown. Such problems typically arise when mechanical constructors want to lighten some parts they are constructing by taking off material, but still want that part to resist some known loadings.

The homogenization method is the only global deterministic method for topological shape optimization. It has been applied to the structural optimization in the domain of linear elastic structural analysis ([BK88, AK93]). The problem is generalized by considering, instead of boolean material/void variables, a continuous density of material in the interval $[0, 1]$. Strong convergence results make the homogenization method robust and secure. But the limits of this method are clearly identified: only one loading case can be considered, no condition on the unknown boundary can be imposed (e.g. pressure or heat transfer conditions), and no extension has yet been proposed regarding other laws (e.g. nonlinear elasticity, elasto-plasticity, ...). Moreover, when it comes to actually design a real part, the continuous density has to be turned into boolean material; this is done in general via a penalization approach, that sometimes fails to converge to pure material.

To overcome these difficulties, stochastic optimization methods have been proposed. In [GMP95], the Simulated Annealing is used to explore the space of polyhedral shapes. A precise definition of this search space is given, together with approximation results. Experimental examples are given in the case of regular discretization grids, on the optimization of the cross section of a beam to maximize its moment of inertia (see section 2.2), and on the optimization of heater systems, in which heat transfer conditions are set on the (unknown) boundary of the shape. In [Jen92, CSJ94], the optimization procedure is GA and the shapes are defined on regular grids, too. The test problems presented there are standard structural optimization problems: the cantilever plate in linear elasticity model. Recent work by the authors [KJS95] presents the first results of topological shape optimization in the domain of nonlinear elasticity (the material is linear but the model is that of large displacements), on the same cantilever plate problems. This latter work also presents an experimental comparison with the homogenization method.

2.2 The test problem

In this section, the experimental environment of the paper is settled: The problem of optimization of the cross section of a beam in order to maximize the inertial momentum is presented.

The main difficulty in stochastic structural shape optimization is that a single fitness computation generally involves performing a finite element analysis on the shape at hand [Jen92, CSJ94, KJS95]. And finite element analyses are very time consuming (0.2-2s per analysis on a middle range HP workstation, even for coarse discretizations). This is why a simpler problem was chosen, the optimization of the cross section of a beam to maximize its moment of inertia in a given direction while minimizing its weight. This problem is introduced and used for the same reasons in [GMP95] to test the Simulated Annealing-based method.

The mechanical hypotheses are that the cross section is constant for the beam, and no specification on how the bending stress is applied to the beam is given. The optimization domain is the unit square $[0, 1] \times [0, 1]$, and the target shape is known to be the "H" shape.

The moment of inertia is given by

$$I_x = \int_{section} (x - x_G) dx$$

where G is the center of gravity of the section. The function to maximize is the ratio between that moment of inertia and the weight of the beam (or the area of the cross section).

3 Representation and GA

The representation of two-dimensional shapes studied in this paper is the bit-array representation induced by a regular discretization on the optimization domain, as in all previous works using stochastic methods for the same kind of problem ([GMP95, Jen92, CSJ94, KJS95]). A reference rectangle is divided into a $N \times M$ regular grid; a shape is discretized according to this grid, and an element of the grid is either majoritary void or material: the shape is thus naturally represented by an array of bits. Figure 1-a shows the bit-array corresponding to the shape of Figure 1-b, where the material is grey and the void is black.

3.1 Fitness computation

Encoding a shape is thus straightforward. But decoding a genotype rises two difficulties:

- In order for the shape to be viable with respect to the mechanical problem, both vertical boundaries of the section must be connected.

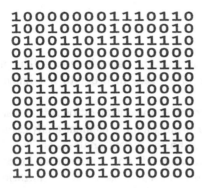
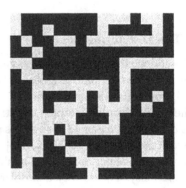

a - *The Bit-array genotype.*   b - *The phenotype.*

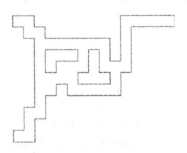

c - *The actual structure.*

Figure 1: *Shape representation: GA applies on the bit-array (a), the fitness is computed on the phenotype (b), but the moment of inertia only depends on the connected component (c).*

- Only material actually connected to both vertical boundaries takes part into the moment of inertia. The disconnected parts have to be removed before the moment of inertia is computed (Figure 1-c).

To cope with the first point, the fitness of a shape which does not connect both vertical edges is set to zero. But there are many possible ways to handle the second difficulty.

- The disconnected parts can be simply removed during the computation of the fitness. But this introduces a fairly high level of degeneracy in the representation (many genotypes correspond to the same phenotype). And, as pointed out in [RS94] and confirmed by experimental results n [Kan95], this is not a desirable feature.
- The disconnected parts can be definitely removed from both the phenotype and the genotype, in a Lamarckian-like way. This is amenable to the so-called *repair* technique used in genetic constraint handling [MJ91]. But these disconnected parts can have the same role as the dominated part of diploid chromosomes, at least in the beginning of the evolution, and this strategy generally leads into local minima (Experimental results account for that statement in [Kan95]).
- The fitness can be modified to favor the disappearing of such disconnected material: the moment of inertia is first computed without the disconnected parts, and some penalty term is added, relative to the amount of disconnected material.

Throughout this paper, the last possibility is used, and the fitness is computed by:

$$\mathcal{F} = \frac{I_x}{weight_{con} + \alpha \ weigth_{dis}}$$

where I_x is the moment of inertia of the connected component, $weight_{con}$ the weight of the connected part $weigth_{dis}$ the weight of the disconnected part and α is some user-supplied positive penalty parameter. The optimal value of $\mathcal{F}$ for the test problem is 1.54, obtained by the perfect "H" shape (without any disconnected material).

3.2 Schema analysis

Using a classical bitstring crossover on bit-arrays rises the following problem. Let H be a schema corresponding to a vertical bar of length L (Fig 2.(a)). In a bitstring representation, H will be encoded as a schema of length $L \times M$ (if the array dimensions are $N \times M$). It follows that H will be disrupted by one-point or two-point crossovers, much more than a schema corresponding to an horizontal bar of same length. More generally, when bit-arrays are handled as bit-strings, n-point crossovers induce a strong geometrical bias *against* vertical building blocks ([Gol89]).

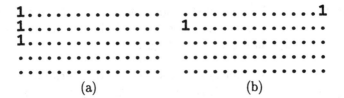

(a)   (b)

Figure 2: *Differences between one- and two-dimensional defining lengths of schemata: Schema (a) has a two-dimensional length of 3, but a one-dimensional length of 31, whereas schema (b) has a two-dimensional length of 15, but a one-dimensional length of 2.*

A quantification of this bias remains to be done, for instance by studying the variance of the fitness on schemas, as a function of the schema length (in the line of what is done in [RS94] for evaluating various representations of the TSP problem).

Nevertheless, the expected inaccuracy of classical n-point crossover (confirmed by experimental results, section 4.1) leads to design specific genetic operators, suited to our two-dimensional problem.

3.3   Experimental settings

The genetic algorithm shell used to obtain the experimental results of the next sections is a home made package. It uses the generational GA scheme (all parents are replaced synchronously by all offsprings) and roulette wheel selection with linear fitness scaling to reach a selective pressure[1] of 2.0. The population size is set to 90. The application of crossover and mutation are governed by user-supplied probabilities (p_{cross} and p_{mut} respectively). The GA stops after 100 generations without improvement of the best fitness, or after 2000 generations [2]. All runs are performed on 15×15 bit-arrays described above, and all results are averages over 20 independent runs.

The test problem presented in section 2.2 is a fairly simple one: Almost all runs did find out a shape very close to the "H" optimal solution. The number of hits of the solution is therefore not the right criterion to use here. So the comparison between different genetic operators in the following sections is made on the plot of the averages of "best fitness vs number of generations", illustrating both the limit values and the dynamics of the runs.

---

[1] ratio between the expected number of offsprings of the best chromosome and that of the average chromosome

[2] The CPU cost of a single run is around 1 hour on a HP712/80 workstation with these parameters.

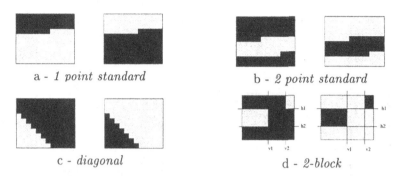

Figure 3: *Examples of "Black and white" offsprings from different crossover operators*

4 Crossover operators

This section is devoted to the study of the crossover operators for the representation of shapes given in section 3. All crossover operators presented in this section are symbolically presented on Figure 3: two offsprings of a black parent and a grey parent are plotted. The color of the cells of the children only tells the parent this cell comes from, regardless of the actual value of the corresponding pixel.

4.1 Standard crossover operators

In order to evaluate the bias induced by handling a bit-array as a bitstring, we first consider the standard crossover operators; the performances of one-point, two-point [Gol89] and uniform [Sys89] crossovers are considered. As pointed out in section. 3.2, one- and two-points crossovers are geometrically biased: only horizontal bands of the parents get exchanged. Experimental results without mutation (Fig. 4) witness the bad influence of this bias. On the opposite, the uniform crossover (easy to imagine, though not presented in Figure 3) does not suffer from such a bias. Two specific two-dimensional operators are introduced in next section to address this issue.

4.2 Specific two-dimensional crossover operators

Diagonal crossover
The basic idea of diagonal crossover is to generalize the popular one-point crossover to the two-dimensional case. As shown on Figure 3-c, a randomly selected straight line separates the rectangle in two parts which are exchanged between both parents.

The block crossover
First introduced in [Jen92], the idea of the block crossover is to cut the whole

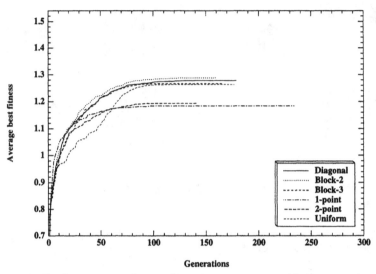

Figure 4: *Performances of recombination operators without mutation*

two-dimensional domain by two horizontal lines and two vertical lines and to exchange some of the large blocks defined by these lines. The values v_1 and v_2 (respectively h_1 and h_2) are chosen uniformly, and the 2 (or 3) blocks to be exchanged are selected randomly. Figure 3-d shows an example of 2-block crossover.

4.3 Comparison without mutation

These operators have been experimentally compared on the problem described in section 2.2. Figure 4 shows the results obtained without any mutation, and with a crossover rate of 1.

The first conclusion that can be drawn from Figure 4 is the ineffectiveness of the one-dimensional crossover operators, similarly outperformed by both uniform and two-dimensional crossovers.

The second remark is that the uniform operator performs poorly in the early stage of the runs, before catching up in the late generations. This fact is confirmed by all other experiments, (see also section 6). The explanation might be the following: in the beginning of a run, the uniform crossover equivalently disrupts schemas of any order, while all other crossovers (even one dimensional) preserve some large areas. These disruptive effects do not occur in the end of evolution, since neither kind of crossover does disturb regions where convergence already occurred: if the bits of the parents at a given position are the same, crossing over has no effect on those bits. In the meantime, the one-dimensional crossovers fail to beneficially exchange schema whose vertical precision is small compared to the horizontal one, like long vertical bars.

A last remark concerns the poor overall results of crossover alone: the optimal shape is almost never reached (remember the optimal fitness is 1.54). The population rapidly becomes uniform, which explains that all curves of Figure 4 stop abruptly between generations 150 and 250, although all runs were allowed 2000 generations.

5 Mutation operators

The standard bit-flip mutation applies on bit-arrays without inducing any geometric bias. Nevertheless, two other mutation operators have been designed and experimented.

5.1 Population-based vs task-oriented

Two directions are explored regarding the mutation operator of bit-array shapes. The first one is problem-independent, and uses statistics on the whole population to keep some genetic diversity. The second mutation operator, purposely devised for the problem of shape optimization, favors small modifications of the boundary of the shape.

Population-based mutation
This mutation operator was first defined in [Cal91] [3], aims at preventing the premature convergence of the population. The probability of flipping a given position is adjusted by considering all bits in that position in the whole population. More precisely, the probability p_i for a given bit i to be flipped depends on the mean value m_i of this bit averaged over the population: if this bit takes a uniform value ($m_i = 0$ or $m_i = 1$), the probability to mutate is set to a high value p_{max}; on the opposite, it is set to a low value p_{min} if there is about the same proportion of 0s and 1s ($m_i = 0.5$). The whole curve is a parabola between these prescribed values (of equation $p_i = 4(p_{max} - p_{min})m_i(m_i - 1) + p_{max}$). This operator thus imposes high values of mutation rate at positions that have already converged.

Boundary mutation
The underlying idea of the boundary mutation is that, for a given topology, i.e. a given number of "holes" in the shape, it is easy to find the optimal design for that topology by slightly moving the boundary. The boundary mutation is defined such that boundary bits, having one edge on the boundary of the connected component of the shape, are given higher probability ($p_{B,max}$) to be flipped than the other bits ($p_{B,min}$).

5.2 Comparison without crossover

Figure 5 shows the results obtained with the three mutation operators, without any crossover. The different mutation rate values have been adjusted by trials and errors to the values given in the plot.

---

[3] where it was improperly termed *epistatic mutation*

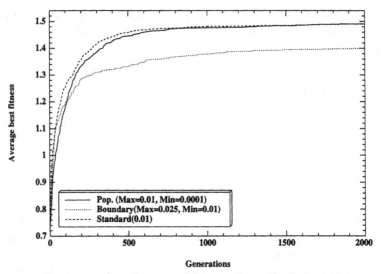

Figure 5: *Performances of mutation operators without recombination*

As expected, boundary mutation alone is outperformed by both the standard and the population-based mutations: since the boundary mutation is oriented toward local optimization, it therefore favors premature convergence. Nevertheless,a closer look at the quality of the results shows that the boundary mutation finds the *exact* optimal solution, when it does find the right local optimum, more quickly then the two other mutations which have great difficulties in adjusting the last pixels. This suggests that the boundary mutation should be used chiefly in the final stages of evolution.

Note that, in the absence of any crossover, no visible difference exists between the standard and the population-based mutation with respect to the best values obtained in 2000 generations: they respectively reach fitness values (and standard deviations) of 1.49083 (0.00772) and 1.4913 (0.01523). Moreover, the population-based mutation clearly slows down the convergence in the early stage of evolution, as could be expected.

6 Cross-comparisons of genetic operators

The experiments presented so far deal with one genetic operator (cross-over or mutation) only. Experiments presented in this section involve both mutation and crossover operators. These results consider only one crossover and one mutation operator with given probability in each run. However, on account of the results of sections 4 and 5, the one-dimensional crossover operators (1-point and 2-points) and the boundary mutation are not considered here.

Table 1 presents the best results obtained on the test problem for pairs of crossover-mutation operators. The crossover rate is 0.7, the mutation rate is 0.001 for the standard mutation, and p_{max} and p_{min} are set to 0.005 and 0.0001 for the population-based mutation (see 5.1).

	Standard mutation	Population-based mutation
Uniform	1.48805 (0.00784)	1.50936 (0.00702)
Diagonal	1.49773 (0.00869)	1.49499 (0.00912)
2-block	1.49203 (0.01138)	1.50914 (0.00872)
3-block	1.50352 (0.01123)	1.5197 (0.0059)

Table 1: *Overall best values (standard deviations) after 2000 generations for all crossover-mutations associations.*

The population-based mutation outperforms the standard mutation in most experiments, and 3-block seems the best choice among crossover operators. Moreover, the association of both operators not only reaches the best value, but also with the smallest standard deviation.

Figure 6 shows the plots corresponding to the values of the population-based mutation in Table 1: once again, the uniform crossover makes a poor start before catching up with the other operators, as in the experiments of section 4. This is confirmed by all other runs, whatever the mutation operator and application rates. On the other side, nothing of that kind can be observed between the population-based and the standard mutation, as it was in section 5 without crossover.

All of the results up to this point strongly suggest that mixed strategies should be used: 3-block crossover and population-based mutation seem to be the best choice; but, on the one hand, the catching-up of the uniform crossover plots indicates it does perform well in the end of the runs; and on the other hand, the locality of the boundary mutation suggests to use it to "polish" the best

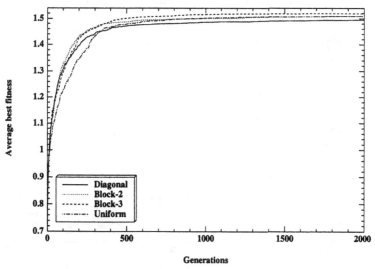

Figure 6: *Overall best results of crossover operators and population-based mutation*

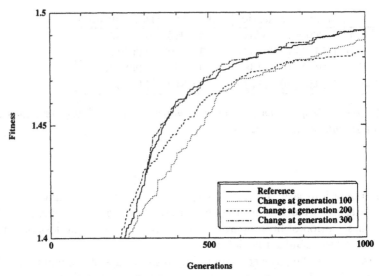

Figure 7: *Too early changes of operators.*

solution found so far in the last generations. Our first attempt toward that goal is to start with the reference results of Figure 5, using 3-block crossover and population-based mutation, and to change to uniform crossover and boundary mutation after a fixed number of generations. Figures 7 and 8 show details of the resulting average plots (beware of the scales).

As can be seen on Figure 7, changing after 100 or 200 generations degrades the results: both curves are below the reference curve, and never catch up. Changing the operators after 300 generations does not seem to change anything. But changing the operators at generation 500 or 600 does indeed increase the performance of the GA, ... at least for some time (see Figure 8).

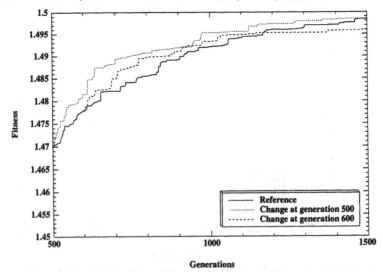

Figure 8: *The benefits of changing the operators does not last*

This improvement is by now not significant: the increase is rather small and the reference curve soon catches up, (all curves ending around similar values). This could be interpreted in the line of Spears experiments [Spe95]: the benefit may result from the changing of operators regardless of the operators themselves.

Experiments on more complex problems are needed to check the significance and interpretation of the above results. However, systematic experiments are not tractable on problems where the fitness computation involves a Finite Element Analysis [KJS95]. So, a tractable test Topological Optimization problem must be devised.

7 Conclusion and perspectives

In this paper, the weaknesses of black-box use of GAs have been demonstrated in one case where the straightforward bitstring representation is not adapted to representing the phenotypes.

The geometrical bias induced by 1- and 2-point crossover badly influences the performances; the isotropy of the uniform crossover allows for good results in any dimension. However, the best results have been obtained with crossover operators purposely devised for 2-dimension problems.

Two mutation operators have been studied in addition to the usual bit-flip mutation. Population-based mutation aims at keeping genetic diversity along generations; it must be tuned with care, as in some cases it might as well prevent any convergence. Second, a boundary mutation has been devised to enable local shape optimization.

The experimental results suggest that the best strategy could be a mixed strategy: two-dimensional crossover and standard or population-based mutation in the beginning, uniform crossover, population-based or boundary mutation in the end. The remaining question is when to switch from one operator to the other. The first experimentations, involving a change after a fixed number of generations demonstrate that the solution should be sought in an evolutionary way. Further research will investigate two evolutionary schemes. The first one consists in encoding into the individual the operators most suited to this individual, in the line of what is done in Evolution Strategies to adjust the mutation on real-valued chromosomes [Sch81], or in GAs to decide whether 2-point or uniform crossover should be applied [Spe91]. Another approach involves off-line experiments, done from time to time along generations; these experiments should enable to decide the operators and parameters most suited to the next few generations, on the basis of a statistical analysis as in [Gre95], or by using Machine Learning techniques as in [SS94].

Another direction of research in shape optimization addresses the representation issue: Almost all real-world applications are three-dimensional problems and the bitstring representation does not scale up. First investigations in the direction of Voronoï diagrams and Delaunay triangulations demonstrate that the cross fertilization of Algorithmic Geometry with Evolutionary Computation can bring a real breakthrough to evolutionary topological shape optimization.

Acknowledgments

The authors are indebted to Michèle Sebag for many helpful discussions and careful proofreading, and to D. Fogel and T. Bäck for accurate comments.

References

[AK93]   G. Allaire and R. V. Kohn. Optimal design for minimum weight and compliance in plane stress using extremal microstructures. *European Journal of Mechanics, A/Solids*, 12(6):839–878, 1993.

[Ant89]   J. Antonisse. A new interpretation of schema notation that overturns the binary encoding constraint. In J. D. Schaffer, editor, *Proceedings of the 3rd International Conference on Genetic Algorithms*, pages 86–91. Morgan Kaufmann, June 1989.

[BK88]   M. Bendsoe and N. Kikushi. Generating optimal topologies in structural design using a homogenization method. *Computer Methods in Applied Mechanics and Engineering*, 71:197–224, 1988.

[Cal91]   D. L. Calloway. Using a genetic algorithm to design binary phase-only filters for pattern recognition. In R. K. Belew and L. B. Booker, editors, *Proceedings of the 4th International Conference on Genetic Algorithms*. Morgan Kaufmann, 1991.

[CS88]   R. A. Caruna and J. D. Schaffer. Representation and hidden bias : Gray vs binary coding for genetic algorithms. In *Proceedings of ICML-88, International Conference on Machine Learning*. Morgan Kaufmann, 1988.

[CSJ94]   C. D. Chapman, K. Saitou, and M. J. Jakiela. Genetic algorithms as an approach to configuration and topology design. *Journal of Mechanical Design*, 116:1005–1012, 1994.

[GMP95]   C. Ghaddar, Y. Maday, and A. T. Patera. Analysis of a part design procedure. *Submitted to Nümerishe Mathematik*, 1995.

[Gol89]   D. E. Goldberg. *Genetic algorithms in search, optimization and machine learning*. Addison Wesley, 1989.

[Gre87]   J. J. Grefenstette. Incorporating problem specific knowledge in genetic algorithms. In Davis L., editor, *Genetic Algorithms and Simulated Annealing*, pages 42–60. Morgan Kaufmann, 1987.

[Gre95]   J. J. Grefenstette. Virtual genetic algorithms: First results. Technical Report AIC-95-013, Navy Center for Applied Research in Artificial Intelligence, February 1995.

[Hol75]   J. Holland. *Adaptation in natural and artificial systems*. University of Michigan Press, Ann Arbor, 1975.

[Jen92]   E. Jensen. *Topological Structural Design using Genetic Algorithms*. PhD thesis, Purdue University, November 1992.

[JM91]   C. Z. Janikow and Z. Michalewicz. An experimental comparison of binary and floating point representations in genetic algorithms. In R. K. Belew and L. B. Booker, editors, *Proceedings of 4th International Conference on Genetic Algorithms*, pages 31–36. Morgan Kaufmann, July 1991.

[Kan95]   C. Kane. *Algorithmes génétiques et Optimisation topologique*. PhD thesis, Université de Paris VI, 1995. Soutenance prévue en 1996.

[KJS95]   C. Kane, F. Jouve, and M. Schoenauer. Structural topology optimization in linear and nonlinear elasticity using genetic algorithms. In *Proceedings of the ASME 21st Design Automation Conference*. ASME, Sept. 1995.

[MJ91]   Z. Michalewicz and C. Z. Janikow. Handling constraints in genetic algorithms. In R. K. Belew and L. B. Booker, editors, *Proceedings of the 4th International Conference on Genetic Algorithms*, pages 151–157, 1991.

[Rad91]   N. J. Radcliffe. Equivalence class analysis of genetic algorithms. *Complex Systems*, 5:183–20, 1991.

[RS94]   N. J. Radcliffe and P. D. Surry. Fitness variance of formae and performance prediction. In D. Whitley and M. Vose, editors, *Foundations of Genetic Algorithms 3*, pages 51–72. Morgan Kaufmann, 1994.

[Sch81] H.-P. Schwefel. *Numerical Optimization of Computer Models*. John Wiley & Sons, New-York, 1981.

[Spe91] W. M. Spears. Adapting crossover in a genetic algorithm. In R. K. Belew and L. B. Booker, editors, *Proceedings of the 4th International Conference on Genetic Algorithms*. Morgan Kaufmann, 1991.

[Spe95] W. M. Spears. Adapting crossover in evolutionary algorithms. In J. R. McDonnell, R. G. Reynolds, and D. B. Fogel, editors, *Proceedings of the 4th Annual Conference on Evolutionary Programming*, pages 367–384. MIT Press, March 1995.

[SS94]   M. Sebag and M. Schoenauer. Controling crossover through inductive learning. In Y. Davidor, H.-P. Schwefel, and R. Manner, editors, *Proceedings of the 3rd Conference on Parallel Problems Solving from Nature*, pages 209–218. Springer-Verlag, LNCS 866, October 1994.

[Sys89] G. Syswerda. Uniform crossover in genetic algorithms. In J. D. Schaffer, editor, *Proceedings of the 3rd International Conference on Genetic Algorithms*, pages 2–9. Morgan Kaufmann, 1989.

[WSF89] D. Whitley, T. Starkweather, and D. Fuquay. Scheduling problems and travelling salesman: The genetic edge recombination operator. In J. D. Schaffer, editor, *Proceedings of the 3rd International Conference on Genetic Algorithms*. Morgan Kaufmann, 1989.

Air Traffic Conflict Resolution by Genetic Algorithms

Frédéric Médioni   Nicolas Durand   Jean-Marc Alliot

ENAC[1]   ENSEEIHT[2]   ENAC

1. Ecole Nationale de l'Aviation Civile
2. Ecole Nationale Supérieure d'Electronique, d'Electrotechnique, d'Informatique, et d'Hydraulique de Toulouse

Abstract. The resolution of Air Traffic Control (ATC) conflicts is a constrained optimization problem: the goal is to propose, for a certain number, n, of aircraft, which might be in conflict in a near future, trajectories that satisfy the separation constraints between aircraft, and minimizes the delays due to the conflict's resolution. The type of conflict resolution trajectories we use allows to split the problem in two steps: first we choose, and freeze, what we call a *configuration* of the problem, i.e. for each aircraft, the direction in which the aircraft is diverted, and for each pair of aircraft, which of the two aircraft passes first at the crossing point of the two aircraft trajectories. We can then compute the optimal trajectories corresponding to this *configuration*, by solving a simple linear optimization problem. Thus we can use an Genetic Algorithm, along with a linear optimization algorithm, such as the *simplex* algorithm: the elements of the population, on which the GA operates, code *configurations* of the problem, and are evaluated using a linear optimization program.The advantage of this approach is that we get, as well as the *fitness* of an element of the population, the local optima corresponding to the *configuration* coded by this element. The GA actually searches for the global optimum among these local optima. We applied this methods to conflicts in which up to 6 aircraft are involved, and obtained really promising results.

1   Introduction

During the first years of civil aviation, the low performances of aircraft, and the fact that they only flew under good visibility conditions, enabled pilots to ensure their own safety. Then, aircraft began to fly faster, and needed to be able to fly even under bad visibility conditions : there was a need for Air Traffic Control. The constantly increasing number of aircraft flying at the same time in one given area led to the division of the airspace in several sectors. The goal of Air Traffic Control is to ensure the aircraft *separation*, *i.e.* the fact that the distance between two aircraft is larger than a given value (the *standard separation*), while minimizing delays due to possible alterations of the aircraft routes. A *conflict* is said to occur when two or more aircraft are not *separated*. Ensuring conflicts resolution while minimizing delays is a complex problem, which is still empirically solved by air traffic controllers. But this way of handling ATC conflicts will probably not be efficient much longer, because the number of aircraft flying, and therefore the number of conflicts to solve simultaneously, is increasing. The automation of ATC would probably allow for a gain in capacity. Several studies have been made in that direction :

- the SAINTEX project [AL92] works like an *expert system*, and tries to minimize delays in some cases.
- Karim Zeghal [Zeg93] presents reactive techniques, which are robust to perturbations, but do not lead to optimal trajectories.
- the AERA 3 project [NFC+83], [Nie89b], [Nie89a], only seeks for optimal trajectories in the case of conflicts involving two aircraft.
- ARC 2000 [K+89], [FMT93], uses priority rules to limit the number of aircraft involved in a conflict, and so be able of optimizing their trajectories. There is no search for a global optimum.
- an ENAC and CENA research group has been using Genetic Algorithms since 1992 to solve *en route* ATC conflicts, and to find optimal trajectories for the aircraft (*cf* [AGS93], [DDAS94]).

We will now present another application a Genetic Algorithm, along with a linear optimization algorithm, to the resolution of ATC conflicts.

2 Background

2.1 Hypotheses

The trajectories generated must meet several criteria: they must be compatible with the aircraft performances; they must be simple enough to be easily transmitted to pilots, either by the controllers, or automatically. This means that the number of changes of heading, altitude, or speed, should be limited.

The hypotheses we will make are actually more restrictive: in the following, we will assume that the aircraft trajectories are in an horizontal plane. We will consider only conflicts involving aircraft flying at the same altitude. Furthermore, we assume that aircraft speeds are constant.

2.2 The optimal collision avoidance trajectory

We will first consider conflicts involving only two aircraft, and alter the trajectory of a single aircraft, and we will then extend the results of this study to the resolution of conflicts involving n aircraft, where $n \geq 2$.

Let us consider two aircraft as two points on the Euclidean plane, a_1 and a_2. At time t, their positions are given by the pairs of coordinates $(x_1(t), y_1(t))$ and $(x_2(t), y_2(t))$. At time t_o aircraft a_1 and a_2 are respectively at points O_1 and O_2, and are supposed to reach points D_1 and D_2 at time t_f. Let C be the point on which the trajectories of the two aircraft would meet if there was no deviation. The collision avoidance trajectories must then satisfy the following constraint:

$$\forall t \in [t_o, t_f], \quad (x_1(t) - x_2(t))^2 + (y_1(t) - y_2(t))^2 \geq d^2 \tag{1}$$

where d is the chosen horizontal standard separation.

We now want to find the optimal collision avoidance trajectory for aircraft a_1 (the trajectory of aircraft a_2 remaining unchanged), *i.e.* the trajectory which leads to the smallest increase of length, compared with the trajectory of a_1 with no deviation, while satisfying constraint (1). It has been shown (*cf* [Dur94]) that in this case, the trajectory of a_1, in the reference system of aircraft a_2, is composed of three phases:

1. a line segment on which constraint (1) is not saturated;
2. an arc of circle on which constraint (1) is saturated;
3. a second line segment on which constraint (1) is not saturated.

Both line segments are tangent to the circle of center C and of radius d (see figure 1). This type of trajectory is difficult to follow precisely for a pilot. This is why we will introduce two other collision avoidance modellings, which will lead to longer but easier trajectories: the *turning point* and the *offset* modellings. These two modellings are presented and compared in [Dur94]. We briefly summarize them here.

2.3   The turning point collision avoidance

The principle of the turning point modelling is simple: the two line segments described above are extended until the arc of circle is reduced to a turning point (*cf* figure 1).

It has been shown that under some additional hypotheses (ensuring that the collision avoidance is started soon enough and that the angle between the initial trajectories of the planes is not too small), the shortest turning point trajectory with no conflict is no more than 5% longer than the optimal collision avoidance trajectory (see [Dur94] for details).

2.4   The offset collision avoidance

The offset modelling for collision avoidance (*cf* figure 2) uses collision avoidance trajectories which are parallel to the initial trajectories, but moved aside. The collision avoidance trajectory is then composed of three linear phases:

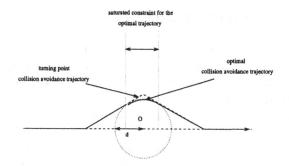

Fig. 1. Comparison optimal trajectory / turning point

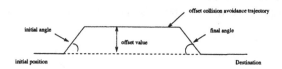

Fig. 2. The offset modelling

1. the *initial phase*: first, the aircraft veers off course with a certain angle (the *initial angle*), until it reaches a certain distance from its original trajectory (this distance is called the *offset value*). The direction to which the aircraft is deviated is called the *offset direction*, the initial angle will always be counted positive.
2. the aircraft then follows a trajectory which is parallel to its initial trajectory.
3. the aircraft finally gets back onto its initial trajectory, with an angle called the *final angle*.

Both the initial angle and the final angle can be fixed. It will be the case in the following. Furthermore, they will be equal, and their common value will be noted β. The duration of the second phase (*i.e.* the time during which the aircraft flies along a line segment which is parallel to its original trajectory) has no influence on the delay due to the collision avoidance. It should however be made as short as possible: while on its deviated trajectory, an aircraft can get in an unpredicted conflict with another one, and the aircraft should anyway be back on its original trajectory when it arrives into a new control sector (about ATC sectors, see [DASF94a], [DASF94b], [DAAS94], [DDAS94]).

When we limit the collision avoidance trajectories to those with an offset, the search for the shortest one is a linear optimization problem, with linear constraints (after some additional simplifications we will describe in the next section). We will then extend these results to the case of a conflict involving n aircraft. The turning point collision avoidance modelling leads to a shorter delay than the offset collision avoidance modelling (*cf* [Dur94]). But the technique we are going to present now uses the fact that the offset modelling leads to linear constraints.

3   A conflict involving n aircraft

Let us consider n aircraft, flying in one ATC sector, from time t_o to time t_f, which may get involved in conflicts. Clearly, the n aircraft are globally separated, if and only if each of them is separated from the others (*i.e.* if and only if constraint (1) is satisfied for each pair of aircraft).

We will now assume that $t_o = 0$. Let us consider two aircraft, a_i and a_j. Let ϕ_{ij} be the angle between their original trajectories (see figure 3), t_{ij}^i (resp. t_{ij}^j) the time at which a_i's original trajectory (resp. a_j's) intersects $a_j's$ (resp. a_i's), and v_i and v_j the norms of the speed vectors of a_i and a_j (which are assumed to remain constant on $[t_o, t_f]$). Let us consider the orthonormal basis (C_{ij}, E_j, E_i), where C_{ij} is the intersection of the original trajectories of aircraft a_i and a_j, and where the x-axis (E_j), is directed by a_j's speed vector (see figure 3).

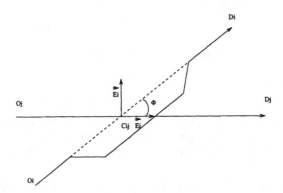

Fig. 3. a conflict involving two aircraft, a single one being deviated.

By writing the coordinates of the aircraft a_i and a_j in this basis, as a function of time $t \in [t_0, t_f]$, we get the following separation constraint, for $t \in [t_o, t_f]$:

$$(v_i - v_j)^2 t^2 + (2v_i v_j \cos(\phi_{ij})(t_{ij}^i + t_{ij}^j) - 2v_i^2 t_{ij}^i - 2v_j^2 t_{ij}^j)\, t$$
$$+ v_i^2 {t_{ij}^i}^2 + v_j^2 {t_{ij}^j}^2 - 2v_i t_{ij}^i v_j t_{ij}^j \cos(\phi_{ij}) - d^2 \geq 0 \qquad (2)$$

It is a second degree inequation in t. Since it is true for a large enough t, it is true for all if and only if its discriminant is negative, which leads to the following inequation:

$$v_i^2 v_j^2 \sin(\phi_{ij})^2 (t_{ij}^i - t_{ij}^j)^2 \geq d^2 (v_i^2 + v_j^2 - 2v_i v_j \cos(\phi_{ij})) \qquad (3)$$

Which yields in turn the two following conditions, depending on which of the two aircraft reach C_{ij} (the intersection of the two trajectories) first:

– If aircraft a_i passes behind aircraft a_j, we get:

$$(t_{ij}^i - t_{ij}^j)v_i v_j \sin(\phi_{ij}) \geq d\sqrt{v_i{}^2 + v_j{}^2 - 2v_i v_j \cos(\phi_{ij})} \tag{4}$$

− If aircraft a_j passes behind aircraft a_i, we get:

$$(t_{ij}^j - t_{ij}^i)v_i v_j \sin(\phi_{ij}) \geq d\sqrt{v_i{}^2 + v_j{}^2 - 2v_i v_j \cos(\phi_{ij})} \tag{5}$$

Note: condition (3) is necessary and sufficient for separating two aircraft on $t \in]-\infty, +\infty[$. We wanted a necessary and sufficient condition for their separation for $t \in [t_o, t_f]$. On this time interval, condition (3) is sufficient but not necessary. The constraints we use are too strict, and we may miss optimal admissible solutions. This can be limited by introducing a superior bound for the offset values: if we know that two aircraft will be separated as long as they are not deviated with an offset value larger to this superior bound, we do not take into account the separation condition for these two aircraft (*cf* [Med94]).

The deviations of aircraft a_i and a_j change the relative positions of their two trajectories. It changes the intersection time. Let us consider the consequences of the offset collision avoidance on the separation constraints for aircraft a_i and a_j, where $1 \leq i < j \leq n$.

The deviation of the two aircraft has two consequences on intersection times $t_{i,j}^i$ and $t_{i,j}^j$. These consequences depend on the offset angle, named β, on the offset direction, and on the offset value. If aircraft a_i is deviated, with its offset value being d_i, we get:

− whatever the offset direction may be, the deviation of the aircraft causes a delay, and $t_{i,j}^i$ increases of $\frac{d_i \tan(\frac{\beta}{2})}{v_i}$.
− When a_i's offset direction is such that a_i's new trajectory is outside angle ϕ $\frac{d_i \cot(\phi_{i,j})}{v_i}$ is added to $t_{i,j}^i$. If a_i's new trajectory is inside this angle, this value is subtracted from $t_{i,j}^i$.
− In the same way, when a_i's offset direction is such that a_i's new trajectory is outside angle ϕ, $t_{i,j}^j$ increases of $\frac{d_i}{v_j \sin(\phi_{i,j})}$.

Thus, by modifying t_{ij}^i and t_{ij}^j in conditions (4) and (5), we get a linear inequation involving d_i and d_j. Its coefficients depend on the offset direction for each aircraft, and for each pair of aircraft, on which one passes behind the other.

For instance, if both aircraft are deviated outside, as shown on figure 4, and if a_i passe behind a_j, we get:

$$\left(t_{ij}^i + \frac{d_i \tan(\frac{\beta}{2})}{v_i} + \frac{d_i \cot(\phi_{ij})}{v_i} + \frac{d_j}{v_i \sin(\phi_{ij})} - t_{ij}^j - \frac{d_j \tan(\frac{\beta}{2})}{v_j}\right.$$
$$\left. - \frac{d_j \cot(\phi_{ij})}{v_j} - \frac{d_i}{v_j \sin(\phi_{ij})}\right)v_i v_j \sin(\phi_{ij}) - d\sqrt{v_i{}^2 + v_j{}^2 - 2v_i v_j \cos\phi_{ij}} \geq 0 \tag{6}$$

This condition, as we mentioned before, is a sufficient condition for the separation of the aircraft, only during the second phase of the offset collision avoidance trajectory (when the aircraft flies along a line segment, parallel to its original trajectory). In the following, we will only consider these constraints. The

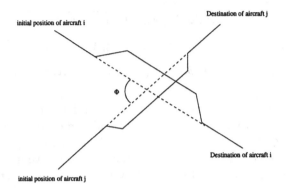

Fig. 4. A conflict involving two aircraft, both of them deviated to the outside

initial and final phases of the offset collision avoidance (when the aircraft leaves its original trajectory to get onto a parallel line segment, and when it gets back onto its original trajectory) lead to non linear constraints, which we will ignore. We will assume that aircraft are deviated soon enough, and will stay on their collision avoidance trajectories long enough, to allow us to ignore these constraints.

We have so far implicitly assume that the original trajectories of aircraft were secant. If the trajectories of two aircraft are parallel (the aircraft may face each other, or one aircraft may fly fast behind a slower one), similar calculi lead to similar linear conditions. What is now important to know is not whether a_j passes behind a_j, or not, but whether a_j passes to the left of a_j, or to the right. For the sake of simplicity, we will assume in the remainder of this article that the trajectories of the planes are secant.

4 A linear, but strongly combinatorial problem

In the previous section, we established that, after some simplifications, the conditions for the separation of each pair of aircraft, lead to a linear constraint. So we have $\frac{n(n-1)}{2}$ linear constraints. We add to these constraints the following ones: $d_i \geq 0$ for $1 \geq i \geq n$, which leads to a total of $\frac{n(n+1)}{2}$ constraints, for a conflict involving n aircraft (or $\frac{n(n+2)}{2}$ constraints if we impose superior bounds for the offset values d_i).

We now want to minimize the global delay, *i.e.* the sum of the delays of aircraft, due to the conflict resolution, under these constraints. The aircraft are delayed only during the initial and final phases of the collision avoidance trajectories. Their delay is a function of the initial angle β. For an aircraft a_i, with offset value d_i this delay is:

$$2\frac{d_i \tan(\frac{\beta}{2})}{v_i}$$

The sum of the delays is then, for the n aircraft:

$$S(d_1, .., d_n) = 2 \sum_{i=1}^{n} \frac{d_i \tan(\frac{\beta}{2})}{v_i}$$

This function is linear in d_i for $1 \le i \le n$.

When the directions of the offsets are fixed for all the aircraft, as well as which aircraft passes behind the other for each pair of aircraft, the offset values leading to the minimal global delay are the solutions of a linear optimization problem, with $\frac{n(n+1)}{2}$ constraints. This problem can be solved, by an algorithm such as the *simplex* algorithm, for instance. But the best we will get can only be a local optimum. We might even get no solution at all.

The linear constraints' coefficients depend on the offset direction for each aircraft and on which aircraft passes behind the other for each pair of aircraft. So these data have to be fixed, before the linear optimization program is run. If n aircraft are involved in the conflict to solve, there are $2^{\frac{n(n+1)}{2}}$ possible combinations for these data. An exhaustive search for a global optimum among the local ones would imply to solve as many linear optimization problems. When n grows, this number quickly becomes too large: 32 768 for $n = 5$, 2 097 152 for $n = 6$.

Two questions naturally arise:

1. Isn't there a fast way to establish that a whole class of such data combinations lead to a linear problem with no solution, and to ignore all the combinations belonging to such a class ?
2. Isn't it possible to group different combinations into one single connected component onto which could be run an optimization program, even with non linear constraints ?

Positively answer to one or both questions would have allowed to reduce the number of combinations to take into account, either by ignoring some of them, or by grouping several ones. Unfortunately, even in the very simple case of a conflict involving only two aircraft, answers to these question are very complex (they mainly depend on the value of the initial angles β, compared to the angle between the trajectories of the aircraft). It gets even more complex for three aircraft, and doesn't lead to a significant combinatorial simplification anyway (*cf* [Med94]).

5 The use of Genetic Algorithms

These combinatorial difficulties lead us to try on this problem a Genetic Algorithm, along with a linear optimization algorithm, such as the "simplex" algorithm.

5.1 The algorithms

To solve the linear problems presented above, we use a program called *lp_solve*, elaborated by the *Design Automation Section, Eindhoven University of Technology*. It uses a *simplex* algorithm, such as the one described in [OH68].

The Genetic Algorithm we use is similar to the ones that are described in [Gol89] and [Mic92]: The elements of the initial population are randomly generated. Then each element is evaluated, by the computation of its *fitness* (the more the element is adapted to the problem to solve the higher, its fitness is). Then, mutation and crossover are randomly applied to the population elements, with probabilities P_c and P_m. At this point, a new population is created, and this process is repeated. At each iteration, a new population is obtained. An iteration is called a *generation*.

To avoid premature convergence toward a local optimum we use the *sharing* technique, described in [GY].

5.2 The encoding

The chromosome of an element of population codes a situation from which the linear optimization will be started: for each aircraft, the offset direction, and for each pair of aircraft, which one passes behind the other. This is coded by an sequence of $\frac{n(n+1)}{2}$ bits, treated as an integer (or as several integers if n is large): the n first bits code the offset directions for the n aircraft (1 for a deviation to the left, 0 for a deviation to the right), and the $\frac{n(n-1)}{2}$ bits, for each of the $\frac{n(n-1)}{2}$ pair of aircraft (a_i, a_j), with $i < j$, which aircraft passes behind the other (1 if a_i passes behind a_j, 0 otherwise).

All the chromosomes built this way do not correspond to solutions of the collision avoidance problem. They code for linear problems, some of which are unfeasible (*i.e.* have no solution), as we've already seen it. But we have no way of knowing *a priori* which element of the population codes an unfeasible linear problems. We will see in the remainder of this article how these elements will be treated.

The mutation operator is classical: first, an element is chosen in the population, with probability P_m ; then, one of the $\frac{n(n+1)}{2}$ bits of its chromosome is randomly chosen, and modified.

The crossover operator is the classical uniform crossover operator. Two element are chosen in the population, with probability P_c, to be the "parents". The parents' bits are randomly distributed between the two "children". To accelerate the algorithm's convergence, we use a technique described by Samir Mahfoud and David Goldberg in [MG92], which is inspired by the Simulated Annealing algorithm. Each child is compared to its best parent. If the child's fitness is higher, it takes its parent's place in the population. If the child's fitness is lower, it takes its poarent's place with probability of law $e^{-\frac{\Delta}{T}}$, where Δ is the difference between the parent's fitness and the child's one, T decreases along the generations.

5.3 Implementation

Each chromosome codes for each aircraft, the offset direction, and for each pair of aircraft, which one passes behind the other. We will say that a chromosome codes for a *configuration* of the collision avoidance problem. As we have already stated, for a fixed configuration, the optimal collision avoidance trajectories are determined by the offset values, which are solutions of a linear optimization problem. We can thus compute, for each element corresponding to a feasible linear problem, the n offset values leading to the shortest delay, with a linear optimization program, and then compute this delay.

It is natural to consider that the shorter this delay is, the better the corresponding configuration will be, and to take for the fitness of an element a decreasing function of the delay given by the linear program applied with the configuration coded by the element's chromosome. However, some configurations might lead to unfeasible linear problems, *i.e.* problems for which all constraints cannot be satisfied at the same time. These configurations might be numerous, and if the fitness of all corresponding elements is set to zero, the genetic algorithm would lose much of its efficiency.

We want to evaluate these elements, to be able to compare one to the other. A linear optimization problem is unfeasible when all of its constraints cannot be satisfied simultaneously. We considered that a chromosome coding an unfeasible problem is better than another one, when more of its constraints can be satisfied simultaneously. So we evaluated these elements the following way: constraints of the linear problem are removed one by one and the linear optimization program is applied after each constraint's removal until the problem is feasible. The constraints to be removed are chosen randomly, to avoid the side effects of possible symmetries. The chromosome's fitness is then given by a decreasing function of the number of constraints that had to be removed to make the problem feasible.

The two different function used for the evaluation of the population's elements have to be adjusted in order to ensure that any elements corresponding to an unfeasible linear problem has a lower fitness than the ones which correspond to feasible problems.

We actually used a slightly different technique to evaluate the elements corresponding to unfeasible problems, which is better adapted to the collision avoidance problem and more efficient. Instead of removing constraints one by one, we remove them "aircraft by aircraft": all the constraints in the coefficient of which a given offset value appears are removed at a time. The results presented in the next section were obtained using this technique.

6 Results

It is difficult to evaluate the results we obtained, as it is difficult to obtain, with another optimization technique a solution known to be optimal we could compare with our best solution. For a conflict involving up to six aircraft however, it is still possible to treat all the possible configurations with a linear optimization program, and to determine which one leads to the global optimum. For a six aircraft conflict, this takes a little more than three hours (there are 2 097 152 different configurations).

We consider the following situation: all the aircraft fly at the same speed (400 knots), at time t_o they are regularly distributed on a semi circle centered in C, of radius 100 nautical miles, and their initial trajectories gather in C (see figure 5). Note that the symmetry of this situation, useful for a man to understand it and make sure there actually is a conflict, do not change the way the machine handles it.

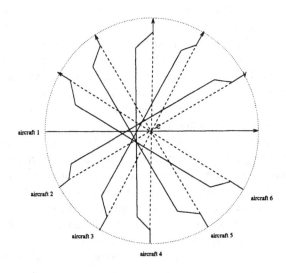

Fig. 5. Optimal collision avoidance system for a six aircraft conflict.

There are two equivalent optimal solutions: either the five first aircraft are deviated to the right, the sixth one goes straight, and for $i < j$, a_i passes behind a_j, or the first aircraft goes straight, the five others are deviated to the left, and for $i < j$, a_j passes behind a_i. We represented this latter solution on figure 5.

Our algorithm found these two optimal solutions. We can compare, for a conflict involving 6 aircraft the time needed to compute the delays corresponding to all the configurations and the time required by our algorithm to reach the optimum.

We ran our program 50 times on the situation presented above. We used populations of 150 elements, and the number of generation was fixed to 98. 39 of the 50 runs lead to one of the two optimal solutions. The distribution of the number of generation before one of the two optimal solution was found, over the 50 runs, is shown on table 1. For instance we found an optimal solution in less than 20 generations 16 times.

The number of generations required to reach an optimal solution is not very meaningful. We present in table 3 the mean value and the standard deviation,

Number of runs	Number of generations needed
16	0 - 20
7	20 -40
7	40 - 60
7	60 - 80
2	80 - 98
11	no optimal solution found in less than 98 generations

Table 1. distribution of the number of generation before an optimal solution is found

Number of runs	Number of calls to the linear optimization program needed
2	0 - 10 000
18	10 000 - 20 000
8	20 000 -30 000
7	30 000 - 40 000
3	40 000 - 50 000
1	50 000 - 55 000
11	no optimal solution found in less than 98 generations

Table 2. Distribution of the number of calls to the linear optimization program needed before an optimal solution is found

over the 39 successful runs, of the CPU time, the number of elements evaluations (Nb of evaluations), and the number of calls to the linear optimization program (Nb of calls), required to reach an optimal solution. Table 2 shows the distribution of this number of calls to the linear optimization program. An exhaustive search would require 2 097 152 calls to this program.

We obtained these results on a HP 720 station.

There are about 250 evaluations at each generation (each of the 150 elements is evaluated, and some other evaluations are needed by the crossover operator).

For each of the first generations, up to as many as 1500 calls to the linear optimization program may be required, whereas this number later decreases to less than 300. Indeed, the evaluation of an element corresponding to an infeasible linear optimization problem requires several calls to the linear optimization program (one after each removal of a group of constraints, the constraints being removed "aircraft by aircraft"), until the problem is feasible. Generation by generation, the elements of population get globally better, the number of elements

	Mean value	Standard deviation
Time (s)	99.0	51.8
Nb of evaluation	9696	6707
Nb of calls	23193	11818

Table 3. Number of calls to the linear optimization program

that correspond to unfeasible problems decreases, and the mean value of the number of constraints to remove also decreases.

With the numerical values we have used, the two optimal solutions led to a 9,6 minute global delay. Each time no optimal solution was found (11 runs out of 50), an under-optimal solution was found, leading to a 10,9 minute global delay.

7 Conclusion

Such a technique is useful to solve only highly combinatorial problems. For a conflict involving three aircraft, only 64 calls to the linear optimization program are required for an exhaustive search, and only 1 024 are required for a conflict involving four aircraft. In these cases an exhaustive search will clearly be faster than the technique presented here.

This technique allows to join the advantages of linear optimization (a quick and efficient search for a local optimum) to those of genetic algorithms, which help finding global optima, and often propose several different solutions to one problem, which is of great interest if we want to help controllers by proposing possible collision avoidance trajectories among which they will be able to choose, rather than completely automate air traffic control.

The work we have presented here is a theoretical approach of the resolution of ATC conflicts. We have made some strong hypotheses (plane trajectories and conflicts, constant speed, separation conditions for all t and not only for $t \in [t_o, t_f]$, etc). If a practical use of this technique is wanted, these hypotheses' strength will have to be reduced (*cf* [Med94]).

These results may probably be made better, by adjusting the coefficient that are used by the algorithm (for instance to evaluate the elements). They are nevertheless already really promising.

References

[AGS93]   Jean-Marc Alliot, Hervé Gruber, and Marc Schoenauer. Using genetic algorithms for solving ATC conflicts. In *Proceedings of the Ninth IEEE Conference on Artificial Intelligence Application*. IEEE, 1993.

[AL92]   Luc Angerand and Hervé LeJeannic. Bilan du projet SAINTEX. Technical report, CENA, 1992. CENA/R92009.

[DAAS94] Nicolas Durand, Nicolas Alech, Jean-Marc Alliot, and Marc Schoenauer. Genetic algorithms for optimal air traffic conflict resolution. In *Submitted to the Second Singapore Conference on Intelligent Systems.* SPICIS, 1994.

[DASF94a] Daniel Delahaye, Jean-Marc Alliot, Marc Schoenauer, and Jean-Loup Farges. Genetic algorithms for partitioning airspace. In *Proceedings of the Tenth Conference on Artificial Intelligence Application.* CAIA, 1994.

[DASF94b] Daniel Delahaye, Jean-Marc Alliot, Marc Schoenauer, and Jean-Loup Farges. Genetic algorithms for air traffic assignment. In *Proceedings of the EuropeanConference on Artificial Intelligence.* ECAI, 1994.

[DDAS94] Daniel Delahaye, Nicolas Durand, Jean-Marc Alliot, and Marc Schoenauer. Genetic algorithms for air traffic control system. *soumis à IEEE Expert '94,* 1994.

[Dur94] Nicolas Durand. Modélisation des trajectoires d'évitement pour la résolution de conflits en route. Technical report, Centre d'Etudes de la Navigation Aérienne, Février 1994.

[FMT93] Xavier Fron, Bernard Maudry, and Jean-Claude Tumelin. Arc 2000 : Automatic radar control. Technical report, Eurocontrol, 1993.

[Gol89] David Goldberg. *Genetic Algorithms.* Addison Wesley, 1989. ISBN: 0-201-15767-5.

[GY] Noel Germay and Xiaodong Yin. A fast genetic algorithm with sharing scheme using cluster analysis methods in multimodal function optimization. Technical report, Université Catholique de Louvain, Laboratoire d'Electronique et d'Instrumentation.

[K+89] Fred Krella et al. Arc 2000 scenario (version 4.3). Technical report, Eurocontrol, April 1989.

[Med94] Frédéric Medioni. Algorithmes génétiques et programmation linéaire appliqués à la résolution de conflits aériens. Mémoire de dea, Ecole Polytechnique, Ecole Nationale de l'Aviation Civile, Juillet 1994.

[MG92] Samir W. Mahfoud and David E. Goldberg. Parallel recombinative simulated annealing: a genetic algorithm. IlliGAL Report 92002, University of Illinois at Urbana-Champaign, 104 South Mathews Avenue Urbana IL 61801, April 1992.

[Mic92] Zbigniew Michalewiicz. *Genetic algorithms + data structures = evolution programs.* Springer-Verlag, 1992. ISBN : 0-387-55387.

[NFC+83] W.P. Niedringhaus, I. Frolow, J.C. Corbin, A.H. Gisch, N.J. Taber, and F.H. Leiber. Automated En Route Air Traffic Control Algorithmic Specifications: Flight Plan Conflict Probe. Technical report, FAA, 1983. DOT/FAA/ES-83/6.

[Nie89a] W.P. Niedringhaus. Automated planning function for AERA3: Manoeuver Option Manager. Technical report, FAA, 1989. DOT/FAA/DS-89/21.

[Nie89b] W.P. Niedringhaus. A mathematical formulation for planning automated aircraft separation for AERA3. Technical report, FAA, 1989. DOT/FAA/DS-89/20.

[OH68] W. Orchard-Hays. *Advanced Linear Programming Computing Techniques.* McGraw-Hill, 1968.

[Zeg93] Karim Zeghal. Techniques réactives pour l'évitement. Technical report, ONERA, June 1993.

Optimizing Hidden Markov Models with a Genetic Algorithm

M. Slimane   G. Venturini   J.-P. Asselin de Beauville
T. Brouard   A. Brandeau

Laboratoire d'Informatique,
Ecole d'Ingénieurs en Informatique pour l'Industrie,
Université de Tours,
64, Avenue Jean Portalis, Technopôle Boîte No 4,
37913 Tours Cedex 9 FRANCE
Tel: (+33)-47-36-14-33, Fax: (+33)-47-36-14-22,
Email : asselin, venturini@univ-tours.fr

Abstract. In this paper is presented the application of genetic algorithms (GAs) to the learning of hidden Markov models (HMMs). The Baum-Welch algorithm (BW), which optimizes the coefficients of a HMM, is improved by the use of a GA. The GA is able to find rapidly a good initial model compared to random generation, and this initial model is optimized further with BW. A representation and adapted genetic operators have been introduced in order to evolve matrix of probabilities. Several tests on artificial data show the interest in using a GA with BW.

1   Introduction

Markov chains are wellknown mathematical tools which are used for instance in pattern recognition. This standard model can be extended to the hidden Markov model which consists in an observable part and a hidden part. To each state of the hidden model is associated a probability of generating an observable symbol. Given a observation of such symbols, it is possible to learn a HMM that has a high probability of generating that observation. In this way, HMMs can model complex stochastic processes and have been applied for instance to speech recognition (Kriouille 1990), language modeling (Gauvain et al. 1994), hand written text recognition (Anigbogu 1992), signal processing (Deng 1992) (Saerens 1993) or image encoding (Mao and Kung 1990).

In order to learn a HMM, several algorithms exist which are similar to gradient search through the space of possible models, like the Baum-Welch algorithm,.

As with gradient search in general, this algorithm is sensitive to the initial point of the search, and may converge to local optima. The idea developped in this paper is to use a genetic algorithm to optimize more precisely HMMs. GAs are known to find rapidly an approximated solution to optimization problems (Holland 1975) (De Jong 1988). Here, the aim of the GA is to find a good initial point for the BW algorithm, in order to find HMMs closer to the optimum. This paper is organized in the following way: section 2 introduces the mathematical definitions and problems relevant to HMMs. Section 3 describes the genetic encoding and the genetic operators chosen that deal with the numeric constraint of the HMM optimization problem. The GA used is described in section 4. Section 5 describes experimental results that show how the GA may improve significantly the BW algorithm on the tested problems.

2 Hidden Markov models and the Baum-Welch algorithm

Markov chains are used to model stochastic processes. They consist of a set of N states denoted by $S = \{S_1, S_2, ..., S_N\}$. At a given time t, the stochastic process is in a given state $q_t \in S$ of the Markov chain. The current state changes according to transition probabilities between the states. A Markov model is thus defined by a state transition matrix $A = [a_{ij}]$ with $1 \leq i \leq N$ and $1 \leq j \leq N$. a_{ij} is the probability to switch from state S_i to state S_j defined by $a_{ij} = P(q_{t+1} = S_j | q_t = S_i)$. For a stationary model of order 1, this probability does depend neither on time and neither on states prior to q_t. In this case, A is a square matrix of dimension N with real coefficients. In addition to A, to each state S_i is associated the probability $\pi_i = P(q_1 = S_i)$ that the stochastic process starts in this state. A vector of such initial probabilities can be defined by $\Pi = \{\pi_i\}$ with $1 \leq i \leq N$. A markov model is thus defined by a model $\lambda = (A, \Pi)$.

This model is useful to model stochastic processes where the current state is known explicitly. Given a series of observed states $\mathcal{O}$, it is possible to learn a Markov model λ that maximises the probability $P(\mathcal{O}|\lambda)$ of generating $\mathcal{O}$ with λ. However, one doesn't always have access to all this information. The observation $\mathcal{O}$ can be only a series of symbols which do not correspond directly to the states of the Markov model. The stochastic process may have hidden states which are not mentioned explicitly in the observation. For instance, one may consider the problem of three urns that contain a different proportion of white and black balls (Rabiner 1989). Balls are chosen one at a time from an urn with replacement. One chooses initially one urn, takes randomly a ball, then replace it in the urn and switch randomly to another urn according to a Markov model. If one observes only the series of "black" or "white" symbols generated in this way, one doesn't know directly which urn has been used. The HMMs have been designed to model such a process.

To define a HMM (Slimane and Asselin de Beauville 1994), let $V = \{v_1, v_2, ...$

| A | | B | | Π |

$$
A = \begin{array}{|ccc|}
\hline
0.2 & 0.3 & 0.5 \\
0.1 & 0.2 & 0.8 \\
0.7 & 0.2 & 0.1 \\
\hline
\end{array}
\qquad
B = \begin{array}{|cc|}
\hline
0.1 & 0.9 \\
0.7 & 0.3 \\
0.4 & 0.6 \\
\hline
\end{array}
\qquad
\Pi = \begin{array}{|ccc|}
\hline
0.1 & 0.5 & 0.4 \\
\hline
\end{array}
$$

Figure 1: A hidden Markov model with 3 states and 2 observation symbols.

$v_M\}$ denote a set of M distinct symbols that will appear in the series $\mathcal{O}$ of observations. At time t, the symbol generated by the model is denoted by $\mathcal{O}_t$ with $\mathcal{O}_t \in V$, and the number of symbols in $\mathcal{O}$ is denoted by T. As with Markov models, a HMM is defined by a transition matrix A and a vector Π of initial probabilities. In addition to A and Π, a third matrix B represents the probabilities of generating each symbol in each state. This matrix is denoted by $B = [b_j(k)]$ where $b_j(k) = P(\mathcal{O}_t = v_k | q_t = S_j)$ with $1 \le j \le N$ and $1 \le k \le M$. A HMM λ is thus defined by $\lambda = (A, B, \Pi)$. For $N = 3$ and $M = 2$, figure 1 gives an example of such a model.

In order to use efficiently a HMM, three basic problems must be solved:

1. Evaluation : Given a series $\mathcal{O}$ of observations and a model λ, how to evaluate the probability $P(\mathcal{O}|\lambda)$ of generating $\mathcal{O}$ with λ? The solution is given by the Forward-Backward algorithm (Rabiner 1989), which computes the exact value of $P(\mathcal{O}|\lambda)$ in polynomial time (Kriouille 1990),

2. Most probable path : this problem corresponds to determining the series Q of hidden states which has the highest probability of generating the series $\mathcal{O}$ of observations. This amounts to finding Q which maximizes $P(Q/\mathcal{O}, \lambda)$. The Viterbi algorithm can find this series of states (Forney 1973),

3. Optimal adaptation of the model : this is the problem of learning a HMM from a series of observations. Given such a series $\mathcal{O}$, how to find a model λ that maximizes $P(\mathcal{O}|\lambda)$? A solution to this problem is given by the Baum-Welch algorithm which improves iteratively an initial model (Baum et al. 1970) (Baum 1972).

In this paper, we only deal with this last point. The Baum-Welch algorithm has the advantage of running fast, but it has also drawbacks inherent to gradient-like algorithms. It may converge to critical points of $P(\mathcal{O}|\lambda)$, like local maxima or inflexion points. It is also sensitive to the initial model. In addition, it creates an "underflow problem": numbers computed by the algorithm can be very small, which raises numerical exceptions and errors on computers. The general idea developped in this paper is to use a GA to find a better initial model to the Baum-Welch algorithm, in order to avoid local maxima and the underflow problem. The GA must be however fast because the execution time of the Baum-Welch algorithm is usually low (see the results in section 5). Thus, the GA is only used to find a better initial model than random generation, and it is not used to optimize completely the model because it would be too long.

3 HMM representation and genetic operators

As with any other GA application, the first problem to solve is to find a representation, an evaluation function and genetic operators.

3.1 Genetic represention of a HMM

As seen in the previous section, a HMM $\lambda = (A, B, \Pi)$ is not directly equivalent to a vector of real parameters because several numerical constraints apply to this vector. For instance, the sum of coefficients of any row in matrix A, B or Π must be equal to 1 (these are stochastic matrix). These constraints must be taken into account in the chosen representation and in the genetic operators in order to avoid generating three matrix (A, B, Π) that would not be a HMM.

The first possible representation is the binary representation which in theory does not require any adaptation of the genetic operators. A stationary HMM of the first order, with N states and M possible symbols, requires $N^2 + N \times M + N$ coefficients (N^2 for matrix A, $N \times M$ for matrix B and N for Π). For example, with $N = 10$ and $M = 5$ which are current values, a HMM requires 160 real parameters. If one considers that a low precision of 8 bits is enough, a chromosome would be represented in this case with $160 \times 8 = 1280$ bits. Studies like (Michalewicz 1992) have shown that this binary representation is less efficient than the real coefficients representation. The "delta-coding" is an interesting way of circumventing this problem by changing the binary representation over time (Whitley et al. 1991). Binary chromosome only encode a move around a given point in this space, which allows the GA to work on much smaller binary representations. However, this technique doesn't solve the problem of constraints.

The chosen representation thus consists in encoding a HMM directly as a chromosome of $N^2 + N \times M + N$ real coefficients. Each coefficient is represented internally with the standard precision of 48 bits.

3.2 Evaluation fonction

The evaluation function of a model λ denoted by $f(\lambda)$ is computed with the Forward-Backward algorithm. Given a model λ and a series $\mathcal{O}$ of observations, this algorithm computes the probability $P(\mathcal{O}|\lambda)$. We thus try to maximize this probability in order to find an initial model with the highest quality before

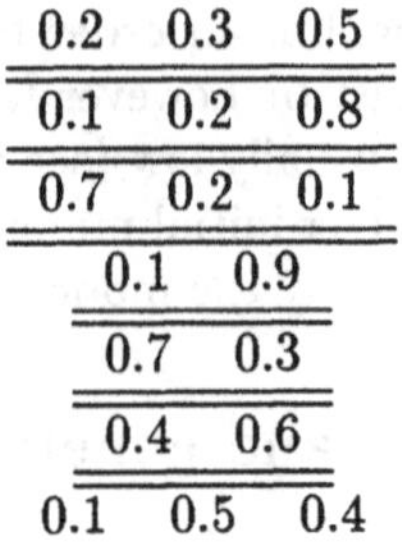

0.2	0.3	0.5
0.1	0.2	0.8
0.7	0.2	0.1
	0.1	0.9
	0.7	0.3
	0.4	0.6
0.1	0.5	0.4

Figure 2: Possible cutting points (double lines) for the 1X crossover operator. These cutting points do not require to normalize the rows of matrix A, B and Π.

using the Baum-Welch algorithm. The evaluation function is thus given by $f(\lambda) = P(\mathcal{O}|\lambda)$.

3.3 Genetic operators

To solve the problem of constraints, one could use for instance techniques developped in (Michalewicz and Janikow 1991) or (Schoenauer and Xanthakis 1993). However, it is possible to define here appropriate genetic operators that will only generate valid models. The constraints are numerous (N for matrix A, N for B and 1 for Π) and involve many variables, but they are all of the same type (sum of coefficients equal to one).

One first possible solution is to use the standard genetic operators for binary or real encoded numbers, like the 1 point crossover that exchanges two parts of two individuals starting from a randomly generated cutting point. With such operators, two parent HMMs are considered as a vector of coefficients. To guarantee that any model generated with this operator is still a HMM that satisfies the constraints, matrix A, B and Π must be normalized by dividing the coefficient of a row by the sum of coefficients on that row. However, this method may modify interesting parts of the model due to the normalization and therefore it has not been used in the following.

A second solution consists in allowing the crossover to cut individuals only between two rows of a matrix. In this way, the crossover operator exchanges only matrix rows or matrix Π between two individuals. The cutting point is chosen randomly between two rows of A, or between A and B, or between B and Π as shown in figure 2. This crossover considers that a row in a matrix is a gene of a single block. The offsprings generated with this operator also satisfy the constraints because these constraints only apply to matrix rows. This crossover may not be able to play its role because it exchanges blocks of at least the size of a row (N or M coefficients). This crossover is denoted 1X in the following and has been used in a single point version only.

A third solution consists in using a linear recombination crossover that computes offsprings as a weighted sum of the parents, like in evolution strategies (Baeck et al. 1991). Given two rows $l = (a_1, a_2, ..., a_N)$ and $l' = (a'_1, a'_2, ..., a'_N)$, if one computes a new row l'' with the formula:

$$l'' = \alpha l + (1 - \alpha)l'$$

$$= (\alpha a_1 + (1 - \alpha)a'_1, \alpha a_2 + (1 - \alpha)a'_2, ..., \alpha a_N + (1 - \alpha)a'_N)$$

with $\alpha \in [0, 1]$, then the sum of coefficients in this row is equal to 1 and thus still satisfies the constraints. A recombination operator can thus be defined as a linear combination of two parent models λ_1 and λ_2. One may set $\alpha = \frac{1}{2}$ for all crossovers. α may also depend on the parents quality by setting:

$$\alpha = \frac{f(\lambda_1)}{f(\lambda_1) + f(\lambda_2)}$$

This technique can be viewed as based on an immune network mechanism (Bersini et Varela 1991), where offsprings must be similar to parents with a high quality. In the following, this operator is denoted by αX.

To define the mutation operator, one may for instance swap coefficients of a given row. This operator avoids using the normalization because the sum of coefficients in the row is still equal to 1. However, it does not introduce new coefficients. Thus, the mutation operator which has been used in this paper modifies randomly one coefficient of the model by generating a new coefficient in $[0, 1]$. Then, any row which has been modified is normalized due to the mutation. The whole row is thus mutated. The probability of mutation used in the following is 1 %.

4 Algorithm

We remind the reader that the aim of the GA is to find rapidly an approximative model and not to optimize completely the model. A GA that would be too slow would be useless here, given the fact that the Baum-Welch algorithm usually runs fast. The algorithm which has been used is the following:

1. $t \leftarrow 1$,
   Generate randomly an initial population $P(t)$ of 60 individuals (one individual = one HMM) and evaluate each individual,

2. Repeat (generate $P(t + 1)$ from $P(t)$)

   (a) Select an intermediate population P_s by keeping only the best 30 individuals of $P(t)$,
   $P_r \leftarrow \emptyset$,

(b) <u>Generate</u> a population P_r of 30 individuals:

- Select randomly and uniformly two individuals (λ_1, λ_2) of P_s,
- Recombine these individuals with the crossover operator to obtain (λ'_1, λ'_2),
- Mutate λ'_1 into λ''_1 with the mutation operator,
- Evaluate $f(\lambda''_1)$
- $P_r \leftarrow P_r \bigcup \{\lambda''_1\}$

(c) $P(t+1) = P_s \bigcup P_r$,
   $t \leftarrow t+1$,

3. <u>Until</u> $t = 200$,

4. Let λ^*_1 be the best model found by the GA,

5. <u>Apply</u> the Baum-Welch algorithm to λ^*_1 to produce a model λ^*_2 and ouput this model.

In order to generate initially individuals that satisfy the constraints, the first generation is created in the following way: for each matrix row, the first coefficient $a_{1,1}$ is generated randomly and uniformly in $[0, 1]$, then the second coefficient $a_{1,2}$ is generated in the interval $[0, 1 - a_{1,1}]$, and so on until the last coefficient of the row is generated.

The selection technique is able to always keep the best individual in the population. All selected individuals have the same chance of reproducing. The stopping criterion is such that the execution time is of the order of 10 minutes on a small PC and for large models (see next section). The GA thus evaluates $30 \times 200 = 6000$ models in one run.

5 Results

The Forward-Backward and Baum-Welch algorithms used in this section have been implemented by ourselves and validated on known examples. In a first experiment, four different methods are compared: a random generation (denoted by RANDOM), a random generation followed by the Baum-Welch algorithm (denoted by BW), a genetic algorithm alone (denoted by GA), and a genetic algorithm followed by the Baum-Welch algorithm (denoted by GA+BW). In this case, the GA uses the 1X operator. In a second experiment, the BW and GA+BW algorithms are compared on the underflow problem. Finally, in a third experiment, the two crossover operators 1X and αX are compared. In all these experiments, the Baum-Welch algorithm has been used with 50 iterations. When it stops because of an underflow exception, the model quality is equal to 0.

$\mathcal{O}_1$	1, 2, 3, 1, 4, 2, 4, 4
$\mathcal{O}_2$	1, 2, 2, 1, 1, 1, 2, 2, 2, 1
$\mathcal{O}_3$	1, 2, 3, 2, 1, 2, 5, 4, 1, 2, 4
$\mathcal{O}_4$	1, 1, 1, 2, 2, 1, 2, 3
$\mathcal{O}_5$	1, 1, 1, 2, 2, 2, 3, 3, 3
$\mathcal{O}_6$	1, 2, 3, 1, 2, 3, 1, 2, 3
$\mathcal{O}_7$	1, 1, 1, 2, 2, 2, 3, 3, 3, 1, 2, 3
$\mathcal{O}_8$	1, 1, 2, 2, 3, 3, 4, 4, 1, 2, 3, 4
$\mathcal{O}_9$	1, 1, 1, 1, 2, 2, 2, 2
$\mathcal{O}_{10}$	1, 2, 3, 4, 5, 6, 6, 5, 4, 3, 2, 1

Table 1: The different series of observations used in the experiments.

5.1 Quality of learned models

An experiment consists in defining artificially a series of observations $\mathcal{O}$. The symbols used in this series are digits as represented in table 1 which contains the 10 series used in this paper. Then, the algorithms mentioned previously are used to learn a HMM λ that maximizes $P(\mathcal{O}/\lambda)$, where the number of states N varies from 2 to 20. The tested algorithm is run for each value of N. The results of each experiment are averaged over 10 runs, and represent directly the $P(\mathcal{O}/\lambda)$ probability.

Table 2 shows the results obtained for the series $\mathcal{O}_1 = 1, 2, 3, 1, 4, 2, 4, 4$ and for different values of the number of states N. The RANDOM search gets the worst results. The GA alone is able to improve those results, but not significantly enough from the problem point of view, since the probabilities are still very low. The GA should probably be run for a much greater number of generations to get better and acceptable performances. However, this improvement in the results is still good enough to give a better starting point to the Baum-Welch algorithm which gets better results than BW for any value of N in this experiment.

Table 3 confirms for the other observations and for $N = 2$ to $N = 20$ that the cooperation between the GA and the Baum-welch algorithm improves significantly the quality of learned models in comparison to BW. Table 3 represents 1900 runs for each algorithm, and the GA+BW results are in average better that BW results and this in an important proportion (in the order of 30 % to 500 %).

To learn a model with 20 states and with a series of observations with 4 different symbols, the search space is $\mathcal{R}^{500}$. The execution time of GA+BW on a PC 486 DX 33 are in the order of 20 seconds for $N = 2$ and 10 minutes for $N = 20$. These performances on large models are acceptable given the fact that a small computer has been used. The execution time of BW alone is less than 30 seconds.

N	RANDOM	BW	GA	GA+BW
2	0.00000	0.00018	0.00017	0.00020
3	0.00001	0.00220	0.00203	0.00320
4	0.00001	0.00763	0.00548	0.01324
5	0.00001	0.03497	0.01796	0.08828
6	0.00000	0.08647	0.01824	0.17314
7	0.00000	0.18561	0.02419	0.42731
8	0.00001	0.18626	0.03555	0.52865
9	0.00000	0.56611	0.02650	0.66371
10	0.00000	0.31816	0.02432	0.54791
11	0.00001	0.50864	0.02547	0.87875
12	0.00000	0.38286	0.03137	0.87585
13	0.00000	0.34342	0.01888	0.78668
14	0.00000	0.44583	0.01865	0.72453
15	0.00000	0.37095	0.01587	0.57936
16	0.00001	0.10914	0.02165	0.73369
17	0.00000	0.12677	0.01195	0.77969
18	0.00000	0.13897	0.01071	0.45478
19	0.00000	0.00000	0.01044	0.82012
20	0.00000	0.00000	0.01293	0.40949
mean	0.00000	0.20075	0.01749	0.49940

Table 2: Results of the different algorithms for the series of observations $\mathcal{O}_1 = 1, 2, 3, 1, 4, 2, 4, 4$. These results represent directly the $P(\mathcal{O}_1/\lambda)$ probability.

Obs.	RANDOM	BW	GA	GA+BW
$\mathcal{O}_1$	0.00000	0.20075	0.01749	0.49940
$\mathcal{O}_2$	0.00057	0.14473	0.05659	0.40208
$\mathcal{O}_3$	0.00000	0.02833	0.00076	0.05738
$\mathcal{O}_4$	0.00015	0.09534	0.03788	0.14531
$\mathcal{O}_5$	0.00002	0.30347	0.02816	0.57107
$\mathcal{O}_6$	0.00002	0.75266	0.30345	0.93553
$\mathcal{O}_7$	0.00000	0.03790	0.00135	0.23589
$\mathcal{O}_8$	0.00000	0.03732	0.00051	0.21510
$\mathcal{O}_9$	0.00268	0.41520	0.26736	0.79877
$\mathcal{O}_{10}$	0.00000	0.01059	0.00001	0.02935

Table 3: Results for the 10 series, with N varying from 2 to 20, averaged over 10 runs. One column in this table thus represents 1900 runs of the corresponding algorithm.

Obs.	% exception BW	% exception AG+BW
$\mathcal{O}_1$	35 %	33 %
$\mathcal{O}_2$	20 %	0 %
$\mathcal{O}_3$	26 %	32 %
$\mathcal{O}_4$	27 %	11 %
$\mathcal{O}_5$	15 %	3 %
$\mathcal{O}_6$	19 %	1 %
$\mathcal{O}_7$	19 %	2 %
$\mathcal{O}_8$	27 %	11 %
$\mathcal{O}_9$	25 %	10 %
$\mathcal{O}_{10}$	50 %	53 %
mean	26 %	16 %

Table 4: Proportion of underflow exceptions for different series of observations.

5.2 The underflow problem

This problem is one drawback of the Baum-Welch algorithm and is due to the fact that numbers computed by the algorithm can become very small and thus difficult to handle with computers. The implementation of the Baum-Welch algorithm used in this paper sometimes stops because of an underflow exception in the 48 bits representation of real numbers. In this case, the quality of the learned model is equal to 0. This is for instance always the case for BW with $N = 19$ and $N = 20$ in table 2. Such kind of errors also occur in GA+BW, since the Baum-Welch algorithm is used also.

It is thus important to know how GA+BW compares to BW on this problem, because the GA could for instance increase the proportion of exceptions. Table 4 shows the proportion of exceptions for each observation. This proportion is simply the number of runs which stopped due to an exception divided by the total number of runs.

One first comment is that even when this proportion is roughly the same for the two algorithms (see for instance $\mathcal{O}_1$ and $\mathcal{O}_3$ in table 4), the results of GA+BW are still really better than those of BW (see table 3). More generally, the starting point found by the GA improves the Baum-Welch algorithm performances, as mentioned in the previous section, but also lowers the proportion of exceptions from 26 % down to 16 %.

5.3 Evaluating the operators

The conclusions derived from a comparaison between two genetic operators on a specific problem may not generalize to other problems, but it may determine

Obs.	1X	αX	1X+BW	αX+BW
$\mathcal{O}_1$	0.01749	0.00098	0.49940	0.57727
$\mathcal{O}_2$	0.05659	0.00983	0.40208	0.30214
$\mathcal{O}_3$	0.00076	0.00002	0.05738	0.06942
$\mathcal{O}_4$	0.03788	0.00381	0.14531	0.18669
$\mathcal{O}_5$	0.30347	0.24892	0.57107	0.61597
$\mathcal{O}_6$	0.75266	0.75266	0.93553	0.94668
$\mathcal{O}_6$	0.03790	0.03224	0.23589	0.23221
$\mathcal{O}_8$	0.03732	0.05279	0.21510	0.22863
$\mathcal{O}_9$	0.41520	0.39715	0.79877	0.77160
$\mathcal{O}_{10}$	0.01059	0.00302	0.02935	0.01408

Table 5: Results with the two genetic operators for the 10 series (averaged over 10 runs).

which operator (namely 1X or αX) is the best for this problem. Table 5 shows the results obtained with these two operators where the αX operator computes α with the strength of the two parents (see section 3.3). The same GA is used with the two operators.

The αX operator makes the population more uniform than 1X. As a consequence, the genetic search converges earlier than for 1X. However, BW gets the same performances for the two operators. This suggests that a higher mutation rate should be used with αX and that BW may use a starting point obtained before 200 generations.

6 Conclusion

In this paper, we have developped a genetic algorithm that improves the Baum-Welch algorithm for learning hidden Markov models. This algorithm finds a better initial model than a random search by using a genetic algorithm, and this initial model can be optimized further by calling the Baum-Welch algorithm. The experiments performed show that this algorithm improves significantly the quality of learned models compared to the Baum-Welch algorithm alone. This approach seems promizing for using hidden Markov models.

Future work consists in trying to use the Baum-Welch algorithm as a local operator in a coarse grain parallel genetic algorithm. The Baum-Welch algorithm could improve an initial population, thus improving the cooperation between the two algorithms. Also, it would be useful to estimate the optimal number of states in order to minimize N while maximizing the probability of generating the observations. Applications will take place in computer vision.

References

Anigbogu J.C. (1992), Reconnaissance de caractères imprimés multifontes à l'aide de modèles stochastiques et métriques, thèse de doctorat, Université de Nancy I, 1992.

Baeck T., Hoffmeister F. and Schwefel H.P. (1991), A survey of evolution strategies, Proceedings of the Fourth International Conference on Genetic Algorithms, 1991, R.K. Belew and L.B. Booker (Eds), Morgan Kaufmann, pp 2-9.

Baum L.E., Petrie T., Soules G. and Weiss N. (1970), A maximization technique occuring in the statistical analysis of probabilistic functions of Markov chains, Ann. Math. Stat. 41(1), pp 164-171, 1970.

Baum L.E. (1972), A inequality and associated maximization technique in statistical estimation for probabilistic functions of Markov processes, Inequalities 3, pp 1-8, 1972.

Bersini H. and Varela F.J. (1991), The immune recruitment mechanism: a selective evolutionnary strategy, Proceedings of the Fourth International Conference on Genetic Algorithms, Morgan Kaufmann, 1991, pp 520-526.

De Jong K. (1988). Learning with Genetic Algorithms: An overview. Machine Learning 3, pp 121-138.

Deng L. (1992), A generalized hidden Markov model with state-conditionned trend functions of time for speech signal, Signal processing, Elsevier n 27, pp 65-78, 1992.

Gauvain J.L., Lamel L.F., Adda G. and Adda-Decker M. (1994), The LIMSI continuous speech dictation system: Evaluation on the ARPA Wall Street Journal Task, ICASSP, 1994, pp 1557-1560.

Holland J.H. (1975), *Adaptation in natural and artificial systems*. Ann Arbor: University of Michigan Press.

Kriouile A. (1990), Reconnaissance automatique de la parole et les modèles markoviens cachés, thèse de doctorat, Université de Nancy I, 1990.

Mao W.D. and Kung S.Y. (1990), An object recognition system using stochastic knowledge source and VLSI architecture, Proceedings of the International Conference on Pattern Recognition, pp 832-836, 1990.

Michalewicz Z. and Janikow C.Z. (1991), Handling constraints in genetic algorithm, Proceedings of the Fourth International Conference on Genetic Algorithms, 1991, R.K. Belew and L.B. Booker (Eds),

Michalewicz Z. (1992), *Genetic algorithms + data structures = Evolution programs*, Springer Verlag, 1992.

Rabiner L.R. (1989), A tutorial on hidden Markov models and selected applications in speech recognition, Proceedings of the IEEE, vol. 77, 1989.

Saerens M. (1993), Hidden Markov models assuming a continuous time dynamic emission of accoustic vectors, Eurospeech 1993.

Schoenauer M. and Xanthakis S. (1993), Constrained GA optimization, Proceedings of the Fifth International Conference on Genetic Algorithms, 1993, S. Forrest (Ed), Morgan Kaufmann, pp 573-580.

Slimane M. and Asselin de Beauville J.P. (1994), Introduction aux modèles de Markov cachés (1ère partie), Internal report, Laboratoire d'Informatique, Université de Tours, Novembre 1994.

Whitley D., Mathias K. and P. Fitzhorn (1991), Delta coding: an iterative search strategy for genetic algorithms, Proceedings of the Fourth International Conference on Genetic Algorithms, Morgan Kaufmann, 1991, pp 77-84.

Springer-Verlag
and the Environment

We at Springer-Verlag firmly believe that an international science publisher has a special obligation to the environment, and our corporate policies consistently reflect this conviction.

We also expect our business partners – paper mills, printers, packaging manufacturers, etc. – to commit themselves to using environmentally friendly materials and production processes.

The paper in this book is made from low- or no-chlorine pulp and is acid free, in conformance with international standards for paper permanency.

Lecture Notes in Computer Science

For information about Vols. 1–987

please contact your bookseller or Springer-Verlag